π-π Scattering – 1973

AIP Conference Proceedings
Series Editor: Hugh C. Wolfe
Number 13

Particles and Fields Subseries No. 5

π-π Scattering – 1973
(Tallahassee Conference)

Editors
P. K. Williams and V. Hagopian
Florida State University

American Institute of Physics
New York   1973

Copyright © 1973 American Institute of Physics, Inc.

This book, or parts thereof, may not be reproduced in any form without permission

L.C. Catalog Card No. 73-81704

ISBN 0-88318-112-6

AEC CONF-730319

American Institute of Physics

335 East 45th Street

New York, N.Y. 10017

Printed in the United States of America

PREFACE

The International Conference on π-π Scattering and Associated Topics was held at The Florida State University in Tallahassee on March 28 to 30, 1973. It was a working, topical conference which was attended by about 85 particle physicists representing most of the groups contributing to the field. In addition to the formal sessions of invited papers, which form the bulk of these proceedings, there was sufficient interest to hold an extra session on the last day of the conference for short contributed papers, many of which are included herein.

The Organizing Committee consisted of G. L. Kane, D. W. G. S. Leith, V. Hagopian, B. Hyams, P. K. Williams and K. W. Lai. Williams was conference chairman, assisted on all matters by Hagopian. The conference drew support from The Florida State University and its Physics Department, the U. S. Atomic Energy Commission, and the National Science Foundation.

We gratefully acknowledge the help of the FSU Office of Continuing Education, particularly Fred Adams, in handling logistics and details. Special thanks are due to members of the Physics Department and their spouses who helped with people, equipment, and meals notably Dr. S. Hagopian, and also D. Pewitt, D. Wilkins, Mrs. D. Pewitt, Mrs. P. Williams, and Mrs. M. Burgess.

P. K. Williams
V. Hagopian

June, 1973

TABLE OF CONTENTS

I. The π-π and K-π Systems   Chairman: J. E. Lannutti

    A. A New Kπ Partial Wave Analysis Below 1 GeV Kπ Mass
       A. Barbaro-Galtieri, M. J. Matison, M. Alston-Garnjost,
       S. M. Flatté, J. H. Friedman, G. R. Lynch, M. S. Rabin,
       and F. T. Solmitz.
       (Presented by A. Barbaro-Galtieri) 1

    B. K_{e4} Decays and Low Energy π-π Phase Shifts
       E. W. Beier 26

    C. π-π Phase Shift Analysis
       P. Estabrooks and A. D. Martin; G. Grayer, B. Hyams,
       C. Jones, P. Weilhammer; W. Blum, H. Dietl, W. Koch,
       E. Lorenz, G. Lütjens, W. Männer, J. Meissburger, and
       U. Stierlin;
       (Presented by A. D. Martin) 37

    D. High Energy π-π Scattering
       W. D. Walker 80

II. π-π Amplitudes and Inelastic Effects   Chairman: D. Morgan

    A. Zeros in ππ Scattering
       M. R. Pennington 89

    B. Coupled Channel Analysis in the $K\bar{K}$ Threshold Region
       G. Grayer, B. Hyams, C. Jones, P. Schlein, P. Weilhammer;
       W. Blum, H. Dietl, W. Koch, D. Lorenz, G. Lütjens,
       W. Männer, J. Meissburger, W. Ochs, and U. Stierlin.
       (Presented by E. Lorenz) 117

    C. π-π Coupled Channels
       P. K. Williams . . . .

    D. Partial Wave Analysis in 2→3 Body Reactions and Comments on Inelastic Final States in ππ and Kπ Scattering
       R. J. Cashmore 144

III. Production Amplitude Analysis and π-π Phase Shifts
                                         Chairman: J. D. Kimel

    A. Amplitude Analyses of Hypercharge Exchange Reactions
       R. D. Field 153

    B. Direct Channel Effects in Nondiffractive Scattering
       J. A. J. Matthews 188

    C. ππ Phase Shift Analysis from 600 to 1900 MeV
        B. Hyams, C. Jones, P. Weilhammer; W. Blum, H. Dietl,
        G. Grayer, W. Koch, E. Lorenz, G. Lütjens, W. Männer,
        J. Meissburger, W. Ochs, U. Stierlin, and F. Wagner.
        (Presented by W. Ochs) 206

    D. Learning About Meson States from Their Production Properties
        G. L. Kane 247

IV. Extrapolation Panel   Chairman: L. J. Gutay

    A. Relative Phases of Single Pion Production Amplitudes
        L. J. Gutay and K. V. Vasavada
        (Presented by L. J. Gutay) 255

    B. Extrapolation Technique Used by the S.L.A.C. Group in Their
    Measurement of the $\pi^+\pi^-$ Cross Section
        P. Baillon 260

    C. Extrapolation with Complex Transformations
        G. Laurens 270

    D. A_2 Exchange and Polarization Effects in the Analysis of ππ
    Scattering
        J. D. Kimel and E. Reya
        (Presented by J. D. Kimel) 274

V. Panel on New Directions   Chairman: W. Selove

    A. ππ and Kπ Scattering Studies with the Argonne Effective
    Mass Spectrometer
        D. S. Ayres, R. Diebold, A. F. Greene, S. L. Kramer,
        A. J. Pawlicki, and A. B. Wicklund.
        (Presented by R. Diebold) 284

    B. New Directions in ππ and Kπ Experiments
    (Status of a 13 GeV/c $K^{\pm}p$ Experiment at SLAC)
        R. K. Carnegie 298

VI. Contributed Papers   Chairman: V. Hagopian

    A. Rho-Omega Interference in the Reaction πN → ππN at 3, 4 and
    6 GeV/c
        D. S. Ayres, R. Diebold, A. F. Greene, S. L. Kramer,
        A. J. Pawlicki, and A. B. Wicklund.
        (Presented by S. L. Kramer) 302

B. $\pi\pi$ Phase Shift Analysis from an Experiment $\pi^-p \to \pi^0\pi^0 n$ at 2 GeV/c Supplemented by $\pi^+\pi^-$, $\pi^-\pi^0$ Final States
G. Villet, M. David, R. Ayed, P. Bareyre, P. Borgeaud, J. Ernwein, J. Feltesse, Y. Lemoigne, P. Marty, and A. V. Stirling.
(Presented by J. Feltesse) 307

C. π-π Phase Shifts in the Energy Region 0.6 to 1.42 GeV
S. Toaff, J. C. Anderson, A. Engler, R. W. Kraemer, F. Weisser; J. Diaz. F., A. DiBianca, W. Fickinger, D. K. Robinson, and C. A. Sullivan.
(Presented by S. Toaff) 312

D. Production of ρ^0 and f^0 in 4 GeV/c π^+d Interactions
J. A. Charlesworth, R. L. Sekulin; M. J. Emms, J. B. Kinson, L. Riddiford, B. J. Stacey, M. F. Votruba, P. L. Woodworth; I. G. Bell, M. Dale, and J. V. Major.
(Presented by R. L. Sekulin) 317

E. Preliminary Results of Chew-Low Extrapolations Using $\pi^-p \to \pi^-\pi^+n$ at 4.5 GeV/c
P. Jacques, S. Barish, J. Bensinger, M. Hearn, W. Kononenko, E. M. O'Neill and W. Selove.
(Presented by P. Jacques) 322

F. A Theoretical Calculation of Low Energy $\pi\pi$ Scattering
K. S. Jhung and R. S. Willey
(Presented by R. S. Willey) 327

G. Relativistic Propagators for π-π Resonances and the N-N Interaction
M. L. Nack, T. Ueda, and A. E. S. Green.
(Presented by M. L. Nack) 332

H. Isopin 2 $\pi\pi$ Phase Shifts from an Experiment $\pi^+p \to \pi^+\pi^+n$ at 12.5 GeV/c
G. Grayer, B. Hyams, C. Jones, P. Weilhammer; W. Blum, H. Dietl, W. Koch, E. Lorenz, G. Lütjens, W. Männer, J. Meissburger, U. Stierlin; and W. Hoogland.
(Presented by W. Hoogland) 337

I. Pion Form Factor and Inelastic π-π Scattering
R. L. Goble 342

J. Phenomenology of $\pi\pi$ Scattering
J. L. Basdevant, C. D. Froggatt, and J. L. Peterson.
(Presented by D. Morgan) 346

K. Comments on "Phenomenology of $\pi\pi$ Scattering"
M. R. Pennington and S. D. Protopopescu 351

L. Structure in the Momentum Transfer Distribution of $\pi^-+p \to \rho + N$ at 2.3 GeV/c
   S. Hagopian, V. Hagopian; and W. Selove 354

M. $\pi^-p \to \pi^-\pi^+n$ Amplitude Analysis and Extrapolation to the π Exchange Pole
   P. Estabrooks and A. D. Martin
   (Presented by A. D. Martin) 357

Chapter 1. The π-π and K-π Systems

A NEW Kπ PARTIAL WAVE ANALYSIS BELOW 1 GeV Kπ MASS*

A. Barbaro-Galtieri, M. J. Matison, M. Alston-Garnjost,
S. M. Flatté,[†] J. H. Friedman,[‡] G. R. Lynch,
M. S. Rabin,[§] and F. T. Solmitz

Lawrence Berkeley Laboratory
University of California
Berkeley, California 94720

ABSTRACT

The Kπ scattering for M(Kπ) < 1 GeV has been studied in the reaction $K^+ p \to \Delta^{++} K^+ \pi^-$ at 12 GeV/c incident K^+ momentum. Both the moments of the Kπ angular distribution and the Kπ cross section are extrapolated to the pion pole, and the results are used in a partial wave analysis. For the s wave we have done both an energy-independent and an energy-dependent partial wave analysis. We find that only the so-called "down" solution is compatible with our data; that is, a phase shift δ_0^1 slowly increasing from 20° at 800 MeV to 60° at 1000 MeV. No evidence for an s-wave resonance near the $K^*(890)$ mass is found, although a resonance with Γ < 7 MeV cannot be excluded by the data. The "down" solution is well represented by an effective range formula with $a_0^1 = -0.33 \pm 0.05$, in good agreement with a current algebra calculation.

I. INTRODUCTION

The interest in Kπ scattering in the energy region of $K^*(890)$ lies in the fact that some theoretical models predict the existence of an s-wave resonance at this mass. Experiments so far have failed to prove or disprove conclusively the existence of this state. Partial wave analyses of Kπ scattering in the $K^*(890)$ region have been done by various authors.[1-5] The most recent analyses found two solutions for the s-wave phase shift in this mass region: a slowly varying "down" solution with δ_0^1 rising from 20° at 800 MeV to about 60° at 1000 MeV, and a rapidly rising "up" solution corresponding to a slowly varying s-wave background plus a narrow resonance ($\Gamma \leq 30$ MeV).

---

*Work done under the auspices of the U. S. Atomic Energy Commission.
†Present address: University of California-Santa Cruz, Santa Cruz, California.
‡Present address: Stanford Linear Accelerator Center, Stanford, California.
§Present address: University of Massachusetts, Amherst, Massachusetts.

A. Barbero-Galtieri

Table I and Fig. 1 summarize the data used and the results obtained in these analyses. Table I shows the total number of events available to each analysis in the $K\pi$ mass interval indicated. The number of events in the $K\pi$ mass region below 1 GeV is much smaller, but this information is not readily available to us. It also shows the events in the incident K^+ momentum region $P_{K^+} > 8$ GeV/c. This is done because events with larger incident K^+ momentum have smaller four-momentum transfer to the $K^+\pi^-$ system and therefore are more valuable in extrapolations to the pion pole. Figure 2 shows the relation t_{min} versus P_K for different values of $K^+\pi^-$ masses for the reaction $K^+p \to \Delta^{++}K^+\pi^-$. Figure 1 shows the δ_0^1 phase shift solutions up to a mass of 1.7 GeV. In the region of $K^*(890)$ the two ambiguous solutions are shown as obtained by the different authors.[2-5] This paper will only deal with $K\pi$ scattering below 1 GeV, therefore the second ambiguity (at $M \approx 1.5$ GeV) will not be discussed here.

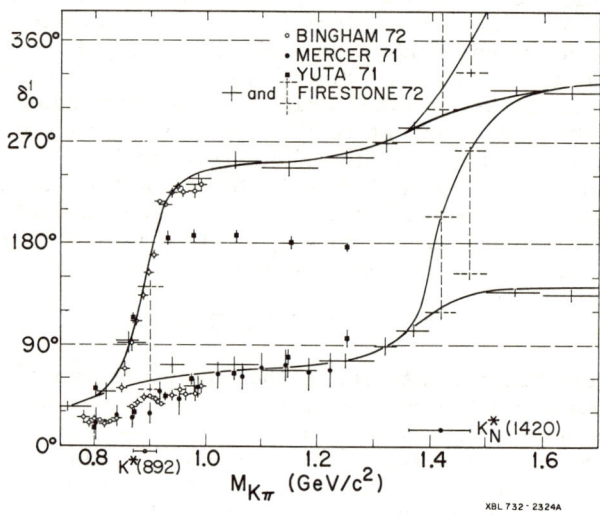

Fig. 1. Phase shift results of Bingham et al.,[3] Mercer et al.,[2] Yuta et al.,[4] and Firestone et al.[5] The curves represent possible paths connecting the ambiguous solutions as drawn by Firestone et al.

Trippe et al.[1] have extrapolated the total cross section to the pion pole and found only the "down" solution in the $K^*(890)$ region; however, they suggested an s-wave resonance at about 1100 MeV. Mercer et al.,[2] who have used the World Data Summary Tape (WDST) compilation,[6] have extrapolated the Y_l^0 moments to the pion pole and found both solutions. However, they eliminated the "up" solution, because it was not in agreement with the extrapolated total cross section. Bingham et al.[3] have also used the WDST compilation[6] when it included a larger number of events. They

Table I. Data used in the partial wave analysis performed by the authors indicated. The column labelled solutions refers to solutions obtained for the s wave in the K*(890) region.

Momentum (GeV/c)	Final states	Number of events	Events in $P_K > 8$ GeV	Interval of K π mass	Solutions	Authors
7.3	$\Delta^{++} K^+ \pi^-$	1363		0.6-2.0	Down	Trippe et al.[1]
	$\Delta^{++} K^0 \pi^0$	219				
3.0-12.7	$\Delta^{++} K^+ \pi^-$	23 244	5449	0.6-3.5	Down	Mercer et al.[2,6]
	$\Delta^{++} K^0 \pi^0$	3967	1148			
2.5-12.7	$\Delta^{++} K^+ K^-$	31 122	12 812	0.6-3.5	Up, down	Bingham et al.[3,6]
	$\Delta^{++} K^0 \pi^0$	4845	2667			
12.0	$\Delta^{++} K^+ \pi^-$	11 073	11 073	0.6-3.5	Up, down	This experiment
5.5	$n K^+ \pi^-$	2875		0.65-1.3	Up, down	Yuta et al.[4]
	$p \bar{K}^0 \pi^-$	2086				
12.0	$K^+ \pi^- p$	2479	2479	0.7-2.0	Up, down	Firestone et al.[5]

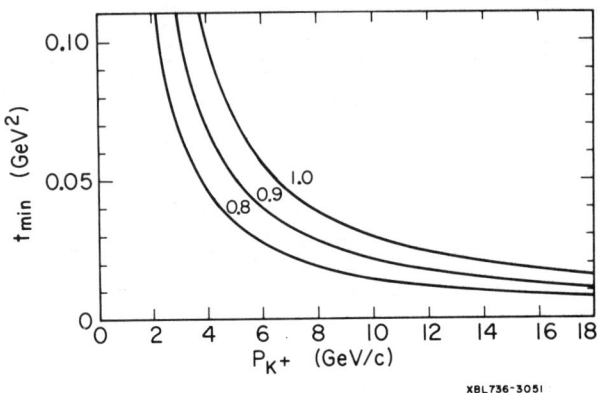

Fig. 2. Minimum momentum transfer, t_{min}, versus incident K^+ momentum for the reaction $K^+ p \to \Delta^{++} K^+ \pi^-$. The three curves refer to the indicated values of $K^+ \pi^-$ invariant mass.

found two solutions but disagreed with Mercer et al.,[2] because they found that with larger statistics the "up" solution could not be eliminated by the total cross section. The next two experiments listed in Table I have used reactions different from those of Refs. 1-3. Yuta et al.[4] used a one meson-exchange model neglecting absorption effects to analyze the off-mass shell $K\pi$ scattering. They found two solutions, one of which required an s-wave resonance at ~ 850 MeV. Firestone et al.[5] extrapolated the moments and the total cross section to the pion pole and performed a partial wave analysis which yielded two solutions as shown in Fig. 1. Recently a new method of analysis has been applied to analyze 9430 events of the reaction $K^- p \to K^- \pi^+ n$ at 4 GeV/c by Chung et al.[7] They studied the mass dependence of each partial wave in the physical region without making any assumption about the nature of the production mechanism. They found that their data can accommodate little if any narrow width daughter state in the $K^*(890)$ region.

II. THE DATA

The analysis described in this paper is based on 11 073 events of the type

$$K^+ p \to \Delta^{++} K^+ \pi^- \qquad (1)$$

with a K^+ incident momentum of 12 GeV/c, obtained in an exposure of the 82-in. Hydrogen Bubble Chamber at SLAC. These events are part of the 4-prong topology of which details have been given elsewhere;[8,9] we only show here some important features of the data.

Figure 3 shows the Kπ invariant mass for events of reaction (1), where Δ^{++} is defined to be $1.16 < M(\pi^+ p) < 1.36$ GeV; it shows that the distribution is dominated by $K^*(890)$ and $K^*(1420)$. Figure 4 shows the momentum transfer distribution $t_{p\Delta}$ for events of reaction (1) in the Kπ mass interval $0.8 < M(K\pi) < 1.0$ GeV. These events are the ones to be used in the partial wave analysis reported in this paper. The t_{min} for these events is about 0.015 (see Fig. 2); that is, they are relatively closer to the pion pole than most of the events used in previous analyses with the same reaction (see Table I and Fig. 2). In particular, a comparison with the sample of Bingham et al.[3] shows that we have a number of events comparable with their sample with $P_{K^+} > 8$ GeV/c; however, t_{min} is smaller (see Fig. 2) for most of the events in our sample. Another advantage of this experiment is that all of the events come from one exposure at one momentum in one bubble chamber, in contrast with the WDST compilation[6] which includes data from 14 different beam momenta, analyzed at 11 different laboratories, so that data may be subject to large uncertainties when combined.

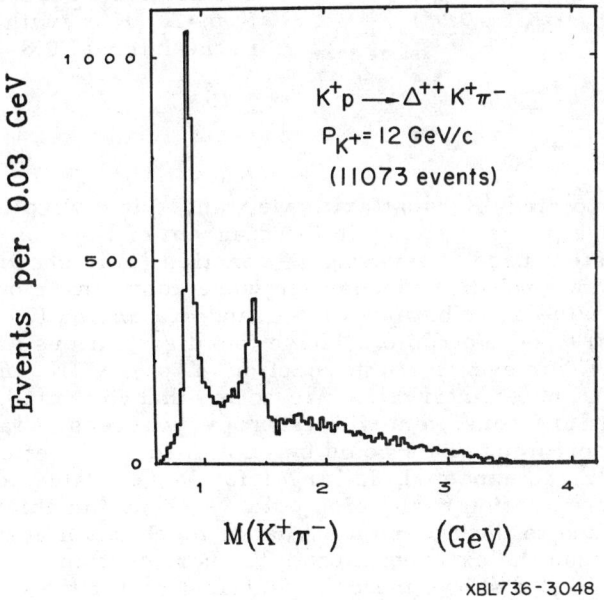

Fig. 3. Invariant mass distribution for the $K^+\pi^-$ system of the reaction $K^+ p \to \Delta^{++} K^+ \pi^-$. The Δ^{++} is defined by $1.16 < M(\pi^+ p) < 1.36$ GeV.

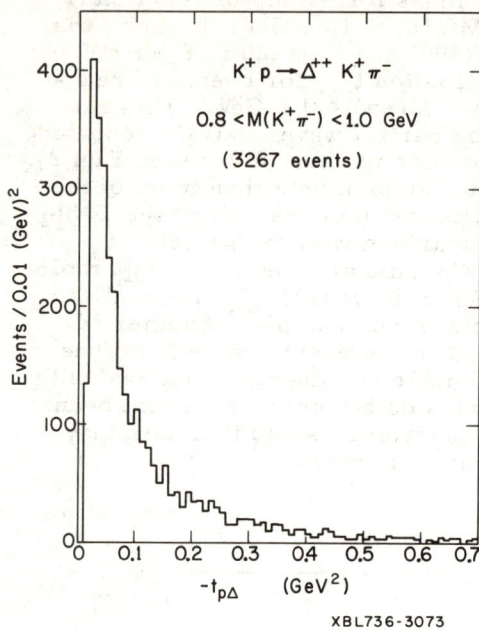

Fig. 4. Distribution of the momentum transfer squared for events of the reaction $K^+ p \to \Delta^{++} K^+ \pi^-$ with $K^+\pi^-$ mass in the interval 0.8-1.0 GeV.

In order to study $K\pi$ scattering we want to investigate the one-pion exchange process shown in the diagram of Fig. 5a. The coordinate system used to investigate reaction (1) is shown in Fig. 5b. If the one-pion exchange mechanism were dominant in our data we would expect the distributions of $\phi_{K\pi}$ and $\phi_{\pi p}$ angles (Treiman and Yang angles) to be isotropic. These two distributions are shown in Figs. 6 and 7 for events of the reaction $K^+ p \to \Delta^{++} K^*$ in various momentum transfer intervals. We notice that for small t the two distributions are consistent with isotropy, whereas at large t they become anisotropic. This could be explained as the effect of background which, as expected, is larger for larger values of $|t|$. By doing an extrapolation to the pion pole we should be able to reduce the background to the one-pion exchange mechanism even further. We will discuss the extrapolation in the next section.

Finally, Fig. 8 shows the $\cos\theta$ distribution for the events that we will use in the partial wave analysis; that is, events of reaction (1) with $0.8 < M(K\pi) < 1.0$ GeV. The moments of the $\cos\theta$ distributions for these events as well as for other $K\pi$ mass intervals and for $|t'| < 0.1$ GeV are shown in Fig. 9. We notice that $Y_3 = Y_4 = 0$ for $M(K\pi) < 1$ GeV, therefore only s and p waves are involved at these energies.

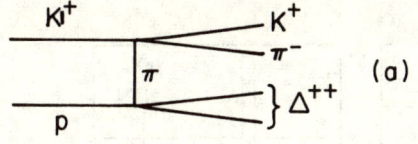

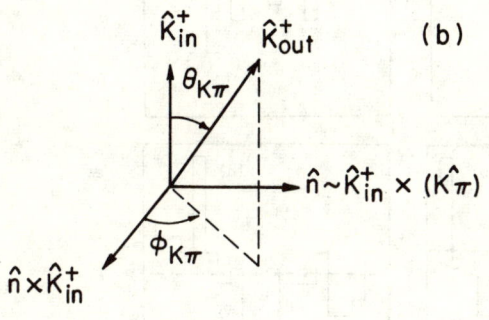

Fig. 5. (a) one-pion exchange diagram, (b) t-channel coordinate system (Jackson frame) for the $K\pi$ vertex. $\phi_{K\pi}$ is the Treiman-Yang angle. An analogous frame can be defined for the πp system.

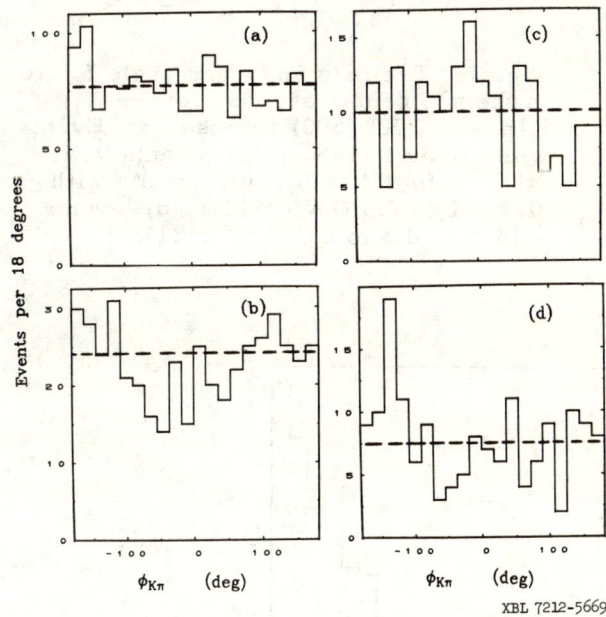

Fig. 6. Treiman and Yang angle $\phi_{K\pi}$ in the $K^+\pi^-$ center of mass for $K^+p \to \Delta^{++}K^*(890)$ events. (a) Events with $|t| < 0.1$ GeV^2 (1551); (b) events with $|t| = 0.1$ to 0.2 GeV^2 (460); (c) events with $|t| = 0.2$ to 0.3 GeV^2 (198); (d) events with $|t| = 0.3$ to 0.5 GeV^2 (156).

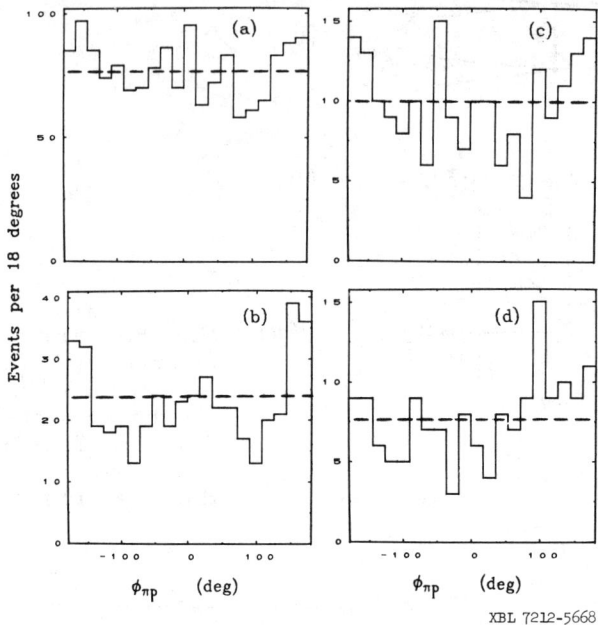

Fig. 7. Treiman and Yang angle for $\phi_{\pi p}$ in the $\pi^+ p$ center of mass for $K^+ p \to \Delta^{++} K^*(890)$ events. (a) Events with $|t| < 0.1$ (1685); (b) events with $|t| = 0.1$ to 0.2 (370); (c) events with $|t| = 0.2$ to 0.3 GeV^2 (185); (d) events with $|t| = 0.3$ to 0.5 GeV^2 (131).

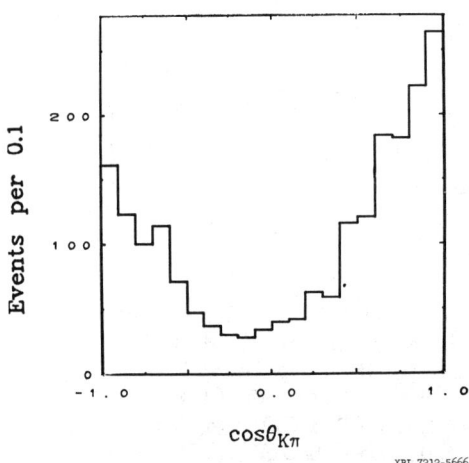

Fig. 8. Angular distribution of the $K\pi$ scattering angle for the reaction $K^+ p \to \Delta^{++} K^+ \pi^-$ for $M(K^+ \pi^-) = 0.8$ to 1.0 GeV and $|t| < 0.1$ GeV^2 (2038 events).

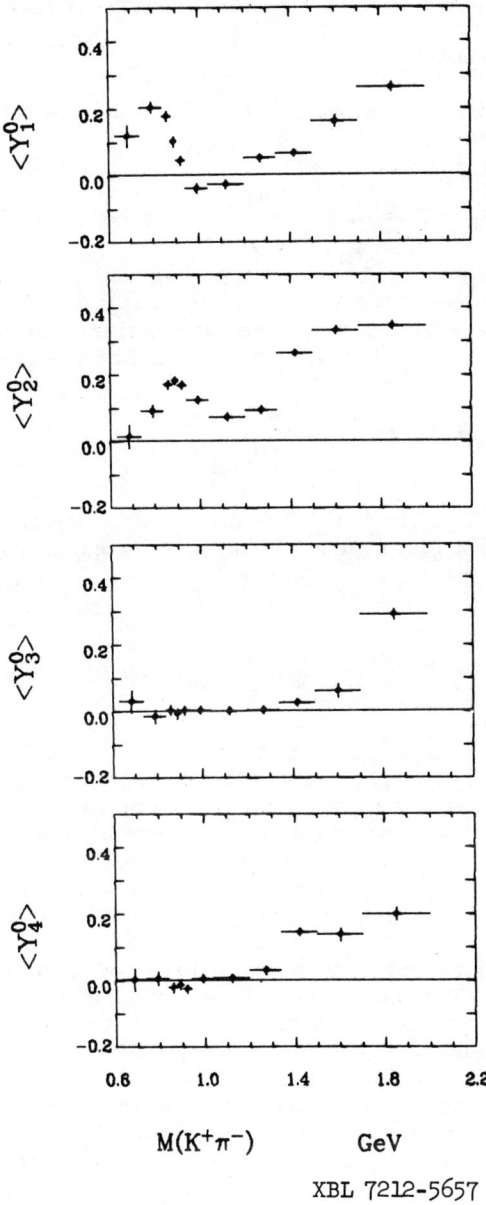

Fig. 9. Moments of the $K^+\pi^-$ angular distribution versus $K\pi$ mass for the reaction $K^+p \rightarrow \Delta^{++}K^+\pi^-$ for events with $|t'| < 0.1$ GeV^2.

III. EXTRAPOLATION TO THE PION POLE

A. Moments Extrapolation

If the differential cross section is expanded in terms of the spherical harmonics as

$$\frac{d\sigma}{d\Omega} = \frac{\sigma}{\sqrt{4\pi}\, a_0} \sum_{\ell=0}^{\ell_{max}} a_\ell Y_\ell^0(\cos\theta),$$

then the expansion coefficients a_ℓ are proportional to the "moments", i.e., the expectation values of the spherical harmonics

$$\langle Y_\ell^0 \rangle = \frac{a_\ell}{\sqrt{4\pi}\, a_0}.$$

We can calculate $\langle Y_\ell^0 \rangle$ for the N events in a given interval of $\pi^+ p$ and $K^+\pi^-$ mass and a given interval of $|t|$ by estimating the expectation value:

$$\langle Y_\ell^0 \rangle = \frac{1}{N} \sum_{i=1}^{N} Y_\ell^0(\cos\theta_i).$$

For a chosen interval in $\pi^+ p$ and $K^+\pi^-$ mass we calculate the values of $\langle Y_\ell^0 \rangle$ in different $|t|$ intervals. We then fit a linear t dependence $a + bt$ for the moments in the physical region and use the parameters a and b to evaluate the moment at $t = \mu^2 = +0.018$, that is, at the pion-pole. The value so obtained will be referred to as the extrapolated $\langle Y_\ell^0 \rangle$ moment. A quadratic extrapolation has also been done and discarded because it was not required by the data.

As a check of the extrapolation procedure we first extrapolate the $\pi^+ p$ moments for the reaction $K^+ p \to \pi^+ p K^*$, of which we have 10 278 events. The extrapolated moments are shown in Fig. 10 with their errors; also shown are the experimental moments for events with $|t'| < 0.1$ GeV.[2] We notice that both fit quite well to the curve calculated from on-shell $\pi^+ p$ scattering[10] in the mass region $M(\pi^+ p) < 1.4$ GeV. Above 1.4 GeV the calculated moments depart considerably from the extrapolated moments. As discussed previously by Schlein,[11] this effect is probably due to background arising from the large $K\pi\pi$ production at $K^*\pi$ and $K\rho$ thresholds, the Q. Figure 11 shows the scatter plot of $K^*\pi$ versus $\pi^+ p$ mass squared in different t intervals.[8] We notice that the Q is produced at every $\pi^+ p$ mass; however, it appears to be less important at small πp masses, especially in the Δ^{++} region. Figure 11c shows the scatter plot for events with $|t'| < 0.05$ GeV^2; we notice that the Q is still present and that outside the Δ^{++} region is dominant. This is again consistent with the hypothesis that the Q is the cause of the discrepancy between calculated and extrapolated moments for

$M(\pi^+p) > 1.4$ GeV. The validity of this hypothesis has been checked in a reaction where there are no strong diffraction phenomena like $pp \to p\pi^+n$ and agreement between the high-mass π^+p moments and the on-shell moments was found to be very good.[12]

For the $K\pi$ moments an analogous situation could arise due to a $\Delta^{++}\pi^-$ threshold enhancement. The Dalitz plot for the $K^+p \to \Delta^{++}K^+\pi^-$ events is shown in Fig. 12 for all events and for different $|t'|$ intervals. Although the $\Delta^{++}\pi^-$ enhancement is less dominant than the Q, we observe the same effect: the $\Delta^{++}\pi^-$ enhancement is less prominent for small $K\pi$ masses and small $|t'|$, and we expect it to produce small distortion of the extrapolated $K\pi$ moments for $K\pi$

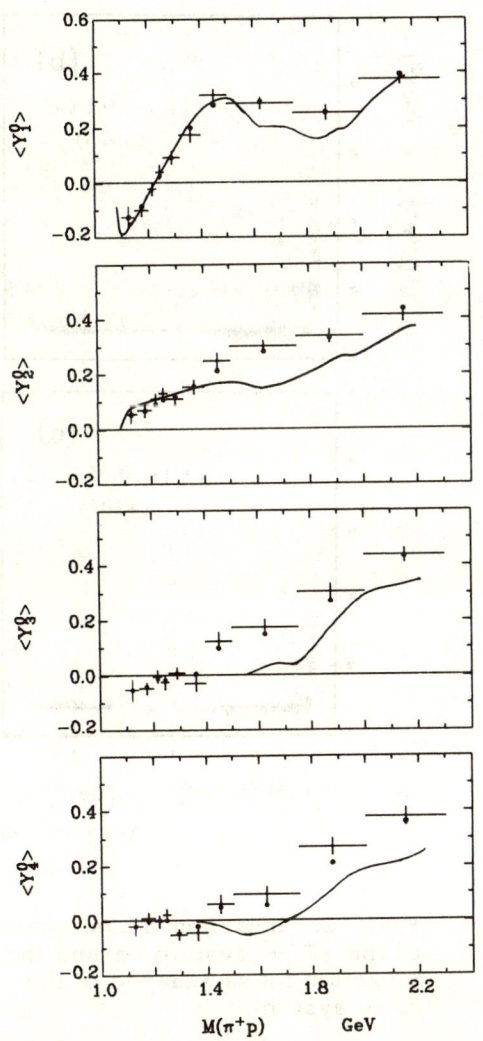

Fig. 10. Extrapolated moments of the π^+p angular distribution versus π^+p mass for events of the reaction $K^+p \to \pi p K^*(890)$. The dots represent the unextrapolated values; the points at the lowest π^+p mass are the unextrapolated values because the statistics were not enough for an extrapolation.

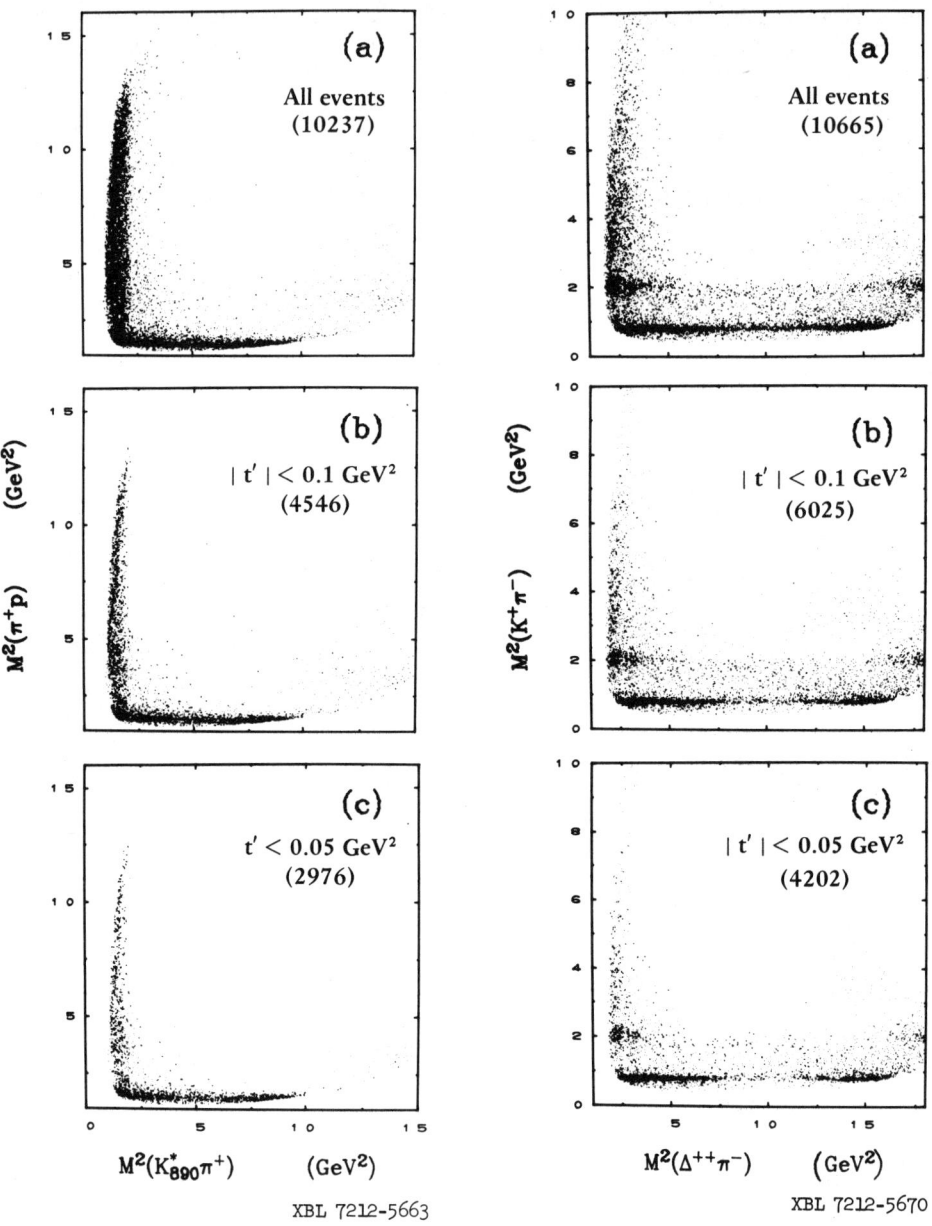

Fig. 11. Invariant mass squared of the $K^*(890) + \pi$ system versus $\pi^+ p$ invariant mass squared.

Fig. 12. Invariant mass squared of the $\Delta^{++}\pi^-$ system versus the invariant mass squared of the $K^+\pi^-$ system.

masses below 1.4 GeV. In this paper we will be considering only
$K\pi$ masses below 1.0 GeV, where only Y_1^0 and Y_2^0 are different from
zero, as noted in Section II. The small $|t|$ and extrapolated $\langle Y_1^0 \rangle$
and $\langle Y_2^0 \rangle$ moments are shown in Figs. 13 and 14 respectively.
More details on the extrapolation can be found in Ref. 9. Note that
in Figs. 13 and 14 we have used overlapping $K\pi$ bins, 20 MeV wide,
whose centers are separated by 10 MeV; therefore, only one-half
of the points are statistically independent.

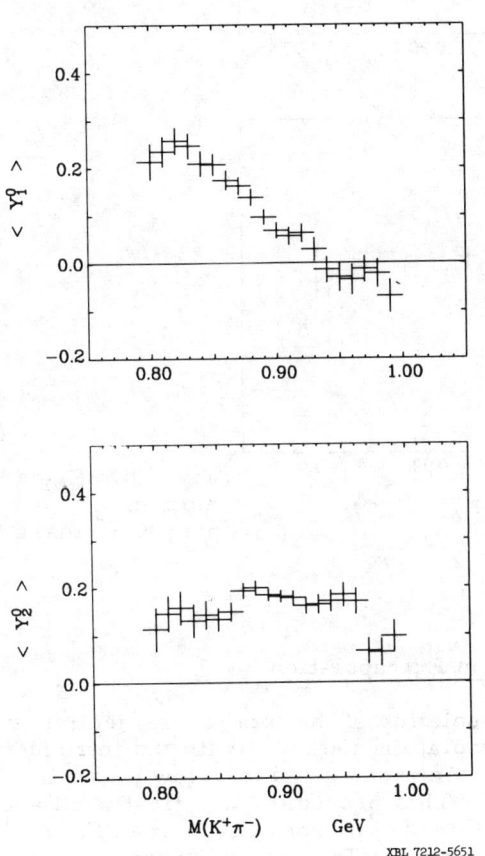

Fig. 13. Moments of the $K^+\pi^-$ angular distribution for events with $|t| < 0.1$ GeV^2. Values for overlapping mass intervals are shown.

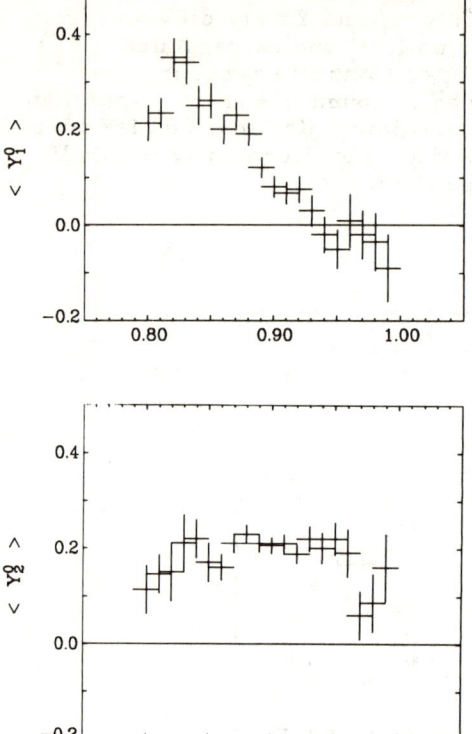

Fig. 14. Extrapolated $K^+\pi^-$ moments. Values for overlapping $K^+\pi^-$ mass bins are shown.

B. Cross-Section Extrapolation

For the extrapolation of the total cross section we have used the Chew-Low extrapolation method[13] with the introduction of Dürr and Pilkuhn[14] (DP) form factors and the slowly varying factor G(t) introduced by Wolf.[15] This procedure was first used successfully by Ma et al.,[16] who obtained π^+p cross sections in the $\Delta^{++}n$ data, in agreement with the measured π^+p cross sections.

For one-pion exchange the differential cross section, modified by Dürr-Pilkuhn and Wolf form factors, is

$$\frac{d^3\sigma}{dm\,dM\,dt} = \frac{1}{4\pi^3 m_p^2 P_L^2} \frac{m^2 q(m)\sigma(m)\, M^2 Q(M)\sigma(M)}{(t-\mu^2)^2} F(m,M,t),$$

(2)

where $F(m, M, t)$ is a form factor which is 1 at the pion pole and has the form

$$F(m, M, t) = (DP)_{\pi^+ p \text{ vertex}} \times (DP)_{K\pi \text{ vertex}} \times G^2(t) \tag{3}$$

with $(DP)_{K\pi} = 1$ for s wave,

$$(DP)_{K\pi} = \left[\frac{q_t(m,t)}{q(m)}\right]^2 \frac{1 + R_{K^*}^2 q^2(m)}{1 + R_{K^*}^2 q_t^2(m,t)} \quad \text{for p wave,} \tag{4}$$

$$(DP)_{\pi^+ p} = \frac{(M+m_p)^2 - t}{(M+m_p)^2 - \mu^2} \left[\frac{Q_t(M,t)}{Q(M)}\right]^2 \frac{1 + R_\Delta^2 Q^2(M)}{1 + R_\Delta^2 Q_t^2(M,t)},$$

$$G(t) = \frac{c - \mu^2}{c - t}.$$

Here the symbol DP is used for the Dürr-Pilkhun form factors and $G(t)$ for the slowly varying additional factor introduced by Wolf.[15] The remaining symbols used in Eqs. (2)-(4) are as follows:

m_p = proton mass $\quad\quad m = M(K^+ \pi^-)$

P_L = lab beam momentum $\quad\quad M = M(\pi^+ p)$

$\sigma(m) = K^+ \pi^-$ cross section $\quad\quad \mu$ = pion mass

$\sigma(M) = \pi^+ p$ cross section

$q(m)$ is the outgoing K^+ momentum in the $K\pi$ c.m.

$Q(M)$ is the outgoing proton momentum in the $\pi^+ p$ c.m.

$q_t(m,t)$ is the virtual π momentum in the $K\pi$ c.m.

$Q_t(m,t)$ is the virtual π momentum in the $\pi^+ p$ c.m.

The values of the numerical constants are taken to be:

$$R_\Delta = 3.97 \pm 0.11 \text{ GeV}^{-1},$$

$$R_{K^*} = 1.25 \pm 0.20 \text{ GeV}^{-1},$$

$$c = 2.29 \pm 0.27 \text{ GeV}^2.$$

R_Δ and c were obtained by Wolf,[15] who fitted many reactions over a large energy range. The value R_{K^*} has been obtained by Trippe et al.[1] by fitting data of the K^* reactions $K^+ p \rightarrow \Delta^{++} K^*$ and $K^- p \rightarrow K^{*0} n$ at various momenta between 3 and 14 GeV/c.[17]

For each $K\pi$ mass interval and t interval we define a quantity

$$"\sigma_{m,t}" = \frac{(d\sigma/dt)_{\text{experimental}}}{(d\sigma/dt)_{\text{DP-OPE}}}, \tag{5}$$

where $(d\sigma/dt)_{\text{DP-OPE}}$ stands for the integration of the right-hand side of Eq. (1) over the Δ^{++} mass region, over the $K\pi$ mass interval and t interval:

$$(d\sigma/dt)_{\text{DP-OPE}} = \frac{1}{4\pi^3 m_p^2 P_L^2} \int dM \int dm \int dt \, \frac{m^2 q(m) M^2 Q(M) \sigma(M)}{(t-\mu^2)^2}$$
$$\times f(m, M, t), \tag{6}$$

with $\sigma(M)$ taken to be the on-shell $\pi^+ p$ cross section and $\sigma(m)$ set equal to one. For each $K\pi$ mass interval we calculate $"\sigma_{m,t}"$ for several t intervals, fit a straight line through these points, and calculate a value of the cross section at $t = \mu^2$. This value, σ_T, should be the on-shell $K\pi$ cross section averaged over the mass interval under consideration, assuming that there are no rapid variations within the interval. The extrapolated cross sections, for $K\pi$ masses below 1 GeV, are shown in Fig. 15. Also shown is the p-wave unitarity limit.

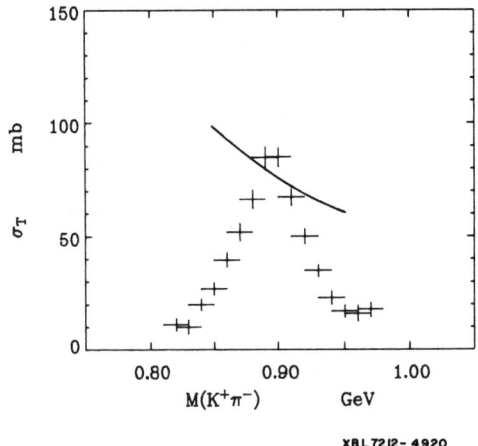

Fig. 15. Extrapolated $K^+\pi^-$ total cross section versus $K\pi$ mass. Values for overlapping $K\pi$ mass bins are shown. The curve is the p-wave unitarity limit.

We have also derived the p-wave cross section from the data. Since there is no evidence for d wave at energies below 1 GeV we can write the total cross section and moments in terms of only s and p waves as follows:

$$\sigma_T = 4\pi \lambda^2 (|s|^2 + 3|p|^2) = \sigma_s + \sigma_p,$$

$$\langle Y_1^0 \rangle = \sqrt{\frac{3}{\pi}} \frac{\text{Re}(sp^*)}{|s|^2 + 3|p|^2} = \sqrt{\frac{3}{\pi}} \frac{|s||p|\cos\phi_{sp}}{|s|^2 + 3|p|^2}, \quad (7)$$

$$\langle Y_2^0 \rangle = \frac{3}{\sqrt{5\pi}} \frac{|p|^2}{|s|^2 + 3|p|^2}.$$

The p-wave cross section is then

$$\sigma_p = \sqrt{5\pi} \ \langle Y_2^0 \rangle \ \sigma_T. \quad (8)$$

We have extrapolated to the pion pole the quantity $\langle Y_2^0 \rangle \sigma_T$, with the method described earlier for σ_T, and obtained the σ_p shown in Fig. 16. The curve shown is a fit with a Breit-Wigner resonance of the type

$$\sigma_p = \frac{16\pi}{3} \lambda^2 \sin^2\delta_1^1, \quad (9)$$

with $\cot \delta_1^1 = (m_R - m)/(\Gamma/2)$

and

$$\Gamma = \Gamma_R \frac{2 m_R}{m_R + m} \frac{q^3(m)}{q_R^3(m_R)} \frac{1 + R^2 q_R^2(m_R)}{1 + R^2 q^2(m)}.$$

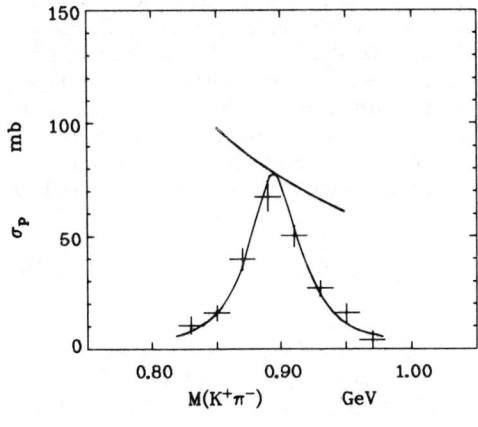

Fig. 16. Extrapolated p-wave cross section versus $K\pi$ mass. The curve is a Breit-Wigner fitted to the data. $M = 896 \pm 2$ MeV and $\Gamma = 47 \pm 3$ MeV. $\chi^2 = 5.5$ for 6 degrees of freedom.

The values obtained for the parameters are

$$m_R = 896 \pm 2 \text{ MeV},$$

$$\Gamma_R = 47 \pm 3 \text{ MeV},$$

$$R = 2 \text{ fermi},$$

in good agreement with the world average[18] values $M = 896.7 \pm 0.7$ MeV and $\Gamma = 51.7 \pm 1.0$ MeV for the neutral K^*.

IV. PHASE SHIFT ANALYSIS

Below 1 GeV the $K\pi$ phase shift analysis requires only s and p waves as already mentioned in Section II. The expression for σ_T, $\langle Y_1^0 \rangle$ and $\langle Y_2^0 \rangle$ in terms of s and p have been given in Eqs. (7). Since we are studying the $K^+\pi^-$ channel, the isospin decomposition gives

$$s = \frac{2}{3} s_{1/2} + \frac{1}{3} s_{3/2},$$
$$p = \frac{2}{3} p_{1/2} + \frac{1}{3} p_{3/2}. \tag{10}$$

In terms of the phase shift δ_ℓ^{2I} each partial wave is written as

$$T_\ell^{2I} = e^{i\delta_\ell^{2I}} \sin \delta_\ell^{2I} \tag{11}$$

Using only data of one charge state we do not have information on both isospin components. The $I = 3/2$ partial waves are best studied in $K^\pm \pi^\pm$ charge states which are pure isotopic spin states. Various authors have measured cross sections for the $I = 3/2$ $K\pi$ system;[19] others have attempted phase shift analyses.[20] We refer to the review by Trippe[19] for a detailed discussion and use here the results of the analyses: the $p_{3/2}$ was found to be very small or consistent with zero for $M(K\pi) < 1$ GeV; the $s_{3/2}$ wave was found to be consistent with a constant cross section, therefore with a phase shift of the form

$$\sigma_0^3 = \frac{4\pi}{q^2} \sin^2 \delta_0^3 = 1.8 \text{ mb}. \tag{12}$$

This is the form used by Bingham et al.[3] and is the form we use.

Using the extrapolated $\langle Y_1^0 \rangle$, $\langle Y_2^0 \rangle$ and σ_T values and Eqs. (7) and (10) we can perform a partial wave analysis. For the four amplitudes involved we assume that:

a) $s_{3/2}$ is given by expression (12) with a negative sign for the phase shift, relative to the $s_{1/2}$ wave, as determined in Refs. 2 and 3;[21]

b) $p_{3/2} = 0$, as discussed above;

c) $p_{1/2}$ is parametrized as the Breit-Wigner form given by Eq. (9), as discussed in Section III-B;

d) $s_{1/2}$ is the only unknown.

A study of other production channels in our experiment[9] shows that below 1 GeV the elasticity can be assumed to be one. This has also been discussed by other authors.[19] For a pure elastic amplitude the phase shifts of Eq. (11) are real, therefore we have only one parameter to determine, δ_0^1. We have made both an energy-independent and energy-dependent phase shift analysis, which will be discussed next.

A. Energy-Independent Partial Wave Analysis

To perform the partial wave analysis of the $K\pi$ system we use as input the extrapolated quantities $\langle Y_1^0 \rangle$, $\langle Y_2^0 \rangle$ and σ_T, shown in Figs. 14 and 15. Therefore at each $K\pi$ mass we have three physical quantities as input and only one unknown, δ_0^1. For a given value of δ_0^1 we calculate a chi-square (χ^2) and search for a minimum of this quantity as a function of δ_0^1. At most $K\pi$ masses only one minimum is found, at others two minima are found. Table II shows the values of δ_0^1 corresponding to these χ^2 minima. More details can be found in Ref. 9.

The results of the analysis are shown in Fig. 17 for all $K\pi$ mass intervals, including overlapping consecutive bins. Here we show the values of δ_0^1 corresponding to all the χ^2 minima, including the points with large χ^2. The values of δ_0^1 for $M(K\pi) \geq 910$ MeV are plotted for the "down" solution as well as for a "down + 180°" solution. Figure 17 shows a continuous slowly increasing "down" solution at every one of the 20 points where the analysis has been done, whereas the "up" solution is present only at two overlapping points at 890 and 900 MeV. We will discuss the two solutions separately.

1. The "down" solution has a smooth behavior in agreement with the solutions found by other authors[1-5] (see Fig. 1). This solution can be parametrized by an effective range formula

$$k \cot \delta_0^1 = \frac{1}{a_0^1} + \frac{1}{2} r_0^1 k^2, \qquad (13)$$

where k is the K^+ momentum in the $K^+\pi^-$ center of mass, a_0^1 is the scattering length, and r_0^1 is the effective range. We have done a fit to the phase shifts of Table II, using every other entry starting at 810 MeV and found

$$\begin{aligned} a_0^1 &= -0.31 \pm 0.05 \text{ fermi}, \\ r_0^1 &= -1.4 \pm 0.5 \text{ fermi}. \end{aligned} \qquad (14)$$

The fit is reasonably good, the chi-square being 10.6 for eight degrees of freedom. The phase shifts and the fitted curve are shown in Fig. 18. The value of the scattering length is in agreement with the current algebra calculation of Griffith,[22] who found $a = -0.22 \pm 0.02$ fermi.

Table II. $K^+\pi^-$ phase shift δ_0^1 fit to extrapolated $\langle Y_1^0 \rangle$, $\langle Y_2^0 \rangle$ and σ_T.

$K\pi$ Mass (GeV)	"Down" solution δ_0^1 (degrees)	χ^2 ($N_D = 2$)	"Up" solution δ_0^1 (degrees)	χ^2 ($N_D = 2$)
0.790-0.810	30 ± 7	1.0		
0.800-0.820	19^{+10}_{-4}	1.0		
0.810-0.830	24 ± 4	4.6		
0.820-0.840	23 ± 4	5.3		
0.830-0.850	29 ± 5	2.4		
0.840-0.860	37 ± 5	0.1		
0.850-0.870	48 ± 6	3.0		
0.860-0.880	43 ± 6	1.7		
0.870-0.890	38 ± 5	0.6	Shoulder at 100	33.7
0.880-0.900	36 ± 4	2.7	133 ± 5	2.3
0.890-0.910	39 ± 5	3.7	151 ± 5	3.2
0.900-0.920	45 ± 4	0.9	163 ± 5	10.3
0.910-0.930	54 ± 5	0.1	164 ± 5	21.3
0.920-0.940	48 ± 5	3.6	179^{+7}_{-6}	21.8
0.930-0.950	40 ± 6	1.3		
0.940-0.960	31^{+8}_{-16}	2.4		
0.950-0.970	45 ± 8	3.5	187^{+8}_{-6}	6.6
0.960-0.980	56 ± 7	0.4		
0.970-0.990	58 ± 9	0.1		
0.980-1.000	56 ± 9	2.5		

We searched the complex energy plane for poles of the scattering matrix and found a pole in sheet II, defined by the convention of Frazer and Hendry,[23] at M = 1062 MeV and Γ = 470 MeV. It is reasonable to expect such a pole, since the phase shift of Fig. 18 would cross 90° if it were to increase with the same energy dependence. However, since we have not used data above 990 MeV, this result is not conclusive.

2. The "up" solution is obtained only at two overlapping points, at 890 and 900 MeV. In this mass region the phase for the p-wave resonance, $K^*(890)$, goes through 90°, and Eqs. (7) show that we expect a phase ambiguity intrinsic to the analysis. In fact both $\vec{s} \cdot \vec{p}$ and $|\vec{s}|$ would be the same for a phase $\delta_0^1 = 90 - \phi_{sp}$ and $(\delta_0^1)' = 90 + \phi_{sp}$. If $s_{3/2}$ were zero this would result in a twofold

ambiguity for $s_{1/2}$. In practice we expect an ambiguity for $s_{1/2}$ because $s_{3/2}$ is small, and the statistical accuracy of our data is limited.

Our results show that the "up" solution is reduced to only two overlapping points where an ambiguity intrinsic to the analysis is expected. In addition the distributions of $\langle Y_1^0 \rangle$, $\langle Y_2^0 \rangle$ and σ_T as a function of $K\pi$ mass do not show any sharp variations, which in general are associated with a narrow resonance. Therefore, there is no evidence in our data for an "up" resonant solution. However, one can still draw a continuous "up" solution by connecting the two points at 890 and 900 MeV with the "down" solution below 890 MeV and the "down + 180°" solution above 900 MeV. This would correspond to a very narrow s-wave resonance at this mass. The resolution of this experiment at the $K^*(890)$ mass is $\Gamma/2 = 5$ MeV;[24] however, we have chosen to analyze the data in 20-MeV intervals in order to have sufficient statistical accuracy for the extrapolation. In order to investigate for what width an s-wave resonance is incompatible with our data, we perform next an energy-dependent partial wave analysis.

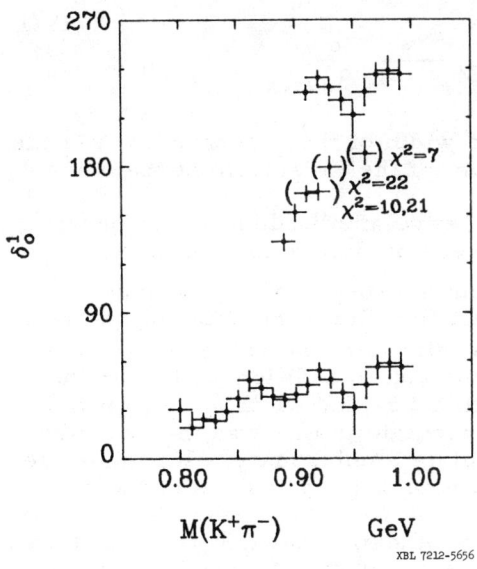

Fig. 17. $I = 1/2$, s-wave phase shift, δ_0^1, from an energy independent fit to the extrapolated $\langle Y_1^0 \rangle$, $\langle Y_2^0 \rangle$, and σ_T.

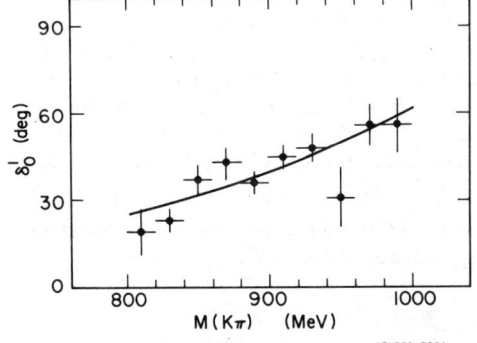

Fig. 18. Effective range fit to the phase shifts of the "down" solution.

B. Energy-Dependent Partial Wave Analysis

We parametrize the $s_{1/2}$ amplitude as

$$s_{1/2} = \frac{1}{\cot\delta_0^1 - i}. \qquad (15)$$

Since the amplitude is elastic, a simple way to combine a background and resonant amplitude preserving unitarity is to add the two phase shifts as follows:[25]

$$\delta_0^1 = \delta_B + \delta_R, \qquad (16)$$

where δ_B is given by Eq. (13), which fits the down solution very well, and δ_R is the phase of an s-wave resonance of the form

$$\cot\delta_R = \frac{M_s - m}{\Gamma/2},$$

$$\Gamma = \Gamma_s \frac{2 M_s}{M_s + m} \frac{q}{q_s}. \qquad (17)$$

Here M_s and Γ_s are the mass and width of the resonance, m is the Kπ mass, and q_s is the momentum of the Kπ system at the mass M_s.

If we include a resonance the s-wave amplitude has four parameters: a_0^1, r_0^1, M_s, and Γ_s. We have 30 data points as input, $\langle Y_1^0 \rangle$, $\langle Y_2^0 \rangle$ and σ_T at 10 different non-overlapping Kπ mass values, which we use for an overall fit. Since the data points are average values over 20-MeV mass intervals, we calculate an average of the function over 20-MeV bins and in addition we fold in the mass resolution as a Gaussian with a ±5-MeV width at half maximum.[24] For each data point we calculate in this way the expected value of the function and then calculate a chi-square. We minimize the sum of the χ^2 over the 30 data points to find values of the parameters.

We find that the non-resonant hypothesis, that is $\delta_R = 0$ in Eq. (16), fits as well as the resonant hypothesis. However, the width of the resonance for the best resonant fit is $\Gamma_s < 1$ MeV, which we cannot detect since we have 20-MeV bins and ±5 MeV resolution. At two standard deviations from the best resonant fit the width is $\Gamma_s = 7$ MeV. The data used in the fit are shown in Fig. 19, where the solid curve represents the scattering length fit, $\delta_R = 0$ in Eq. (16), and the dashed curve represents the fit for $\Gamma_s = 7$ MeV. A resonance with this width could produce a detectable effect especially in the Y_1^0 and σ_T distributions. The non-resonant fit gives $a_0^1 = -0.33$, $r_0^1 = -1.1$, $\chi^2 = 36.0$ for 26 degrees of freedom, with parameters in agreement with the ones obtained in the energy-independent fit [Eq. (14)].

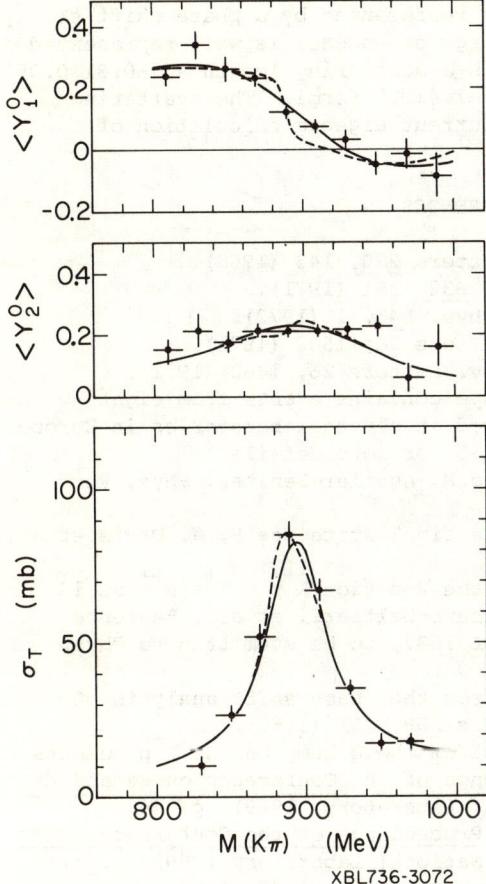

Fig. 19. Extrapolated $\langle Y_1^0 \rangle$, $\langle Y_2^0 \rangle$ and σ_T. The curves are the results of energy-dependent fits. The solid curve represents the best fit for the non-resonant hypothesis; the dashed curve is the fit for an s-wave resonance with $\Gamma = 7$ MeV added to an effective range background.

V. CONCLUSIONS

We find no evidence in our data for an s-wave resonance near the $K^*(890)$. In both the energy-independent and energy-dependent partial wave analysis we find that the "down" solution fits our data adequately. However, since we have limited statistics and a mass resolution of ±5 MeV we cannot exclude an $s_{1/2}$ resonance with $\Gamma < 7$ MeV.

The analysis of Bingham et al.,[3] who used the WDST compilation data,[6] found two solutions that fitted the data equally well: a "down" solution similar to ours and an "up" solution corresponding to a resonance added to background with $\Gamma \lesssim 30$ MeV. In our experiment we have a better mass resolution, and in addition we have included the total cross-section measurements in the fit, thus adding constraints in the fit. We found no "up" solution, but could not exclude one corresponding to a resonance with $\Gamma_s < 7$ MeV. The other analyses listed in Table I with two solutions had fewer statistics than our analysis. Chung et al., who used a different method of analysis, agree with our conclusions.[7]

In conclusion, we find that the s-wave $K\pi$ scattering in the 0.8- to 1.0 GeV mass region is adequately represented by a phase shift slowly varying from 20° to 70°. Its energy dependence is well represented by an effective range formula with a scattering length $a_0 = -0.31 \pm 0.05$ fermi and an effective range $r_0 = -1.4 \pm 0.5$ fermi. The scattering length is in agreement with the current algebra calculation of Griffith:[22] $a = -0.22 \pm 0.02$ fermi.

REFERENCES

1. T. G. Trippe et al., Phys. Letters **28B**, 143 (1968).
2. R. Mercer et al., Nucl. Phys. **B32**, 381 (1971).
3. H. H. Bingham et al., Nucl. Phys. **B41**, 1 (1972).
4. H. Yuta et al., Phys. Rev. Letters **26**, 1502 (1971).
5. A. Firestone et al., Phys. Rev. Letters **26**, 1460 (1971).
6. The K^+p World Data Summary Tape contains events from eight different experiments conducted at eleven laboratories in Europe and the USA. See Refs. 4 and 5 for more details.
7. S. U. Chung, R. L. Eisner, and M. Aguilar-Benitez, Phys. Rev. Letters **29**, 1570 (1972).
8. For further discussion of this final state see P. J. Davis et al., Phys. Rev. D **5**, 2688 (1972).
9. "A Study of $K\pi$ Scattering in the Reaction $K^+p \to K^+\pi^-\Delta^{++}$ at 12 GeV/c"; by M. Matison, A. Barbaro-Galtieri, et al., Lawrence Berkeley Laboratory Report LBL-1537, to be submitted to Phys. Rev. (1973).
10. The moments were calculated from the phase shift analysis of Donnachie et al., Phys. Letters **26B**, 161 (1968).
11. For a discussion of the effect of the Q bump on the π^+p moments see P. E. Schlein in _Proceedings of the Conference on $\pi\pi$ and $K\pi$ Interactions_ (Argonne National Laboratory, 1969), p. 446.
12. E. Colton and P. Schlein, in _Proceedings of the Conference on $\pi\pi$ and $K\pi$ Interactions_ (Argonne National Laboratory, 1969), p. 1.
13. G. F. Chew and F. E. Low, Phys. Rev. **113**, 1640 (1959).
14. H. Dürr and H. Pilkuhn, Nuovo Cimento **40A**, 899 (1965).
15. G. Wolf, Phys. Rev. Letters **19**, 925 (1967).
16. Z. Ming Ma et al., Phys. Rev. Letters **23**, 342 (1969).
17. P. E. Schlein, in _Meson Spectroscopy_, edited by C. Baltay and A. H. Rosenfeld (Benjamin, 1968), p. 161.
18. Review of Particle Properties, Particle Data Group, Rev. Mod. Phys. **45**, S1 (1973).
19. For a recent review see T. G. Trippe, Recent Experimental Studies of the $K\pi$ Interactions, in _Zero Gradient Synchrotron Workshops, Summer 1971_, ANL/HEP 7208, Vol. I, page 6 (1971).
20. A. M. Bakker et al., Nucl. Phys. **B24**, 211 (1970), and _Meson Resonances and Related Electromagnetic Phenomena_, edited by R. H. Dalitz and A. Zichichi (International Physics Series, Bologna, 1971), p. 53; B. Jongejans and K. Voorthius, ibid., page 57.

21. The sign can be determined only in experiments which fit both $K^+\pi^-$ and $K^0\pi^0$ simultaneously.
22. R. W. Griffith, Phys. Rev. $\underline{176}$, 1705 (1968). The values and errors for the coupling constants used to calculate this prediction have been taken from the review of G. Ebel et al., Nucl. Phys. $\underline{B83}$, 317 (1971).
23. W. R. Frazer and A. W. Hendry, Phys. Rev. $\underline{134}$, B1307 (1964).
24. P. J. Davis et al., Phys. Rev. Letters $\underline{23}$, $\underline{1701}$ (1969).
25. See, for example, similar fits made for the Δ^{++}(1236) in A. Barbaro-Galtieri, "Selected Topics in Baryon Resonances," Lectures for the 1971 Erice Summer School, Lawrence Berkeley Laboratory LBL-555 (1972).

K_{e4} DECAYS AND LOW ENERGY π-π PHASE SHIFTS[*]

Eugene W. Beier

Department of Physics, University of Pennsylvania,

Philadelphia, Pennsylvania, 19174

ABSTRACT

The determination of the low energy π-π phase shifts from the decays $K^{\pm} \to \pi^{+}\pi^{-}e^{\pm}\nu$ is discussed. Results of the University of Pennsylvania K_{e4} experiment are presented. These results are compared with higher energy data on the π-π phase shifts and with the results of another K_{e4} experiment.

In this contribution to the conference I intend to demonstrate how the low energy π-π phase shifts are determined in the decays $K^{\pm} \to \pi^{+}\pi^{-}e^{\pm}\nu$. I will then make a few remarks about the University of Pennsylvania K_{e4} experiment [1,2] and compare the results of this experiment with some other knowledge of the low energy π-π phase shifts.

The decays $K^{\pm} \to \pi^{+}\pi^{-}e^{\pm}\nu$ provide a way to determine the low energy pion-pion interaction which is not complicated by strong interactions of the pions with the other particles which appear in the final state[3]. We believe that we understand the way in which the leptons contribute to the weak interaction, i.e., that the interaction is V-A and that the leptons are created at the same point in space-time. It is then possible to parameterize the hadronic contribution in terms of form factors which have phases specified by the pion-pion interaction.

To describe a four body final state when we are not interested in spin correlations requires five independent kinematic variables. For K_{e4} decays these variables are customarily chosen by considering the K to decay to a dipion system and a dilepton system, each of which then decays into two particles. Thus, the five kinematic variables are chosen to be $M_{e\nu} = \sqrt{s_{\ell}}$ and $M_{\pi\pi} = \sqrt{s_{\pi}}$, the invariant masses of the dilepton and dipion; θ_e and θ_π, the polar angles of the $e^{+}(e^{-})$ and $\pi^{+}(\pi^{-})$ in $K^{+}(K^{-})$ decay, defined in the dilepton and dipion center of momentum, respectively; and φ, the azimuthal angle between the dipion and dilepton planes. These coordinates are shown schematically in Figure 1.

---

[*] Work supported in part by the United States Atomic Energy Commission

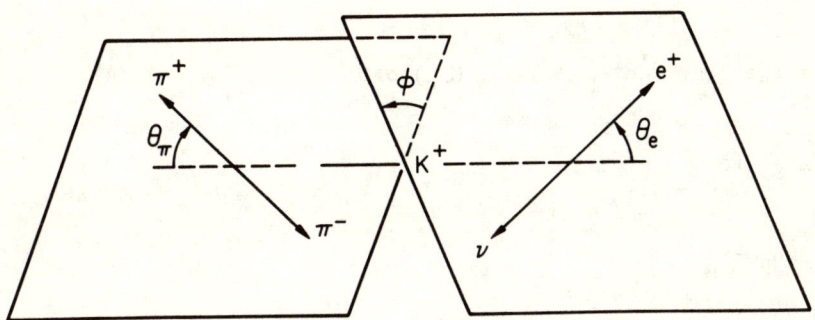

Fig. 1 Coordinate System describing K_{e4} Decay

The phenomenological form of the invariant amplitude is specified by Lorentz invariance, the V-A interaction, lepton locality, energy conservation, and the approximation $m_e = 0$, to be (for K^+ decay, for example)

$$M = \frac{G}{\sqrt{2}} \sin\theta_c <\pi^+\pi^-|A_\alpha + V_\alpha|K^+> \bar{u}(p_\nu)\gamma_\alpha(1+\gamma_5)v(p_e) \quad (1)$$

The hadronic axial vector and vector currents are written

$$A_\alpha = f\frac{(p_+ + p_-)_\alpha}{M_k} + g\frac{(p_+ - p_-)_\alpha}{M_k} \quad (2)$$

$$V_\alpha = h\varepsilon_{\alpha\beta\nu\delta}\frac{p_\beta^k(p_+ + p_-)_\nu(p_+ - p_-)_\delta}{M_k^3} \quad (3)$$

where p_+ and p_- are the π^+ and π^- four momenta and f, g, and h are dimensionless form factors which depend only on $M_{\pi\pi}$, $M_{e\nu}$, and $\cos\theta_\pi$.

A partial wave expansion of the form factors makes the $\cos\theta_\pi$ dependence explicit. We retain terms with $\ell \leq 1$ as required by the data[2], and assume the validity of the semileptonic $\Delta I = \frac{1}{2}$ rule in order to eliminate $I = 2$ states. The assumption of time reversal invariance for the reaction specifies the phases of the form factors through the Fermi-Watson theorem. Thus:

$$f = f_s e^{i\delta_s} + f_p' e^{i\delta_p} b(M_{e\nu}, M_{\pi\pi}) \cos\theta_\pi \quad (4)$$

$$g = g_p e^{i\delta_p}$$

$$h = h_p e^{i\delta_p}$$

where f_s, f_p', g_p and h_p are real and b is a kinematic factor. The form factor f_s induces transitions to dipion states with $\ell = 0$, $I = 0$ and f_p', g_p and h_p induce transitions to dipion states with $\ell = 1$, $I = 1$.

The decay distribution function is obtained from the invariant amplitude in the standard way. The distribution is a function of the five kinematic variables and the five parameters f_s, f_p', g_p, h_p, and $\delta_s - \delta_p$. The phase shift difference $\delta_s - \delta_p$ appears in the s-wave, p-wave interference terms.

There are three methods generally used to obtain the parameters of the distribution function. The first method is to project the five one-dimensional distributions out of the distribution function and then to fit the one-dimensional distributions to the form factors and $\delta_s - \delta_p$ by minimizing the total chi-squared. This method has the drawback that information contained in correlations between variables is lost upon integration over variables.

An elegant method for obtaining $\tan(\delta_s - \delta_p)$ independent of the form factors was suggested by Pais and Treiman[3]. In this method one considers the two dimensional distribution in $\cos\theta_e$ and φ. The ratios of certain coefficients in this two dimensional distribution then give $\tan(\delta_s - \delta_p)$ directly. The problems which arise in applying this technique are twofold. Radiative corrections can distort the form of the $\cos\theta_e$ distribution dictated by lepton locality at the level of several per cent, although such corrections do not significantly affect the other distributions. Furthermore, the correlations in φ turn out to be quite weak, thereby requiring a large number of events to obtain a significant fit.

The third method is to perform a multi-dimensional analysis, thereby including as much of the information in the correlations between variables as possible. Such an analysis is generally performed using the maximum likelihood method.

We concentrate here on the first method of analysis, since this is the method our group at Pennsylvania has completed at this time. The information contained in each of the one dimensional distributions is shown in Table I.

Table I. Information in the one dimensional distributions

One dimensional distribution	Parameters determined
$M_{\pi\pi}$	f_s, f_p', g_p
$M_{e\nu}$	f_s, f_p', g_p
$\cos\theta_\pi$	f_s, f_p', g_p, $\cos(\delta_s - \delta_p)$
$\cos\theta_e$	f_s, g_p
φ	f_s, g_p, h_p, $\sin(\delta_s - \delta_p)$

We note that if the absolute rate for the decay is not determined, it is only possible to obtain the ratios of the p wave form factors to f_s. Also, the strongest constraints on g_p are provided by the two mass distributions and $\cos\theta_\pi$ distribution. If the $\cos\theta_e$ distribution is omitted from the fit, there is essentially no information lost.

It is evident that the φ distribution is important in determining $\delta_s - \delta_p$. The explicit form of the φ distribution is, for K^+ decay:

$$\frac{dN_+}{d\varphi} = \text{Const} \times \left\{1 + \beta \sin\varphi + \gamma \cos\varphi + \delta \cos^2\varphi\right\} \quad (5)$$

The factor $\sin(\delta_s - \delta_p)$ appears only in the coefficient of $\sin\varphi$;

$$\beta = K f_s g_p \sin(\delta_s - \delta_p) \quad (6)$$

where K is a kinematic factor. The $\sin \varphi$ correlation is a parity violation which measures the up-down asymmetry of the electron with respect to the dipion plane.

We know from experimental tests of discrete symmetries[1] that the K_{e4} decay interaction not only violates parity P, but also violates particle-antiparticle conjugation C. Under the combined operation CP the decay interaction is invariant. The parity operation has the effect that $\varphi \to -\varphi$ and other variables remain unchanged. Under C, $K^+ \to K^-$, so that the requirement of CP invariance is that the distribution for K^+ at φ be the same as the distribution for K^- at $-\varphi$. This leads to the explicit form of the φ distribution for K^- decay:

$$\frac{dN_-}{d\varphi} = \text{Const} \times \left\{ 1 - \beta \sin \varphi + \gamma \cos \varphi + \delta \cos^2 \varphi \right\} \tag{7}$$

The distribution is the same as that for K^+ decay except that the sign of the $\sin \varphi$ correlation is reversed.

The reason for discussing weak interaction symmetries at a π-π conference can now be explained. Let the experimental apparatus efficiency function be

$$\epsilon(\varphi) = \text{Const.} \times \left\{ 1 + b \sin \varphi + c \cos \varphi + d \sin 2\varphi + e \cos 2\varphi + \ldots \right\}$$

where b, c, d, e, etc. are small compared to unity. If one has comparable samples of K^+_{e4} and K^-_{e4} events, the value of the corrected $\sin \varphi$ correlation relative to the isotropic term is equal to half the difference of the uncorrected $\sin \varphi$ correlations for K^+_{e4} and K^-_{e4} events, respectively. This statement is valid up to corrections of order b^2. If the corrections make the coefficient of $\sin \varphi$ more positive, for example, the magnitude of the coefficient for one sample will increase while the magnitude of the coefficient for the other sample will decrease. Within the approximation, the difference of the coefficients will be the same in the corrected and uncorrected samples.

As we shall see later, the $\sin \varphi$ correlation is small relative to the isotropic term. If one hopes to extract accurate π-π phase shifts from a high statistics K_{e4} experiment, it is necessary to know the corrections to the φ distribution to better than .01, or to have a measurement of the $\sin \varphi$ correlation which is independent of the corrections. This can be most easily accomplished by studying both K^+ and K^- decays and exploiting the consequences of C violation and CP invariance.

The University of Pennsylvania K_{e4} experiment (collaborators: E. Beier, D. Buchholz, A. K. Mann, S. H. Parker, J. B. Roberts, W. K. McFarlane) performed at the Brookhaven AGS has so far yielded a sample of 8141 K_{e4} events. The sample includes 4800 K^+ events and 3341 K^- events and has less than one per cent contamination. Details of the experimental arrangement and event reconstruction can be found in references 1 and 2.

The five one dimensional distributions extracted from this sample of events are displayed in Figure 2. The dashed lines are the one dimensional projections of the efficiency function ε.

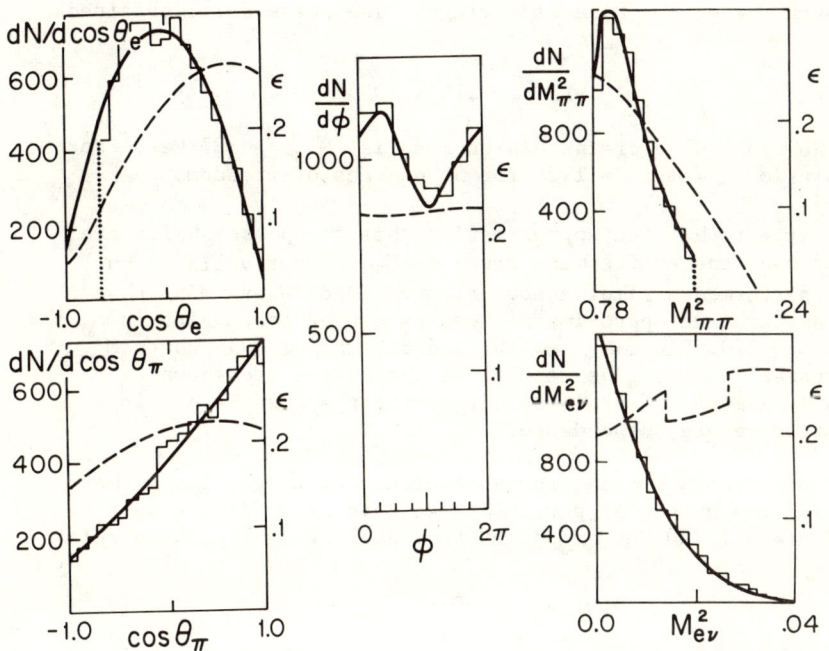

Fig. 2 The observed one dimensional distributions in the five independent variables. The histograms are the data corrected for experimental detection efficiency. Dashed lines are the Monte Carlo calculated detection efficiency (ε). Solid lines are calculated from a simultaneous fit to the one dimensional distributions (excluding the $\cos \theta_e$ distribution) assuming all parameters are constant.

We have eliminated events with $\cos \theta_e < -.6$ and $M_{\pi\pi} \geq 395$ MeV where the detection efficiency is low. Note that the efficiency function is almost uniform in φ. The corrections to the $\sin \varphi$ coefficient are typically $< .01$ and are small compared to the statistical errors. The φ distribution has significant correlations at the level of five to ten per cent.

The smooth curves through the data are calculated from a simultaneous fit to the distributions in $M_{\pi\pi}$, $M_{e\nu}$, $\cos\theta_\pi$, and φ. The $\cos\theta_e$ distribution was omitted from the fit because radiative corrections distort this distribution at the level of several per cent.

In the fit in Figure 2, all parameters are assumed to be constant over the entire kinematic range. The phase shift obtained is

$$<\delta_s - \delta_p> = .19 \pm .05$$

where the mean value of the dipion mass is $<M_{\pi\pi}> = 327$ MeV. For this solution $\chi^2/DF = 1.24$ for 54 degrees of freedom.

We expect that the approximation that the phase shifts are constant over the entire mass range of $M_{\pi\pi}$ is not valid. Thus, we have performed a simultaneous fit to the data assuming that the parameters are approximately constant over three bins in $M_{\pi\pi}$: $280 < M_{\pi\pi} < 310$, $310 < M_{\pi\pi} < 340$, and $340 < M_{\pi\pi} < 395$ where $M_{\pi\pi}$ is expressed in MeV. The results of these fits are shown as circles in Figure 3. It is seen that the phase shift displays an important energy dependence.

In order to build an energy dependence into the fit without increasing the number of parameters we have also fit the data in the interval 280 MeV to 353 MeV to a scattering length formula:

$$\beta \cot \delta_s = \frac{1}{a_s} \qquad (8)$$

where

$$\beta = \left\{ \frac{M_{\pi\pi}^2 - 4m_\pi^2}{M_{\pi\pi}^2} \right\}^{\frac{1}{2}} \qquad (9)$$

Over this interval we can neglect δ_p and the contribution to the $\pi-\pi$ scattering amplitude from terms in $q^2 = \frac{1}{4}(M_{\pi\pi}^2 - 4m_\pi^2)$. The fit yields $a_s = (.17 \pm .13)m_\pi^{-1}$ with $\chi^2/DF = 1.15$ for 50 degrees of freedom.

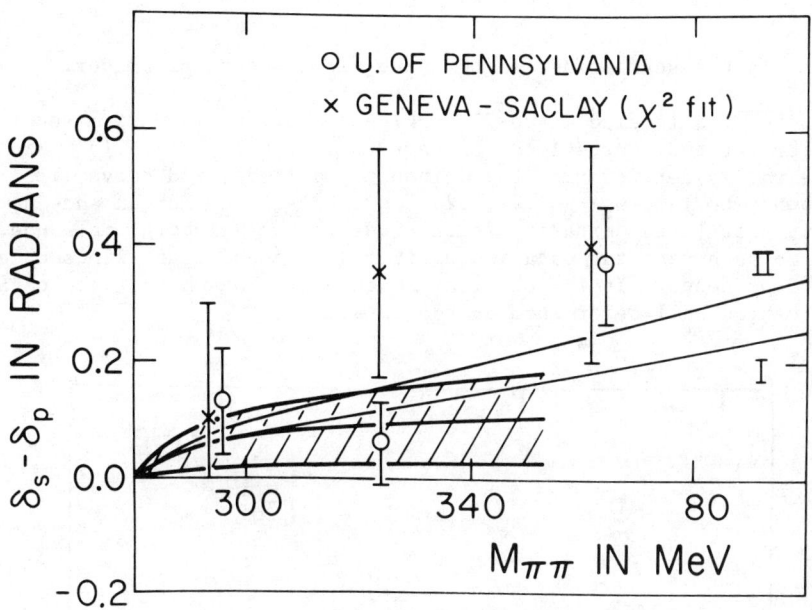

Fig. 3 Dependence of $\delta_s - \delta_p$ on $M_{\pi\pi}$. The circles represent a fit to the data assuming the parameters are constant over these intervals in $M_{\pi\pi}$. The cross hatched area is a fit (including the error) of $\beta \cot \delta = a_s^{-1}$ to the data in the interval $280 < M_{\pi\pi} < 353$ MeV. The fit yields $a_s = (0.17 \pm 0.13) m_\pi$. The solid lines labelled I and II are theoretical predictions obtained from Refs. 4 and 5. The data of the Geneva-Saclay experiment (Ref. 7) are represented by ×.

The presence of a zero below threshold in the I = 0 s-wave amplitude as suggested, for example, by the Weinberg derivation[4] of the scattering length a_s, leads to an effective range which is negative and large. The effective range formula thus does not lead to a good approximation to the π-π scattering phase shifts in the interval 280 MeV $< M_{\pi\pi} <$ 400 MeV. A more appropriate parameterization follows from assuming, as did Weinberg, that the amplitude is a linear function of q^2. Just above threshold the amplitude is, to a good approximation, real, so that the amplitude

$$A_o^o = \frac{\omega}{q} e^{i\delta_s} \sin \delta_s \tag{10}$$

can be approximated by its real part

$$\text{Re}A^o_o = \frac{\omega}{q} \frac{1}{2} \sin 2\delta_s \cong a_s + b_s q^2 \qquad (11)$$

where a_s is the scattering length and b_s is a second parameter.

The curves labeled I and II in Figure 3 are derived from such a parameterization. Curve I corresponds to $(a_s, b_s) = (.16, .18)$ which are the values derived from the Weinberg amplitude, and curve II corresponds to $(a_s, b_s) = (.21, .25)$ as suggested by Morgan and Shaw[5], who have calculated unitarity corrections to the Weinberg amplitude. It can be seen that the data is qualitatively consistent with such an energy dependence. It is not clear at this writing whether two parameters can be well determined by the data.

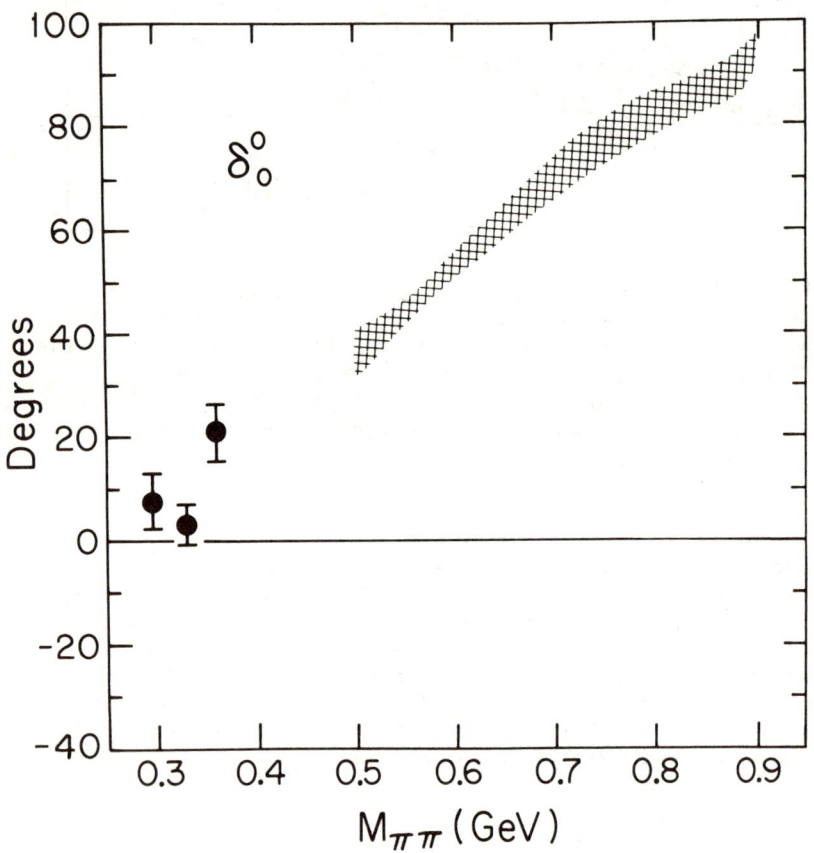

Fig. 4 Comparison of the results of the Pennsylvania K_{e4} experiment with higher energy data of Ref. 6.

The s-wave, $I = 0$ phase shifts obtained from this K_{e4} experiment are displayed in Figure 4 with the higher energy data of Protopopescu, et. al.[6] from the reaction $\pi^+ p \to \pi^+ \pi^- \Delta^{++}$. It is seen that the K_{e4} data connect smoothly with the data at higher invariant dipion mass.

A Geneva-Saclay collaboration[7] has obtained 1609 K_{e4}^+ events in a spark chamber experiment at the CERN PS. The form factors obtained in this experiment, are in good agreement with those from the Pennsylvania experiment, but the average value of $<\delta_s - \delta_p> = .34 \pm .13$ is somewhat higher. The energy dependence of $\delta_s - \delta_p$ for the Geneva-Saclay data is given by the symbol ✗ in Figure 3. The comparison is made for the same method of analysis, that is, the fit to the one dimensional distributions[8]. It is seen that the difference in the average value of $\delta_s - \delta_p$ arises from a different value for the central bin in $M_{\pi\pi}$.

In conclusion, let me emphasize that the $K_{e4}^\pm$ decays provide a <u>direct</u> measurement of the low energy pion-pion interaction. These measurements show that the interaction is weak, and that the energy dependence of the phase shift is significant.

REFERENCES

1. E. W. Beier, D. A. Buchholz, A. K. Mann, W. K. McFarlane, S. H. Parker, and J. B. Roberts, Phys. Rev. Lett. $\underline{29}$, 511 (1972).

2. E. W. Beier, D. A. Buchholz, A. K. Mann, S. H. Parker, and J. B. Roberts, Phys. Rev. Lett. $\underline{30}$, 399 (1973).

3. Phenomenological discussions of K_{e4} decay are given by N. Cabibbo and A. Maksymowicz, Phys. Rev. 137, B348 (1965); T. D. Lee and C. S. Wu, Ann. Rev. Rev. Nucl. Sci, $\underline{16}$, 471 (1966); A. Pais and S. B. Treiman, Phys. Rev. $\underline{168}$, 1858 (1968); and F. A. Berends, A. Donnachie, and G. C. Oades, Phys. Rev. $\underline{171}$, 1457 (1968).

4. S. Weinberg, Phys. Rev. Lett. 17, 616 (1966).

5. D. Morgan and G. Shaw, Nucl. Phys. $\underline{B43}$, 365 (1972).

6. S. D. Protopopescu, et al, *Experimental Meson Spectroscopy* - 1972 A. H. Rosenfeld and K-W. Lai editors (American Institute of Physics, New York, N. Y. 1972) p. 17 and S. D. Protopopescu, et al, preprint, Lawrence Berkeley Laboratory - 970.

7. P. Basile et. al., Phys. Lett. $\underline{36B}$ 619 (1972) and A. Zylberztejn, Phys. Lett. $\underline{38B}$ 457 (1972).

8. The Geneva-Saclay experiment obtains $<\delta_s - \delta_p> = .39 \pm .08$ from a maximum likelihood calculation. See Ref. 7 for the energy dependence found using this analysis.

ππ PHASE SHIFT ANALYSIS*

P. Estabrooks[+] and A.D. Martin["]
Theory Division, CERN, Geneva

G. Grayer[Θ], B. Hyams, C. Jones and P. Weilhammer
Nuclear Physics Division, CERN, Geneva

W. Blum, H. Dietl, W. Koch, E. Lorenz, G. Lütjens,
W. Männer, J. Meissburger and U. Stierlin
Max-Planck-Institut für Physik und Astrophysik,
Munich

ABSTRACT

We perform an energy independent ππ phase shift analysis in the energy range $440 < M_{\pi\pi} < 1400$ MeV using high statistics $\pi^-p \to \pi^-\pi^+n$ data at 17.2 GeV/c. The method is based on an amplitude analysis of the production process and the extrapolation of the dominant π exchange amplitudes to the π exchange pole. We consider the phase shift ambiguities and study ways of selecting the physical solution. We determine the parameters of the ρ and f resonances. We find an $I = 0$ S wave resonance under the f and we comment on the properties of the S^* resonance near the $K\bar{K}$ threshold.

1. INTRODUCTION

It was originally proposed by Goebel[1] and by Chew and Low[2] that the $\pi N \to \pi\pi N$ cross-section suitably extrapolated from the physical region to the π exchange pole $(t = \mu^2)$ would provide a valuable means of determining the ππ differential cross-section. Many attempts to extract ππ phases have been based on Chew-Low extrapolations, until now agreement has been reached on the general picture of the phases in the ρ region[3]. However, with the recent increase in experimental statistics of the $\pi^-p \to \pi^-\pi^+n$ data we are confronted with the problem of finding the best way to account for the other exchange mechanisms which

---

* Invited talk by A.D. Martin at the International Conference on ππ Scattering and Associated Topics, Tallahassee, Fl. March 28-30, 1973.

[+] Supported by the National Research Council of Canada.

["] On leave of absence from the University of Durham, England.

[Θ] Now at the Max-Planck-Institut, Munich.

are seen to occur in addition to π exchange. The method we propose is to use the observed moments of the $\pi^-\pi^+$ angular distribution to perform an amplitude analysis of the production process. In this way we can isolate the dominant π exchange pole. That we are able to perform such an amplitude analysis without knowledge of the nucleon polarization observables is a fortunate circumstance of the nature of the exchanges (see Section 3).

We use this method to extract $\pi\pi$ phase shifts from the high statistics $\pi^-p \to \pi^-\pi^+ n$ data at 17.2 GeV/c[4]. We discuss separately the $\pi\pi$ phase shift analysis below and above the $K\bar{K}$ threshold. The former is described in Sections 4-6 and the latter in Section 7. In Section 8 we comment on the behaviour of the $I=0$ $\pi\pi$ S wave near the $K\bar{K}$ threshold.

2. PION EXCHANGE

Suppose that the reaction $\pi^-p \to (\pi^-\pi^+)n$ were mediated entirely by π exchange. Then the differential cross-section is

$$\frac{d^3\sigma}{dt\, dM_{\pi\pi}\, d\Omega} = \frac{2}{4\pi^2 m^2 p_L^2} \left(\frac{g^2}{4\pi}\right) \frac{-t}{(t-\mu^2)^2} |F(t)|^2 q\, M_{\pi\pi}^2 \frac{d\sigma_{\pi\pi}}{d\Omega} \quad (1)$$

where m is the nucleon mass, p_L is the incident π^- laboratory momentum, t is the momentum transfer at the nucleon vertex, F(t) a form factor satisfying $F(\mu^2) = 1$, and the πNN coupling $g^2/4\pi = 14.4$. q, $M_{\pi\pi}$ and $d\sigma_{\pi\pi}/d\Omega$ are the $\pi^-\pi^+$ momentum, mass and differential cross-section in the $\pi^-\pi^+$ c.m. frame.

The experimental observables are the moments $\langle Y_M^J \rangle$ of the $\pi^-\pi^+$ angular distribution as a function of t and $M_{\pi\pi}$

$$\frac{d^3\sigma}{dt\, dM_{\pi\pi}\, d\Omega} = N \sum_J \sum_{M=-J}^{J} \langle Y_M^J \rangle \operatorname{Re} Y_M^J(\theta,\phi) \quad (2)$$

where N is the number of events in the element $dt\, dM_{\pi\pi}$ and where we have chosen the y axis normal to the $\pi^-p \to (\pi\pi)n$ reaction plane. We use $\langle Y_M^J \rangle$ to abbreviate $\operatorname{Re} \langle Y_M^J \rangle$.

π exchange produces only (t channel) helicity zero $\pi^-\pi^+$ systems, and in this simplified situation only the $M=0$ moments, $\langle Y_0^J \rangle$, would be non-zero. We may express these moments in terms of the $(\pi^-p \to \pi^-\pi^+n)$ amplitudes for the production of S, P,... wave helicity zero $\pi^-\pi^+$ states

$$\sqrt{4\pi} \, N \langle Y_0^0 \rangle = |S_0|^2 + |P_0|^2 + ...$$

$$\sqrt{4\pi} \, N \langle Y_0^1 \rangle = 2\text{Re}(S_0 P_0^*) + ...$$

$$\sqrt{4\pi} \, N \langle Y_0^2 \rangle = \frac{2}{\sqrt{5}} |P_0|^2 + ...$$

(3)

where the omitted terms involve D_0, F_0,... Up to a normalization constant these amplitudes are given by

$$L_0 = \frac{\sqrt{-t}}{t-\mu^2} \, F_L(t) \, \frac{M_{\pi\pi}}{\sqrt{q}} \, \sqrt{2L+1} \, f_L \qquad (4)$$

where f_L are the $\pi^-\pi^+$ partial wave amplitudes at c.m. energy $M_{\pi\pi}$

$$f_L = f_L^{I=1} \qquad \text{for } L \text{ odd}$$

$$f_L = \frac{2}{3} f_L^{I=0} + \frac{1}{3} f_L^{I=2} \qquad \text{for } L \text{ even.}$$

(5)

The f_L are defined so that in the $\pi\pi$ elastic region

$$f_L^I = \sin \delta_L^I \, e^{i\delta_L^I} \qquad (6)$$

Thus $\pi\pi$ phases can be obtained by extrapolating the production amplitudes, or rather $(t-\mu^2)L_0/\sqrt{-t}$, from the physical region $(t<0)$ to $t=\mu^2$.

In Fig. 1 we show the mass spectrum up to $M_{\pi\pi} = 2$ GeV of the unnormalized t channel moments integrated over the interval $0 < -t < 0.15$ GeV^2 obtained in $\pi^-p \to \pi^-\pi^+n$ at 17.2 GeV/c. From these moments we see:

a) the presence of the $\rho(770)$, $f(1260)$, $g(1700)$ mesons with spins 1, 2, 3 respectively; to establish[6,4] spin 3 for the g meson requires the additional knowledge that the $J=7$ and higher moments are small near 1700 MeV;

b) from $\langle Y_0^1 \rangle$ the presence of a large S wave under the ρ meson;

c) from $\langle Y_0^2 \rangle$ the presence of a large S wave under the f meson[7];

d) sharp structure near $M_{\pi\pi} = 1$ and 1.45 GeV which Odorico[8] associates with the double pole killing zeros propagating linearly into the $\pi^-\pi^+$ physical region from the forward

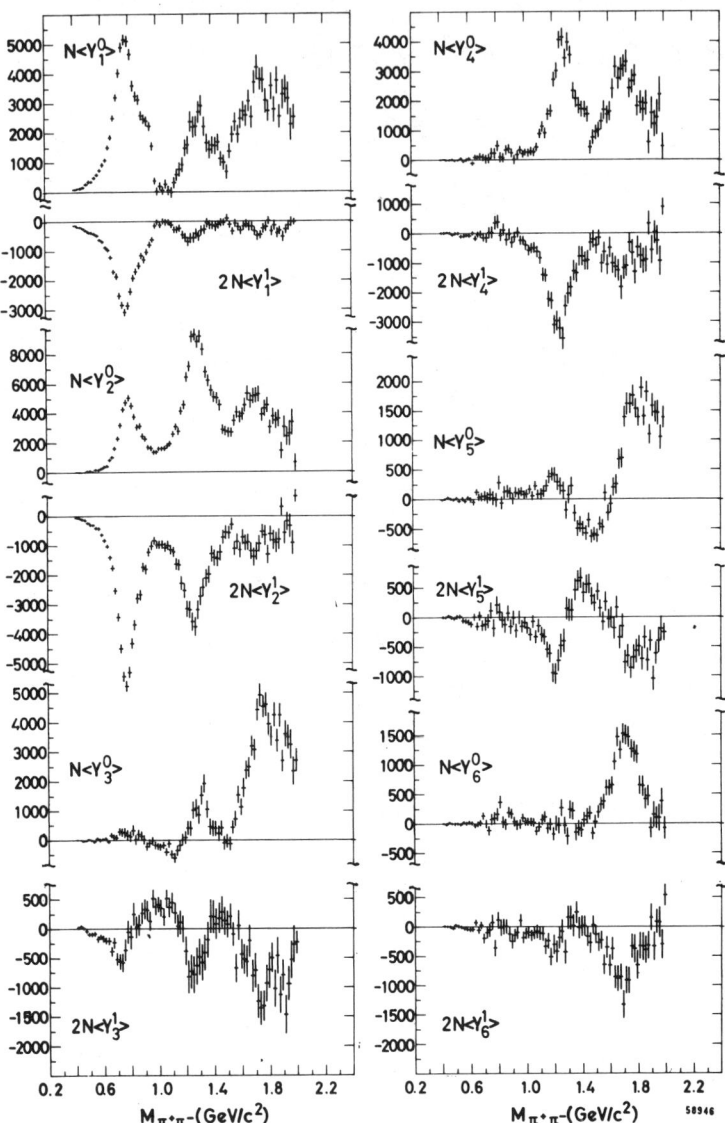

Figure 1   The corrected unnormalized t channel moments, $N < Y_J^M >$, as a function of $\pi^+\pi^-$ mass for the interval $0 < -t < 0.15$ GeV^2, taken from Ref. 4. A factor of 2 should be included in the $M \neq 0$ moments shown in Ref. 5.

direction ; the effect at 1 GeV is complicated by the opening of the $K\bar{K}$ channel with the cross-section at its S wave unitarity limit, suggesting the nearby presence of the S^* meson[9] ;

e) from $< Y_1^J >$ moments the non-negligible presence of helicity one $\pi^-\pi^+$ production.

This last observation indicates that there are other exchange mechanisms in addition to π exchange, such as A_2 exchange or absorptive corrections. Thus a $\pi\pi$ phase shift analysis based on a straightforward extrapolation of the $< Y_0^J >$ moments can be very misleading[10,11]. Additional terms occur on the right-hand sides of Eqs. (3) involving amplitudes describing non-zero helicity $\pi\pi$ production. To allow for these exchanges we perform a production amplitude analysis of all the observed moments as a function of t and $M_{\pi\pi}$.

3. AMPLITUDES AND EXCHANGE MECHANISMS FOR $\pi^-p \to \pi^-\pi^+n$

To describe the reaction $\pi^-p \to \pi^-\pi^+n$ we use the variables shown in Fig. 2. The production of a $\pi^-\pi^+$ system of spin L is described by helicity amplitudes $H^{L,\lambda}(s,t,M_{\pi\pi}^2)$ with $\pi\pi$ helicity $\lambda = 0, \pm 1, \ldots, \pm L$. For the moment we omit the nucleon helicity labels. This simplifies the discussion and will be corrected for later.

It is convenient to introduce the combinations of helicity amplitudes

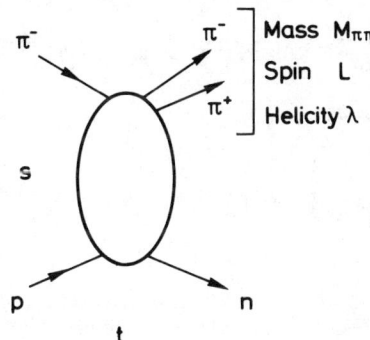

Figure 2 Variables for the process $\pi^-p \to \pi^-\pi^+n$.

$$L_{\lambda\pm} = \frac{1}{\sqrt{2}} \left(H^{L,\lambda} \mp (-1)^\lambda H^{L,-\lambda} \right)$$

(7a)

At high energies (that is, to order $1/s$) the amplitudes $L_{\lambda+}$ and $L_{\lambda-}$ describe the production of a $\pi\pi$ system of spin L, helicity λ by natural and unnatural parity exchange, respectively. We see that $L_{\lambda+} = 0$ for $\lambda = 0$ $\pi\pi$ production, that is, a zero helicity $\pi\pi$ system cannot be produced at high energies by natural parity exchange. In this case we have only an unnatural parity exchange amplitude, which we define as

$$L_0 \equiv H^{L,0}$$

(7b)

The observables $<Y_M^J>$ may be expressed in terms of the amplitudes $(S_0, P_0, P_{1\pm}, D_0, D_{1\pm}, D_{2\pm}, \ldots)$ of Eqs. (7). Each moment is a sum over bilinear terms of the form $\text{Re}(L'_\lambda, L^*_\lambda)$. A given moment $<Y_M^J>$ will only contain terms with $L' + L \geq J$ and $|\lambda' - \lambda| = M$. Furthermore $L' + L$ must be even (odd) if J is even (odd). These restrictions are embodied in the Clebsch-Gordan coefficients $<LL'\lambda-\lambda'|JM>$ and $<LL'00|J0>$ which occur when the density matrix is expressed in terms of the moments[12]. Moreover, the moments contain no interference terms between $L_{\lambda+}$ and $L_{\lambda-}$ amplitudes.

For example, in a region of $M_{\pi\pi}$ where only S and P wave $\pi\pi$ production is appreciable, the observables can be expressed in terms of the production amplitudes $S_0, P_0, P_{1\pm}$ as follows

$$\sqrt{4\pi} \, N \langle Y_0^0 \rangle = |S|^2 + |P_0|^2 + |P_+|^2 + |P_-|^2$$

$$\sqrt{4\pi} \, N \langle Y_0^1 \rangle = 2 \text{Re}(SP_0^*)$$

$$\sqrt{4\pi} \, N \langle Y_1^1 \rangle = \sqrt{2} \, \text{Re}(SP_-^*)$$

$$\sqrt{4\pi} \, N \langle Y_0^2 \rangle = \frac{1}{\sqrt{5}} \left(2|P_0|^2 - |P_+|^2 - |P_-|^2 \right) \quad (8)$$

$$\sqrt{4\pi} \, N \langle Y_1^2 \rangle = \sqrt{\frac{6}{5}} \, \text{Re}(P_0 P_-^*)$$

$$\sqrt{4\pi} \, N \langle Y_2^2 \rangle = -\sqrt{\frac{3}{10}} \left(|P_+|^2 - |P_-|^2 \right)$$

So far we have simplified the discussion by disregarding the nucleon helicities. Each amplitude is really two independent amplitudes, a nucleon helicity flip and a non-flip amplitude, $H_{+\frac{1}{2}}^{L\lambda}$ and $H_{-\frac{1}{2}}^{L\lambda}$ respectively. The combinations of Eqs. (7) are to be formed for both the nucleon flip and non-flip amplitudes. For an experiment involving unpolarized nucleons, Eqs. (8) are correct provided it is understood that the nucleon helicities are summed over as follows

$$|L|^2 \equiv |L_{++}|^2 + |L_{+-}|^2$$

$$\text{Re}(L'L^*) \equiv \text{Re}(L'_{++} L_{++}^* + L'_{+-} L_{+-}^*) \quad (9)$$

Here we have omitted the $\pi\pi$ helicity label.

The absence of nucleon polarization data prevents a model independent determination of the amplitudes for $\pi N \rightarrow (\pi\pi)N$. However, the unnatural parity exchanges have the simplifying property* that π exchange contributes only to nucleon flip amplitudes, whereas the amplitudes with the quantum numbers of A_1 exchange have nucleon non-flip (cf. Table I). We shall call the latter A_1 exchange contributions regardless of whether they arise from A_1 exchange, absorption, etc., with the exception of the order $1/s$ π exchange contribution in the s channel which we include explicitly. A study of the eigenvalues of the density matrix within the positivity domain[13] indicates that the A_1 contributions are small. Here we shall neglect these contributions. This should be a good approximation, particularly as the neglected quantities only enter quadratically in the expressions for the observables $< Y_M^J >$, that is, there are no π-A_1 interference terms.

The most direct check of this assumption will be nucleon polarization measurements for $\pi N \rightarrow \pi\pi N$; the polarization associated with unnatural parity exchange is due to π-A_1 interference. Also we can check the small t dependence of the observable $(3\rho_{00}^P + \rho_{00}^S)d\sigma/dt$ in a $\pi\pi$ mass region where S and P waves are dominant. The π exchange contribution vanishes like t in contrast to the non-flip A_1 contribution. In practice this test[14] is difficult, requiring very high statistics and depending mainly on the extreme forward data points.

With the assumption of negligible A_1 exchange contributions it follows, for example, that the relative phases

$$\varphi = \arg(P_-) - \arg(P_0)$$
$$\Delta = \arg(S_0) - \arg(P_0) \tag{10}$$

determine the phase between S_0 and P_-. Thus in a region of $M_{\pi\pi}$ where only S and P wave $\pi^-\pi^+$ states are important we can use the six observable moments, Eqs. (8), to determine[15] $|P_+|$, $|P_0|$, $|P_-|$, $|S|$, φ and Δ as functions of $M_{\pi\pi}$ and t. In Section 4 we discuss the uniqueness of the solution and also how we include the small D wave contribution.

---

* This is exactly true for the π pole in the t channel; in the s channel we have order $1/s$ π exchange contributions to the nucleon non-flip amplitudes.

s ch. hel. amp. $H^{L,\lambda}_{\lambda_n, \lambda_p}$	n, x	ang. mom. $(\sqrt{-t'})^n$	Regge pole exchange		$(\sqrt{-t'})^{n+x}$
			nat. p. $(L_{\lambda+})$	unnat. p. $(L_{\lambda-})$	
$H^{L,0}_{+-}$	1 0	$\sqrt{-t'}$		π	$\sqrt{-t'}$
$H^{L,0}_{++}$	0 0	const		A_1	const
$H^{L,1}_{+-}$	0 2	const	$A_2 + \pi$		$-t'$
$H^{L,-1}_{+-}$	2 0	t'	$A_2 - \pi$		$-t'$
$H^{L,1}_{++}$	1 0	$\sqrt{-t'}$	$A_2 + A_1$		$\sqrt{-t'}$
$H^{L,-1}_{++}$	1 0	$\sqrt{-t'}$	$A_2 - A_1$		$\sqrt{-t'}$
$H^{L,2}_{+-}$	1 2	$\sqrt{-t'}$	$A_2 + \pi$		$(\sqrt{-t'})^3$
$H^{L,-2}_{+-}$	3 0	$(\sqrt{-t'})^3$	$-A_2 + \pi$		$(\sqrt{-t'})^3$
⋮					

TABLE I : Regge exchange contributions to the s channel helicity amplitudes for $\pi^- p \to (\pi^-\pi^+)n$ and their behaviour in the forward direction, $t' \equiv t - t_{min} = 0$. λ is the helicity of the $\pi^-\pi^+$ system. The amplitudes $L_{\lambda+}$ and $L_{\lambda-}$ are defined in Eqs. (7). To leading order in s, only the exchanges listed contribute to $L_{\lambda+}$ and $L_{\lambda-}$.

Choice of frame and absorptive corrections :

In order to extract $\pi\pi$ phase shifts we must isolate the $\pi N \to (\pi\pi)N$ amplitudes which are dominated by π pole exchange and suitably extrapolate them from the physical region to $t = \mu^2$. Clearly $S_0, P_0, D_0,\ldots$ are the desired amplitudes. Now the amplitude analysis can be done equally well using either the s or t channel moments of the $\pi^+\pi^-$ angular distribution. However, we argue that it is appropriate to extrapolate s channel amplitudes. The reason is that we believe the absorptive corrections to the exchange pole contributions are simpler in the s channel[16]. At present we do not have a reliable prescription for determining these corrections. The indications are that they interfere destructively with the pole contributions and that, to a good approximation, they conserve s channel helicities. Moreover they are expected to be largest in $x \neq 0$ s channel amplitudes, and, for $x = 0$ amplitudes, to decrease with increasing net helicity flip n (the n,x notation is that of Ref. 16). For an s channel helicity amplitude the net helicity flip, $n = |\lambda + \lambda_p - \lambda_n|$, specifies the forward behaviour arising from angular momentum conservation, and $n + x = |\lambda| + |\lambda_p - \lambda_n|$, specifies the behaviour for definite parity (Regge pole) exchange. This behaviour, together with the values of n and x, is listed in Table I.

Consider the $x = 2$ $H_{+-}^{L,1}$ s channel helicity amplitude. The pole contributions, which are required to vanish as t', are expected to be modified by destructive interference with a non-vanishing (absorptive) background. The cross-over zeros in the s channel $< Y_1^l >$ moments near $-t = \mu^2$ are experimental support for this picture. This absorptive correction to the s channel $P_{1-}, D_{1-},\ldots$ amplitudes will, when crossed, affect the t channel $P_0, D_0,\ldots$ helicity amplitudes. Of course, the s channel $S_0, P_0,\ldots$ amplitudes may themselves have absorptive corrections, but as these are helicity flip amplitudes these modifications should be relatively small. In either channel the absorptive modifications to $S_0, P_0, D_0,\ldots$ do not in principle cause a problem since they should disappear on appropriate extrapolation to the π exchange pole. However, to determine $\pi\pi$ phases it is desirable to extrapolate what are believed to be the "purest" π exchange amplitudes and for this reason we shall use the s channel $S_0, P_0, D_0,\ldots$ amplitudes.

One slight complication of this choice is that the π pole contribution, which in the t channel contributes only to $S_0, P_0,\ldots$, is distributed among all the $L_{\lambda-}$ s channel amplitudes. For example, for P wave $\pi^+\pi^-$ production we have, to leading order in s,

$$(P_0)_{+-} \equiv H_{+-}^{1,0} = -\frac{g_\pi}{\sqrt{q}} \frac{t + M_{\pi\pi}^2 - \mu^2}{J} \frac{\sqrt{-t'}}{t - \mu^2} \qquad (11)$$

$$(P_-)_{+-} = \frac{g_\pi}{\sqrt{q}} \frac{2M_{\pi\pi}}{\mathcal{J}} \frac{t'}{t-\mu^2}$$

$$(P_0)_{++} \equiv H_{++}^{1,0} = \tau H_{+-}^{1,0} \quad \text{with} \quad \tau = \sqrt{\frac{t_{min}}{t'}} \quad (11)$$

where $\mathcal{J}^2 \equiv [t - (M_{\pi\pi} - \mu)^2][t - (M_{\pi\pi} + \mu)^2] \simeq M_{\pi\pi}^2$.

The last amplitude is only relevant at very small t. To include its contribution we multiply each product of $\lambda = 0$ amplitudes (e.g., $|P_0|^2$, $Re(S_0 P_0^*)$) occurring in the expressions for the observable moments, Eqs. (8), by $1 + \tau^2$ before solving for the amplitudes[15]. Thus from now on by $S_0, P_0, \ldots$ we mean only the helicity flip amplitudes $H_{+-}^{L,0}$.

4. PRODUCTION AMPLITUDE ANALYSIS FOR $M_{\pi\pi}$ BELOW 1 GeV

We have seen that the neglect of A_1 exchange amplitudes permits the determination of the magnitudes and relative phases of the amplitudes (S, P_0, P_-) and the magnitude of P_+ directly from the data. Instead of using the relative phases φ and Δ of Eq. (10) it is convenient to project S and P_- into components parallel and perpendicular to P_0 on the Argand plot as illustrated in Fig. 3. In terms of these amplitude components the moments of the $\pi^-\pi^+$ angular distribution become

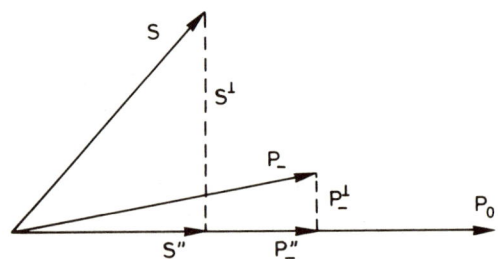

Figure 3 Vectors representing the unnatural parity exchange amplitudes. The components $S^\parallel$, $P_-^\parallel$ and P_0 are well determined by the data, whereas essentially only the product of the perpendicular components, $S^\perp P_-^\perp$, is measured.

$$\sqrt{4\pi}\, N \langle Y_0^0 \rangle = |S''|^2 + |S^\perp|^2 + |P_0|^2 + |P_+|^2 + |P_-''|^2 + |P_-^\perp|^2$$

$$\sqrt{4\pi}\, N \langle Y_0^1 \rangle = 2 S'' |P_0|$$

$$\sqrt{4\pi}\, N \langle Y_1^1 \rangle = \sqrt{2}\, (S'' P_-'' + S^\perp P_-^\perp)$$

$$\sqrt{4\pi}\, N \langle Y_0^2 \rangle = \frac{1}{\sqrt{5}} (2|P_0|^2 - |P_+|^2 - |P_-''|^2 - |P_-^\perp|^2)$$

$$\sqrt{4\pi}\, N \langle Y_1^2 \rangle = \sqrt{\frac{6}{5}}\, P_-'' |P_0|$$

$$\sqrt{4\pi}\, N \langle Y_2^2 \rangle = -\sqrt{\frac{3}{10}}\, (|P_+|^2 - |P_-''|^2 - |P_-^\perp|^2)$$

(12)

Eliminating all amplitude components in favour of $|P_0|$ we obtain a cubic equation for $|P_0|^2$. From the observed moments it turns out that one solution is unphysical, $|P_0|^2 < 0$, and that the remaining two solutions are both physical with similar values of $|P_0|^2$.

We are considering a region of $M_{\pi\pi}$ where D waves are relatively small. Although in Eqs. (12) we have omitted the terms depending on the D wave amplitudes we do, in fact, allow for these small contributions. In the first place we solve analytically for the two solutions using $\langle Y_0^1 \rangle - \sqrt{28/27} \langle Y_0^3 \rangle$ instead of $\langle Y_0^1 \rangle$ since, unlike $\langle Y_0^1 \rangle$, this combination does not contain the dominant D wave interference term $\mathrm{Re}(P_0 D_0^*)$. Moreover using the $\langle Y_0^3 \rangle$ moment we estimate D_0, $D_{1\pm}$ as described in Section 5.b. We allow for these small D wave contributions in the S and P wave amplitude analysis by iteration starting from the two exact solutions.

As an example we show in Fig. 4 the two solutions found at the different t values from the s channel moments in the mass bin $700 < M_{\pi\pi} < 720$ MeV. The amplitudes S and P_0 have similar t behaviour and so we show $\gamma_S^\parallel = S^\parallel/|P_0|$ and $\gamma_S^\perp = S^\perp/|P_0^\perp|$. By inspection of Eqs. (12) we notice that the component $S^\perp$ is less constrained than $S^\parallel$ and this is reflected in the resulting errors. Similarly the component $P_-^\parallel$ is better determined than $P_-^\perp$. Moreover, the data do not determine the absolute signs of $S^\perp$ and $P_-^\perp$, but only their relative sign. In the ambiguous cases we have chosen $P_-^\perp$ to be positive in Fig. 4.

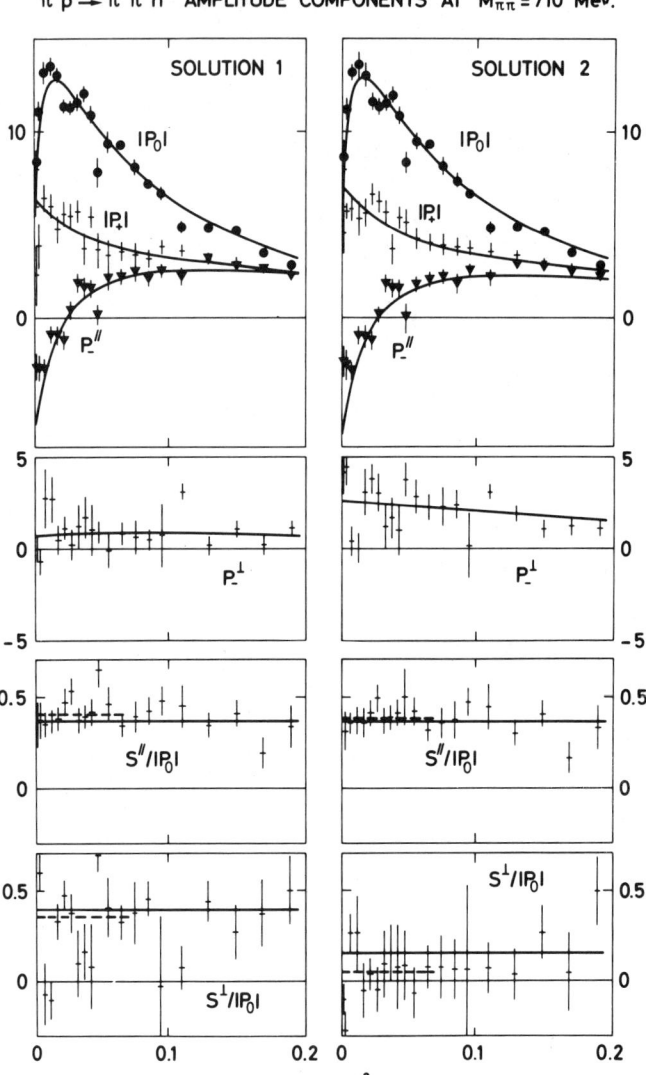

Figure 4 The two solutions for the s channel amplitude components calculated from the data in the mass bin $700 < M_{\pi\pi} < 720$ MeV. The points are the solutions obtained at the different t values and the dashed lines represent the resulting average values of S/P_0. The continuous lines are obtained from the parametric fit to the data that is described in Section 5.

We may compare the structure of these amplitude components with the behaviour anticipated from the contributions of π exchange to P_0 and P_-, and A_2 exchange to P_+ (cf. Section 3). Up to slope factors of the form $\exp[b(t-\mu^2)]$, the P wave amplitudes are expected to have the following t structure

$$P_0 = -g_\pi \mathcal{S}_\pi M_{\pi\pi} \frac{\sqrt{-t'}}{t-\mu^2}$$

$$P_- = g_\pi \left[\mathcal{S}_\pi \frac{2t'}{t-\mu^2} - C(t) \right]$$

$$|P_+|^2 = \left| -t' g_A^F \mathcal{S}_A - g_\pi C(t) \right|^2 - t' \left| g_A^{NF} \mathcal{S}_A \right|^2$$

(13)

to leading order in s, where $\mathcal{S}_i$ are signature factors $\frac{1}{2}+\frac{1}{2}\exp(-i\pi\alpha_i)$ with $\alpha_\pi \approx t-\mu^2$ and $\alpha_A \approx 0.5+t$, and g_A^F and g_A^{NF} are the A_2 exchange couplings to nucleon helicity flip and non-flip respectively. Studies[17] of ρ and A_2 exchange in spin 0 - spin $\frac{1}{2}$ processes indicate that there $g_A^F/g_A^{NF} \approx 4$. The additional contribution $g_\pi C(t)$, which is non-vanishing at $t' = 0$, can be regarded as the absorptive correction to π and A_2 exchange in the (evasive) s channel H_{+-}^1 amplitude. At $t' = 0$ we have $P_+ = P_-$. The Williams' model[18] is a special case of Eqs. (13), namely that with $g_A^{F,NF} = 0$, $\mathcal{S}_\pi = 1$ and $C = 1$.

For the amplitudes obtained from the data in the ρ mass band, $730 < M_{\pi\pi} < 810$ MeV, such a breakdown has been discussed in Ref. 15 (see also Refs. 19, 20). The actual interpretation of the various contributions to $P_\pm$ is not important as far as the extraction of $\pi\pi$ phases is concerned. However, $P_\pm$ give sizeable contributions to all the moments and it is crucial to allow for their presence in the Chew-Low extrapolation.

The two allowed solutions for the amplitude components are distinguished mainly by their differing values of $S^\perp$ and this leads to an ambiguity in the determination of the $\pi\pi$ S wave phase.

Connection with $\pi\pi$ phases :

For each $M_{\pi\pi}$ bin the dominant π exchange amplitudes, S and P_0, are the appropriate quantities to extrapolate in t to $t = \mu^2$ to determine $\pi\pi$ phase shifts. We discuss first the P wave and then the S wave extrapolation.

To extrapolate $|P_0|$ to the π exchange pole we fit the calculated amplitudes for $-t < 0.2$ to the form [cf. Eq. (4)]

$$P_0 = A \frac{\sqrt{-t'}}{\mu^2 - t} e^{b(t-\mu^2)} \sqrt{3} f_P, \tag{14}$$

where for $M_{\pi\pi}$ in the $\pi^-\pi^+$ elastic region $f_P = \sin \delta_P e^{i\delta_P}$. In other words from $|P_0|$ we determine $A|f_P|$ and thus, knowing the normalization A, we obtain δ_P, the P wave $\pi\pi$ phase shift. The normalization factor A has an $M_{\pi\pi}$ dependence

$$A^2 = \mathcal{N} \frac{M_{\pi\pi}^2}{q} \left[\frac{M_{\pi\pi}^2}{M_{\pi\pi}^2 - 4\mu^2} \right], \tag{15}$$

where $M_{\pi\pi}^2/q$ arises from the Chew-Low formula and the factor in brackets is due to crossing P_0 from the t to the s channel at $t = \mu^2$. It remains to fix the over-all normalization constant, $\mathcal{N}$, of the $\pi^-p \to \pi^-\pi^+n$ cross-section $d\sigma/dM_{\pi\pi}$. To do this we extrapolate $|P_0|$ for each $M_{\pi\pi}$ bin in the region of the ρ resonance, and adjust the constant $\mathcal{N}$ until the resulting δ_P goes smoothly through the resonance. Knowing the constant $\mathcal{N}$, and therefore A, we can calculate δ_P as a function of $M_{\pi\pi}$ in the $\pi^-\pi^+$ elastic region.

Consider now the extrapolation of the S wave amplitudes. To a good approximation the values of S/P_0 are constant in t. Therefore, to obtain the value at $t = \mu^2$, we simply fit the values for $-t < 0.2$ to a constant

$$\frac{S}{P_0} = \frac{\frac{2}{3} f_S^0 + \frac{1}{3} f_S^2}{\sqrt{3} f_P} \tag{16}$$

At the sample energy, $M_{\pi\pi} = 710$ MeV, the resulting extrapolations for $S^{\parallel}/P_0$ and $S^{\perp}/P_0$ are indicated by dashed lines on Fig. 4.

It is illuminating to view the results in terms of the unitarity circles for the $\pi\pi$ partial wave amplitudes, f_L^I. We cannot determine both the $I = 0$ and $I = 2$ S wave $\pi\pi$ phase shifts and so we input f_S^2 using the values obtained in analyses [3,21] of $\pi^+p \to \pi^+\pi^+n$ data. The values we use for δ_S^2 are listed in Table II. The situation at 710 MeV is shown in Fig. 5. The larger unitarity circle corresponds to the P wave which we assume to be elastic. Then, as described above, $|P_0|$ determines δ_P. The P wave results for the two solutions are almost identical and are shown by a single line on Fig. 5. Also we show the unitarity circle for the $I = 0$ S wave, scaled down by the factor $2/3\sqrt{3}$ arising from $\sqrt{2L+1}$ and isospin.

51

$M_{\pi\pi}$ (MeV)	Solution 1				Solution 2				Input
	δ_S^0	δ_P	δ_D^0	χ^2	δ_S^0	δ_P	δ_D^0	χ^2	δ_S^2
450	39.1 ± 2.7	4.8 ± 0.9	0.4	101	21.0 ± 3.4	7.6 ± 0.5	0.4	96	-5.2
470	43.1 ± 1.8	6.2 ± 0.6	0.6	104	16.7 ± 2.4	9.9 ± 0.4	0.6	96	-6.0
490	48.5 ± 2.0	7.9 ± 0.4	0.8	117	20.8 ± 2.6	11.5 ± 0.5	0.8	118	-6.6
510	51.8 ± 2.2	9.3 ± 0.5	1.1	136	25.2 ± 2.7	12.2 ± 0.5	1.1	140	-7.2
530	52.0 ± 2.2	10.0 ± 0.6	1.5	149	26.6 ± 3.3	13.1 ± 0.5	1.5	157	-7.8
550	55.9 ± 2.3	11.5 ± 0.5	1.9	94	27.4 ± 3.5	14.6 ± 0.5	1.9	109	-8.5
570	59.1 ± 2.7	12.5 ± 0.5	2.4	141	28.0 ± 3.5	15.3 ± 0.5	2.4	131	-9.1
590	61.3 ± 2.2	15.4 ± 0.5	3.0	90	27.7 ± 2.8	18.0 ± 0.5	3.0	153	-9.7
610	67.8 ± 2.2	17.5 ± 0.4	3.7	104	27.5 ± 2.5	20.8 ± 0.4	3.7	131	-10.3
630	71.6 ± 2.2	19.8 ± 0.4	4.5	139	28.3 ± 2.4	22.6 ± 0.5	4.5	107	-10.9
650	74.0 ± 2.3	24.2 ± 0.4	4.5	140	34.1 ± 2.6	26.5 ± 0.5	4.5	139	-11.5
670	67.3 ± 3.7	31.0 ± 0.4	4.5	98	47.8 ± 3.0	31.8 ± 0.5	4.5	99	-12.2
690	76.9 ± 2.9	36.9 ± 0.5	4.5	126	45.6 ± 2.4	38.2 ± 0.5	4.5	126	-12.8
713	80.6 ± 3.0	43.6 ± 0.5	4.5	141	54.4 ± 3.4	44.2 ± 0.6	4.5	146	-13.4
730	87.5 ± 3.3	56.5 ± 0.6	4.5	112	59.6 ± 3.3	57.2 ± 0.8	4.5	117	-14.0
750	89.0 ± 5.3	70.0 ± 1.2	4.5	149	72.7 ± 8.8	70.4 ± 1.5	4.5	150	-14.7
770	83.8 ± 3.2	106.1 ± 1.7	4.5	134	113.2 ± 3.5	102.5 ± 1.8	4.5	140	-15.9
790	94.5 ± 4.3	118.6 ± 0.9	4.5	131	105.2 ± 5.8	117.5 ± 0.9	4.5	136	-16.5
810	90.3 ± 2.9	127.7 ± 0.8	4.5	133	122.2 ± 4.0	127.3 ± 1.3	4.5	134	-17.2
830	93.5 ± 3.1	135.0 ± 0.8	4.5	96	131.3 ± 3.0	133.8 ± 0.9	4.5	95	-17.8
850	92.0 ± 2.7	137.2 ± 0.7	4.5	110	140.7 ± 2.4	134.1 ± 0.8	4.5	112	-18.4
870	89.2 ± 2.7	137.2 ± 0.7	4.5	110	140.7 ± 2.4	134.1 ± 0.8	4.5	112	-18.4
890	99.8 ± 4.1	143.6 ± 0.6	4.5	102	137.1 ± 2.4	141.5 ± 0.8	4.5	104	-19.1
910	99.1 ± 4.1	147.4 ± 1.1	2.6 ± 0.7	234	144.7 ± 2.6	145.2 ± 1.1	2.6 ± 0.6	235	-19.4
930	105.6 ± 2.5	151.7 ± 0.5	3.4 ± 0.5	256	132.5 ± 3.9	149.7 ± 0.5	5.3 ± 0.6	256	-20.0
950	108.7 ± 5.0	152.8 ± 1.1	3.7 ± 0.8	282	157.5 ± 4.2	150.6 ± 1.0	4.5 ± 0.7	278	-20.5
970	112.4 ± 10.0	156.3 ± 1.2	6.1 ± 1.2	183	143.3 ± 3.3	153.8 ± 0.9	6.5 ± 0.6	175	-20.9

TABLE II : $\pi\pi$ phase shifts, δ_L^I in degrees determined from $\pi^- p \to \pi^- \pi^+ n$ data at 17.2 GeV/c in 20 MeV $\pi\pi$ mass bins. In each mass bin below 900 MeV six moments were fitted at 20 t values (0.0025 < -t < 0.2 GeV^2). Above 900 MeV, 15 moments were used at the 20 t values. The $\pi^0\pi^0$ mass distribution selects Solution 1 as the physical solution.

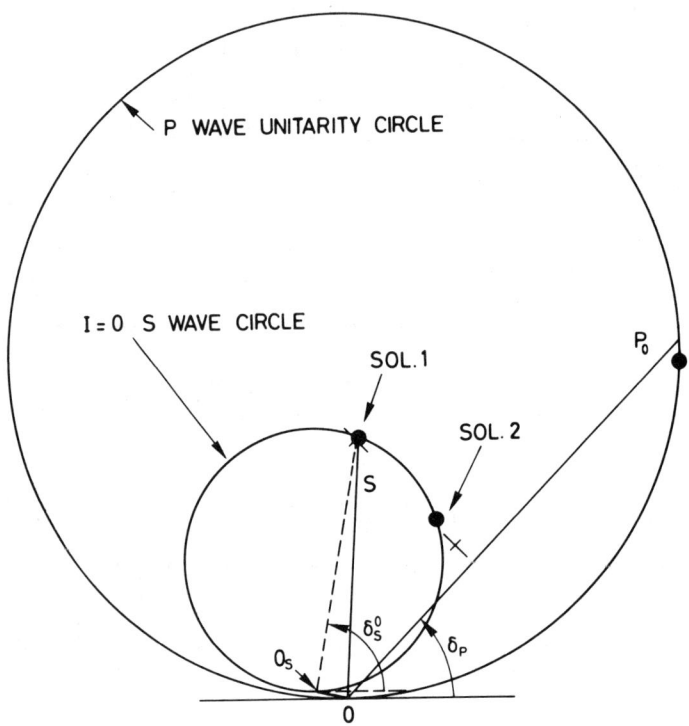

Figure 5 The S and P wave ππ phases at $M_{\pi\pi}$ = 710 MeV obtained by extrapolating the amplitude solutions of Fig. 4. The scale of the unitarity circles (1 : 2/3√3) represents the relative size of the amplitudes in the production process. The length OO_S is the input I = 2 S wave. The crosses are the S wave results obtained from the dashed lines of Fig. 4. The black dots are the results of the parametric fit with elastic unitarity imposed.

The shift of origin from O to O_S is due to the input $I=2$ S wave amplitude f_S^2. Knowing $S^{\parallel}/P_0$ and $S^{\perp}/P_0$ we may plot the S wave amplitude on Fig. 5. The two solutions are indicated by crosses which represent their error bars. If the S wave is elastic, as we expect, then the cross representing the physical solution should lie on the unitarity circle. We notice that the well determined component $S^{\parallel}$ is similar for the two solutions, whereas the poorly known component $S^{\perp}$ distinguishes the two solutions. It is apparent that it is not going to be easy to select the physical solution for δ_S^0 simply from an analysis of $\pi^-p \to \pi^-\pi^+n$ data alone. As in previous analyses [3,5,22-24] we will have to take care to keep track of both solutions as a function of $M_{\pi\pi}$. Essentially the $M=0$ moments determine $|P_0|$, $S^{\parallel}$ and $|S|^2 + |P_0|^2$ and so the best chance of getting a unique δ_S^0 appears to be in a region away from the ρ where $|S|$ is significantly different for the two solutions and leads to different extrapolated cross-sections. The inclusion of the $M \neq 0$ moments is necessary to allow a reliable determination of P_0 and S. Moreover, in principle, from a knowledge of the sign of $P^{\perp}$, they also allow the sign of $S^{\perp}$ to be determined (cf. $<Y_1^1>$). In practice $P^{\perp}$ is small and poorly determined and so is not decisive.

So far S wave unitarity has not been imposed. At first sight it appears that this could select the physical solution - perhaps one solution is always nearer to the circle than the other solution. However, $S^{\perp}$, like $P^{\perp}$, is badly determined and it would be misleading to select the solution in this way. Rather at the outset we should impose unitarity (at $t = \mu^2$) on the analysis and then see if one solution is preferred to the other. We describe such an analysis below.

5. $\pi\pi$ PHASE SHIFT ANALYSIS FOR $M_{\pi\pi}$ BELOW 1 GeV

The analysis is based on the high statistics $\pi^-p \to \pi^-\pi^+n$ data obtained[4] at a laboratory momentum of 17.2 GeV/c. S channel moments of the $\pi^-\pi^+$ angular distribution are used in 20 MeV $\pi\pi$ mass bins from $M_{\pi\pi} = 440$ MeV upwards. In each mass bin we determine the structure of the production amplitudes in the range $0 < -t < 0.2$ GeV^2 by fitting the moments to parametric forms based on Eqs. (13). For the results that we present, the slopes of the π, C, A_2 contributions, the complex $C(\mu^2)$, and g_A^F of Eqs. (13) are taken as parameters in each $M_{\pi\pi}$ bin, in addition to δ_S^0 and δ_P; the signature factors $\mathcal{S}_i$ are included and $g_A^{NF} = 0.25 \, g_A^F$. We include D waves as outlined in Section 4. We impose elastic unitarity, $f_L = \sin\delta_L \, e^{i\delta_L}$, through Eqs. (14)-(16), except that, above the onset of the $\pi\omega$ channel, $M_{\pi\pi} = 920$ MeV, we allow the P wave to be inelastic.

The unitarity constraint is only true at $t = \mu^2$. However, we included a term $[1 + a(t - \mu^2)]$ on the right-hand side of Eq. (16). Since the unitarity phase is preserved[25] we took the parameter a to be real. Such a term could also arise from differing amounts of absorption in the π pole contributions to S and P_0, or from a t dependence associated with crossing from the t to the s channel. The values found for the parameter a in the different mass bins were distributed about zero, and typically $a \approx \pm 0.5$ GeV^2. The results we present have $a = 0$. Values of $a = \pm 0.5$ lead to changes in δ_S^0 and δ_P of about $\pm 2°$ and $\pm 0.5°$, respectively.

We tried several different forms of amplitude parametrization based on Eqs. (13), allowing different slope factors $\exp[b(t - \mu^2)]$ on individual contributions, using different input values of g_A^F/g_A^{NF}, etc. The phase shift results were extremely stable to such changes of parametrization. We also repeated the analysis using only data for $-t < 0.1$. Again the results were essentially unaltered. The curves shown in Fig. 4 are the form of the amplitudes at 710 MeV. We see that they are a good description of the amplitude components determined t by t indicating that the chosen parametric form is adequate. The fit to the observed s channel moments is shown in Fig. 6.

a) S and P wave ππ phases :

The phase shifts obtained by the method outlined above are shown in Fig. 7 and listed in Table II. There are two solutions mainly differing in the values of δ_S^0. Solution 1 is characterized by a small $P^\perp$ and solution 2 by a small $S^\perp$. By this means or by following the Barrelet zeros (see Section 7, Fig. 16) it is possible to keep track of the two solutions through the ρ mass region. The solutions are stable to changes of parametrization and to changes of the t region over which the data are fitted. With the exception of solution 2 below $M_{\pi\pi} = 650$ MeV, the solutions are also stable to reasonable variations of the input values of δ_S^2 and δ_D^0. The black dots and open circles of Fig. 7 denote solutions 1 and 2 respectively. For comparison we show by a dashed curve the solution obtained by Protopopescu et al.[24] from an analysis of $\pi^+ p \to \pi^+\pi^- \Delta^{++}$ data.

In contrast to recent analyses[6,22,24], we obtain two acceptable solutions below (as well as above) the ρ mass. At low $M_{\pi\pi}$ the solutions have very different $|S|$, and therefore lead to different extrapolated ππ cross-sections. However, both values are compatible with the physical region data since the non π exchange background differs for the two solutions in such a way as to give good fits to the data in each case. We discuss this further in the next section.

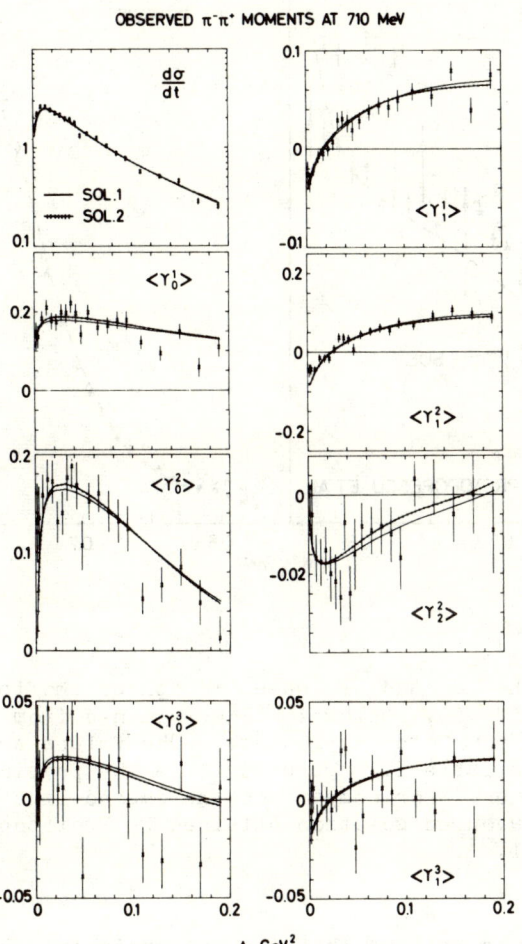

Figure 6 The fits to the s channel moments with $J \leq 2$ at $M_{\pi\pi} = 710$ MeV corresponding to solutions 1 and 2 of Table II. The description of the $J = 3$ moments is also shown ($\delta_0^0 = 4.5°$).

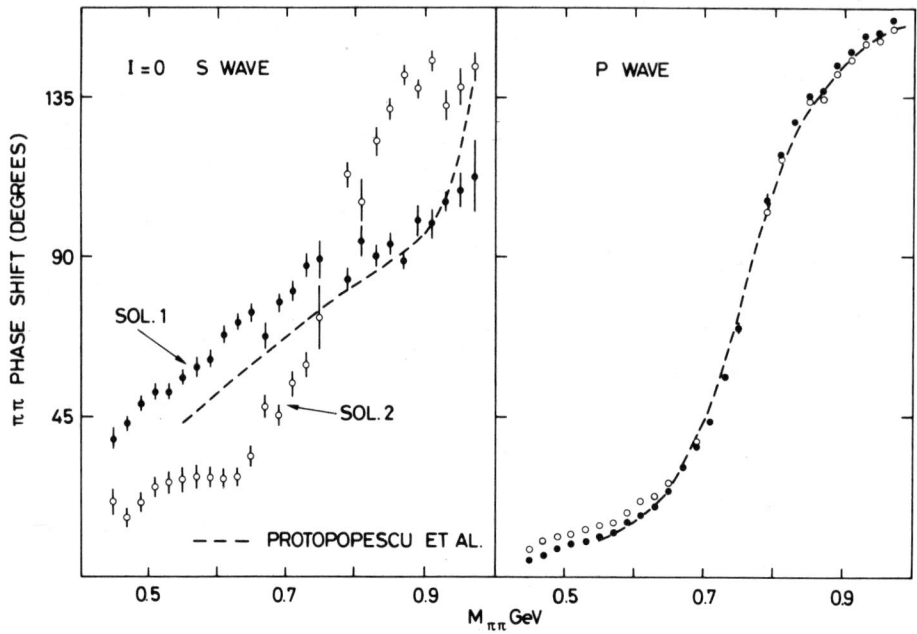

Figure 7 The S and P wave $\pi\pi$ phase shifts, δ_S^0 and δ_P, below 1 GeV determined from the $\pi^- p \to \pi^-\pi^+ n$ amplitudes. The values are listed in Table II. Solution 1 is the physical solution. For comparison the dashed line is the favoured solution obtained by Protopopescu et al.[24]

If we were to believe that the non-vanishing absorptive background [$C(t)$ of Eq. (13)] is dominantly real relative to π exchange then this appears to favour the solution with the smaller $P_\perp^i$, that is, solution 1. On the other hand, although the results of the phase coherent analysis ($P_\perp^i = 0$) described in the next section do in general prefer solution 1, we find even there an acceptable solution 2 at $M_{\pi\pi}$ values below and above the ρ mass region. From the analysis above it is clear that the $\pi^-\pi^+ n$ data alone do not resolve the S wave ambiguity in the elastic region.

The most direct way to select the physical solution is to study the $\pi^0\pi^0$ mass distribution[26,27], since here only even L $\pi\pi$ partial waves can contribute. The histogram in Fig. 8 is the $\pi^0\pi^0$ mass spectrum for $2\mu^2 < -t < 8\mu^2$ obtained from a $\pi^- p \to \pi^0\pi^0 n$ experiment[26] at 8 GeV/c. In terms of $\pi\pi$ phase shifts this spectrum is, to a good approximation, proportional to

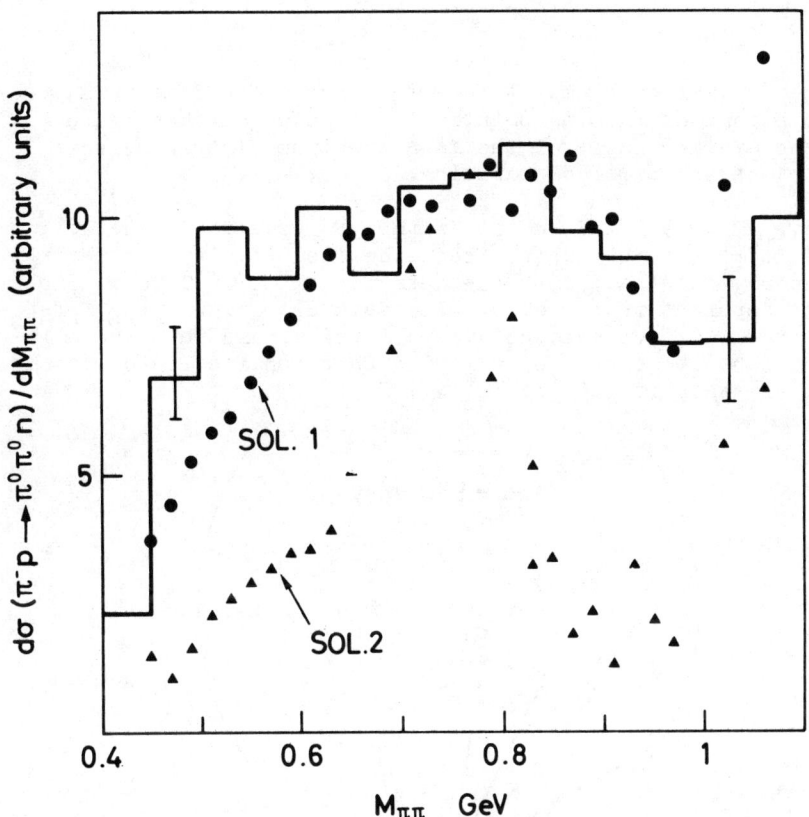

Figure 8 The histogram is the $\pi^0\pi^0$ mass spectrum for $2\mu^2 < -t < 8\mu^2$ from $\pi^-p \to \pi^0\pi^0 n$ at 8 GeV/c[26]. The circles (triangles) are the shape of the spectrum calculated from the $\pi\pi$ phases of solution 1 (solution 2) respectively. The scale is arbitrary.

$$\frac{d\sigma(\pi^0\pi^0)}{dM_{\pi\pi}} \propto \frac{M_{\pi\pi}^2}{q} \sum_{L=0,2} (2L+1) \left| \frac{1}{3} f_L^0 - \frac{1}{3} f_L^2 \right|^2$$

where the partial wave amplitudes f_L^I, are defined as in Eq. (6). The predictions of the two solutions are shown on the figure. A comparison of the shapes of the mass spectrum clearly selects solution 1 as the physical solution.

Above $M_{\pi\pi} = 920$ MeV we allow the P wave to be inelastic. Although $|f_P|$ is well determined, the inelasticity parameter η_P is poorly constrained by the data for $M_{\pi\pi} \sim 950$ MeV. The reason is apparent from Fig. 9. The data determine $|f_P|$, $|f_S|$ and $\cos(\delta_S - \delta_P)$ but not the over-all phase, and thus the solutions shown have comparable χ^2. The phase shifts listed assume that the P wave is elastic below 1 GeV.

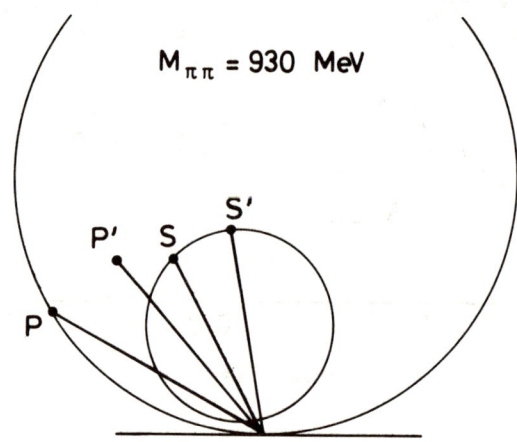

Figure 9 S, P denote solution 1 at $M_{\pi\pi} = 930$ MeV. The solution S', P', with an inelastic P wave gives a comparable fit to the data.

To determine resonance parameters we use the form[28]

$$f_L = \frac{x M_R \Gamma}{M_R^2 - M_{\pi\pi}^2 - iM_R \Gamma},$$ (17)

with

$$\Gamma = \left(\frac{q}{q_R}\right)^{2L+1} \frac{D_L(q_R r)}{D_L(q r)} \Gamma_R$$

For the ρ we use $D_1(y) = 1 + y$ and fit to the P wave phase shift of solution 1 in the range $650 \leq M_{\pi\pi} \leq 890$ MeV. With this parametrization, we find for the ρ

$$M_\rho = 772.2 \pm 0.6 \text{ MeV}, \qquad \Gamma_\rho = 143.1 \pm 1.1 \text{ MeV},$$

$$r_\rho = 0.83 \pm 0.08 \, f.$$

b) D wave in the elastic region :

In the ρ mass region the observed s channel $<Y_0^3>$ moment has a systematic behaviour versus both $M_{\pi\pi}$ and t which is consistent with P-D interference. For a fixed $t \sim -0.05$ GeV^2, the normalized $<Y_0^3>$ moment decreases from around 0.04 for $M_{\pi\pi} \sim 600$ MeV, through zero in the region of the ρ mass, to about -0.03 for $M_{\pi\pi} \sim 900$ MeV. For fixed $M_{\pi\pi}$, $<Y_0^3>$ reaches a maximum size for $-t \sim 0.05$ and then decreases, changing sign for $-t \sim 0.15$ GeV^2. Examples of the observed s channel $<Y_0^3>$ moment are shown in Fig. 10.

In terms of production amplitudes

$$\sqrt{4\pi}\, N \langle Y_0^3 \rangle = \frac{6}{\sqrt{35}} \text{Re}\left(\sqrt{3}\, P_0 D_0^* - P_- D_{1-}^* - P_+ D_{1+}^*\right)$$

$$\simeq \frac{6\sqrt{3}}{\sqrt{35}} \left|\frac{D_0}{P_0}\right| \left[|P_0|^2 - |P_-|^2 - |P_+|^2\right] \cos(\delta_P - \delta_D).$$ (18)

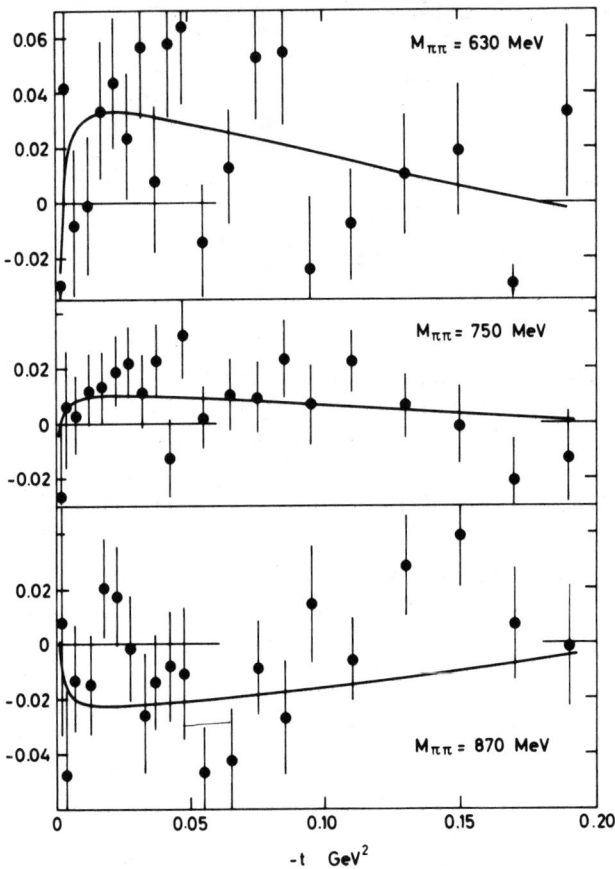

Figure 10 The t dependence of the s channel $<Y_0^3>$ moment in three typical 20 MeV $M_{\pi\pi}$ bins. The curves are the description of the data for $\delta_D^0 = 4.5°$.

The last equality is obtained assuming the proportionality relation

$$D_{1\pm} = \sqrt{3}\, D_o (P_\pm/P_o), \tag{19}$$

where the $\sqrt{3}$ arises from crossing the π exchange contribution to the s channel. For instance such a relation is implied by the Williams model. In Eq. (18) the $M_{\pi\pi}$ behaviour of $<Y_0^3>$ arises mainly from the factor $\cos(\delta_P - \delta_D)$, while the t behaviour of $<Y_0^3>$ is due to the term in square brackets.

Knowing the P wave amplitudes and taking the D wave to be elastic below $M_{\pi\pi} = 900$ MeV, and dominantly $I = 0$, we calculate δ_D^0 for each $M_{\pi\pi}$ bin by comparing Eq. (18) with the $<Y_0^3>$ data. Over the entire range 620 MeV $< M_{\pi\pi} < 900$ MeV we find δ_D^0 is essentially constant with a value of $\delta_D^0 = 4.5°$. For mass values close to $M_{\pi\pi} = 780$ MeV δ_D^0 cannot be reliably determined since the P and D amplitudes are about $\pi/2$ out of phase. That δ_D^0 should be so large for $M_{\pi\pi} \sim 650$ MeV is puzzling. The curves in Fig. 10 are calculated from Eq. (18) using $\delta_D^0 = 4.5°$. Equation (18) gives a good (one parameter) description of the t and $M_{\pi\pi}$ behaviour of the $<Y_0^3>$ moment. Versus $M_{\pi\pi}$ it predicts a $<Y_0^3>$ sign change at $M_{\pi\pi} \sim 780$ MeV (the data crossover is at $M_{\pi\pi} \sim 800$ MeV) and versus t a sign change at $-t \sim 0.2$ (compared to 0.15 in the data).

In the S and P wave analysis in the region $620 < M_{\pi\pi} < 900$ MeV we took $\delta_D^0 = 4.5°$. Below 620 MeV the estimates of δ_D^0 may not be reliable, since for $M_{\pi\pi} \lesssim 500$ MeV an anomalous behaviour is observed in some higher moments[4] and, therefore, we assumed a q^5 threshold behaviour of δ_D^0. For $M_{\pi\pi} > 900$ MeV we included the $J = 3, 4$ moments and determined $\delta_D^{0\pi\pi}$ in each mass bin.

6. PHASE COHERENT ANALYSIS

We repeated the $\pi\pi$ phase shift analysis using essentially a Williams' model parametrization of the production amplitudes. In place of Eqs. (13) we use the simplified forms

$$P_0 = -\frac{g_\pi}{\sqrt{q}} M_{\pi\pi} \frac{\sqrt{-t'}}{t - \mu^2} e^{b(t - \mu^2)}$$

$$P_- = \frac{g_\pi}{\sqrt{q}} \left[\frac{2t'}{t - \mu^2} - C \right] e^{b(t - \mu^2)} \qquad (20)$$

$$P_+ = -\frac{g_\pi}{\sqrt{q}} C\, e^{b(t - \mu^2)}$$

where C, which specifies the absorptive background, is assumed to be real and independent of t. Strictly speaking the Williams' model has $C = 1$, however, here we take C as a parameter to be determined in each mass bin. Departures from the above simple parametrization occur for $-t \gtrsim 0.15$ GeV^2 due to the neglect of A_2 exchange contributions, etc.[15,4]. Therefore we restrict the analysis to the data in the region $-t < 0.1$ GeV^2. The points shown in Fig. 11 for $M_{\pi\pi} < 1$ GeV are the results of this analysis. The values of δ_S^0 are in excellent agreement with solution 1 of Section 5 and are a demonstration of the stability of the phase shifts to a change of parametrization.

Since phase coherence between P_0 and P_- is an input assumption here, it is not surprising that we obtain solution 1. However, for mass bins where the two solutions of Section 5 are dissimilar (i.e., away from the region of the ρ mass) we also find solution 2, and with comparable χ^2. Moreover the values of C are almost identical for the two solutions, whereas for $M_{\pi\pi} \sim 500$ MeV we have already remarked (cf. Section 5) that the background has to be different for the two solutions. This apparent contradiction is resolved when we note that δ_P is significantly different for the two solutions in this mass region.

7. PHASE SHIFT ANALYSIS IN THE INELASTIC REGION

Above the $K\bar{K}$ threshold we cannot impose elastic unitarity. Indeed, the $\pi\pi \to K\bar{K}$ cross-section is observed[24,29] to rise rapidly to its S wave unitarity limit. Further we can no longer regard the D wave as a small correction. On the other hand we still want to perform a phase shift analysis at each $M_{\pi\pi}$ independently. We use a similar method to that described in Sections 4 and 5.

a) Production amplitude analysis :

From the observed s channel moments with J, $M \leq 4$ we determine the production amplitudes $L_{\lambda\pm}$ with L, $\lambda \leq 2$, and extrapolate S, P_0, D_0 to the π exchange pole. The data determine the magnitudes and relative phases of S, P_0, D_0, but not the over-all phase. In the elastic region unitarity determined the over-all phase, but in the inelastic region the unitarity constraint is weaker. For example a solution, such as shown in Fig. 12, can be rotated through any angle provided that the partial waves lie within their unitarity circles.

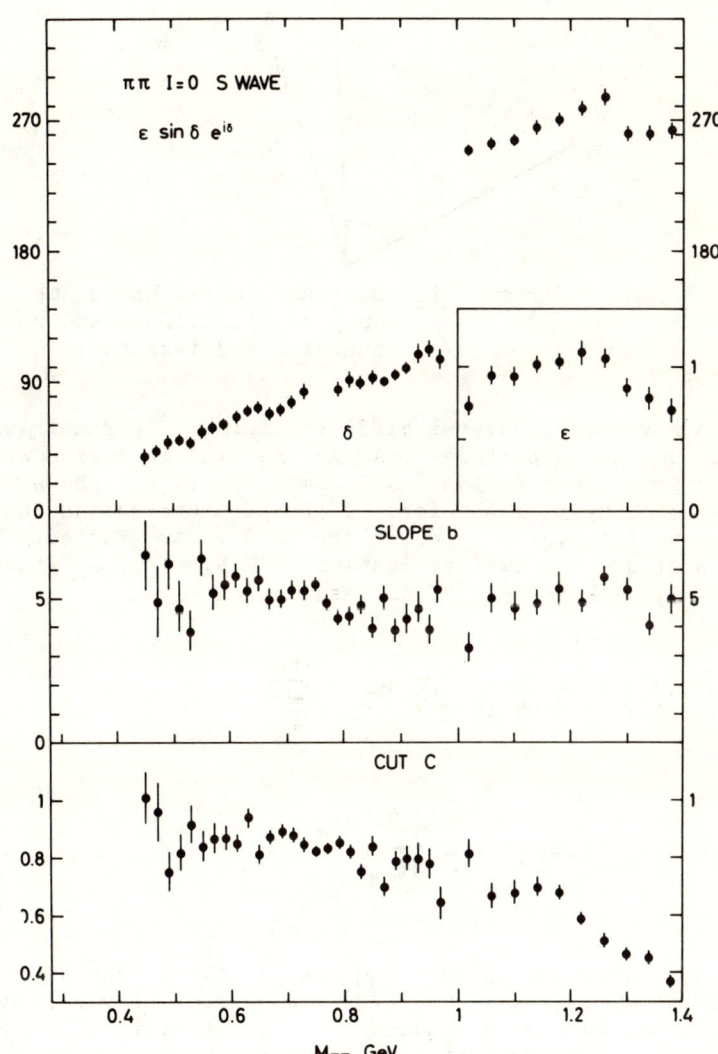

Figure 11 Some results of a phase shift analysis using a simplified parametrization, Eqs. (20). The results above 1 GeV are discussed in Section 7.d.

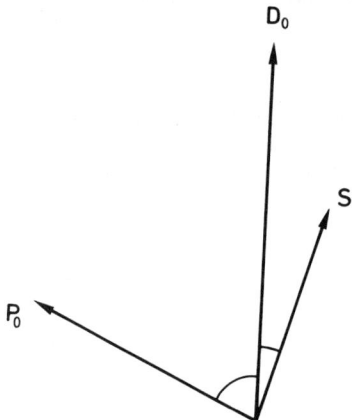

Figure 12 Dominant π exchange amplitudes. Only the magnitudes and relative phases are determined.

We have tried several different forms of parametrization of the s channel amplitudes. As in the elastic region we find that, as long as we include the $\lambda \neq 0$ amplitudes, the phase shifts are stable to changes of the form of the parametrization and to changes of the t interval over which the moments are fitted. The results we present use the parametrization of Eqs. (13), as stated in Section 5, with the additional assumptions

$$D_{1\pm} = \sqrt{3}\, P_{\pm}\, \frac{D_0}{P_0} \qquad (21)$$

$$D_{2\pm} = \frac{\sqrt{-t'}}{M_{\pi\pi}}\, D_{1\pm}, \qquad (22)$$

which are motivated by studying the t to s channel crossing matrix. Equations (21) and (22) are correct provided the main contribution to the s channel $\lambda = 1,2$ amplitudes is due to π exchange and its absorptive correction. This is expected to be a good approximation for $-t < 0.2$ GeV^2, particularly as A_2 exchange decreases* relative to π exchange with increasing $M_{\pi\pi}$.

---

* For instance, for $-t \gtrsim 0.4$ GeV^2 the moments indicate that the natural parity exchange contribution is less dominant in the f region than in the ρ region.

In the region $1.0 < M_{\pi\pi} < 1.4$ GeV we used data in 40 MeV mass bins. In each mass bin we fitted the s channel moments with $J \leq 4$ in the interval $0.005 < -t < 0.2$ GeV^2. A typical fit is shown in Fig. 13. The two solutions have comparable χ^2 but have different magnitudes and relative phases of S, P_0 and D_0. We parametrize the partial wave amplitudes, f_L of Eq. (4) in the form

$$f_L = \begin{cases} r_L e^{i\delta_L} & \text{for } L=1 \\ \frac{2}{3} r_L e^{i\delta_L} & \text{for } L=0,2 \end{cases} \quad (23)$$

and fix the over-all phase by the choice $\delta_D = 90°$. The 2/3 is inserted for even L so that, if there were no I = 2 $\pi\pi$ amplitude, unitarity would require $r_L \leq \sin\delta_L$. This bound is not imposed in the fit to the data. The results for each 40 MeV mass bin are listed in Table III. In Fig. 14 they are shown in the form $\sqrt{2L+1}\, f_L$, which represents their relative strength in the production process. We must now explain why we have shown two solutions.

b) Zero contours :

In addition to the continuum ambiguity of the over-all phase, are there discrete ambiguities in the phase shift analysis ? Yes, for S, P, D waves there are four solutions giving identical $d\sigma_{\pi\pi}/d\Omega$. A useful way[30,31] of seeing this and of keeping track of the four solutions is to study the zeros of the $\pi^+\pi^-$ scattering amplitude in the complex $z \equiv \cos\theta_{\pi\pi}$ plane. These have been called Barrelet zeros[32]. For S, P, D waves the $\pi\pi$ amplitude, $A(z)$, will have two such zeros, say at $z = z_1$ and $z = z_2$. Thus

$$\frac{d\sigma_{\pi\pi}}{d\Omega} = A(z) A^*(z^*) = c(z-z_1)(z-z_2)(z-z_1^*)(z-z_2^*), \quad (24)$$

and for each z_i there is a twofold ambiguity. Is $z = z_i$ or $z = z_i^*$ the zero of A ? For example, at $M_{\pi\pi} = 1.18$ GeV from solution 1 of Fig. 15 we calculate the zeros and predict the other three solutions that are shown. Having obtained one solution the procedure is to use these three predictions as starting values in the analysis of the observed moments. In this way we obtain four solutions at each $M_{\pi\pi}$. Two of the solutions (those denoted solutions 3 and 4 in the example shown in Fig. 15) are clearly ruled out by studying continuity of the partial waves with $M_{\pi\pi}$. The real and imaginary parts of the positions of the zeros for the other two solutions are shown in Fig. 16. By following the zero contours we can keep track of the solutions. However, two similar

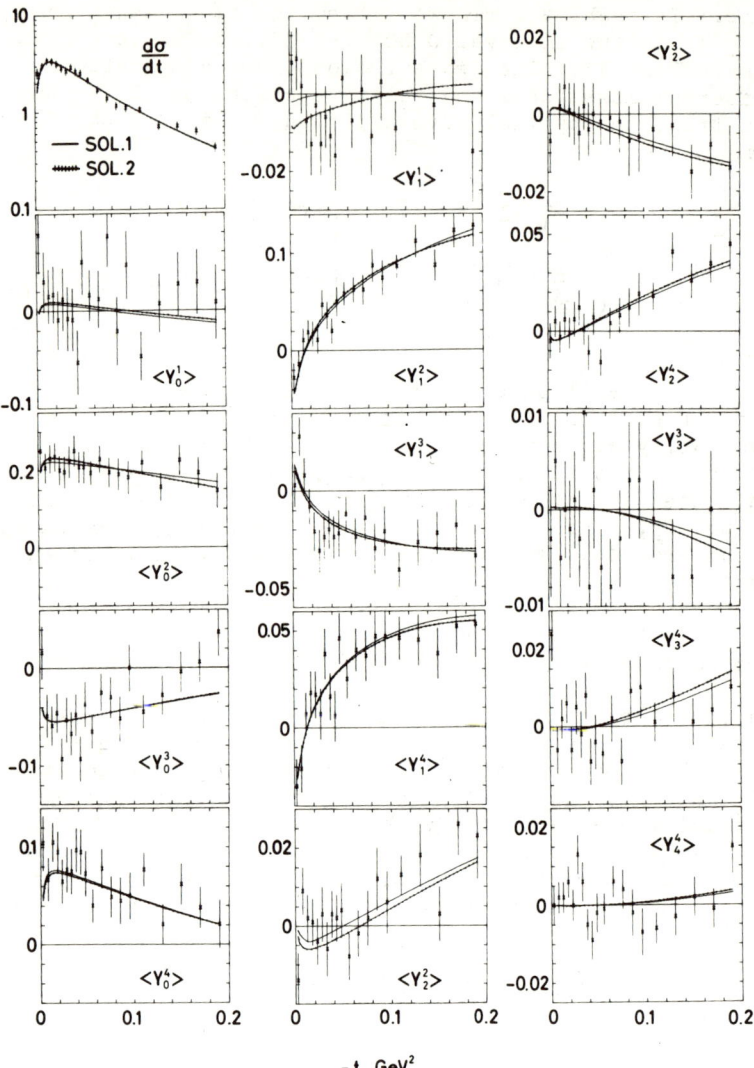

Figure 13 The fits to the s channel moments in a typical mass bin, $1.12 < M_{\pi\pi} < 1.16$ GeV. The partial wave parameters are given in Table III.

$M_{\pi\pi}$ (GeV)	Solution 1					Solution 2				
	f_S	δ_S	f_P	δ_P	χ^2	f_S	δ_S	f_P	δ_D	χ^2
1.02	0.74±0.10	146±2	0.32±0.02	212±2	324	0.16±0.03	193±24	0.42±0.01	0.21±0.02	327
1.06	0.81±0.08	146±2	0.33±0.02	213±2	286	0.22±0.05	190±22	0.45±0.01	0.24±0.02	296
1.10	1.01±0.04	149±2	0.27±0.01	218±2	348	0.28±0.04	168±8	0.45±0.01	0.32±0.01	364
1.14	1.08±0.05	148±2	0.22±0.02	211±2	273	0.32±0.04	157±5	0.45±0.01	0.39±0.01	277
1.18	1.13±0.06	145±2	0.27±0.03	207±2	268	0.62±0.10	140±3	0.44±0.04	0.54±0.02	274
1.22	1.23±0.06	136±2	0.22±0.03	195±2	266	0.86±0.10	127±4	0.40±0.04	0.69±0.01	272
1.26	1.11±0.06	121±3	0.23±0.03	171±5	353	1.21±0.05	126±3	0.15±0.03	0.81±0.02	348
1.30	1.02±0.06	99±5	0.32±0.03	163±2	327	1.29±0.03	122±3	0.10±0.03	0.78±0.02	319
1.34	0.94±0.08	83±6	0.32±0.03	150±4	292	1.17±0.05	116±5	0.19±0.03	0.66±0.01	286
1.38	0.93±0.09	81±10	0.31±0.04	144±6	307	1.14±0.04	118±5	0.20±0.02	0.56±0.02	311

TABLE III : The $\pi\pi$ partial wave amplitudes, f_L of eqn. (23), obtained in the analysis of reaction $\pi^-p \to \pi^-\pi^+n$ in the region $1.0 < M_{\pi\pi} < 1.4$ GeV. The over-all phase is fixed by the choice $\delta_D = 90°$. In each 40 MeV mass bin we fit 15 moments at 19 t values in the range $0 < -t < 0.2$ GeV^2. Solution 1 is the physical solution.

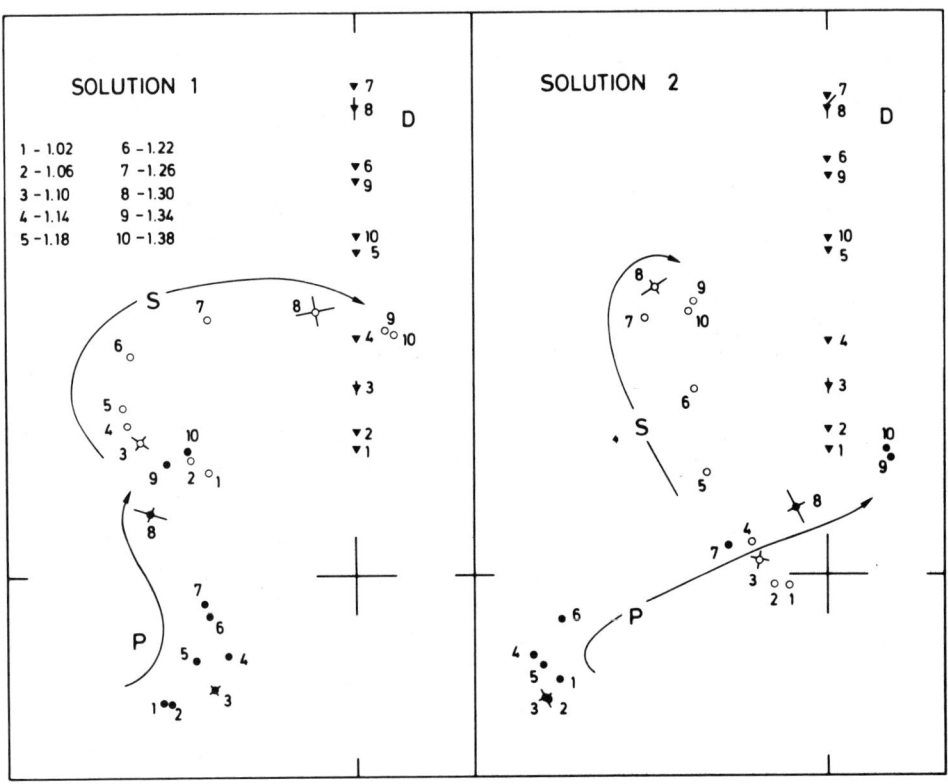

Figure 14 Two solutions for the $\pi^+\pi^-$ partial wave amplitudes f_L (scaled by $\sqrt{2L+1}$) found by analyzing $\pi^-p \to \pi^-\pi^+n$ data in 40 MeV mass bins in the range $1.0 < M_{\pi\pi} < 1.4$ GeV. We choose $\delta_D = 90°$. Typical errors are shown at $M_{\pi\pi} = 1.1$ and 1.3 GeV. The results are listed in Table III.

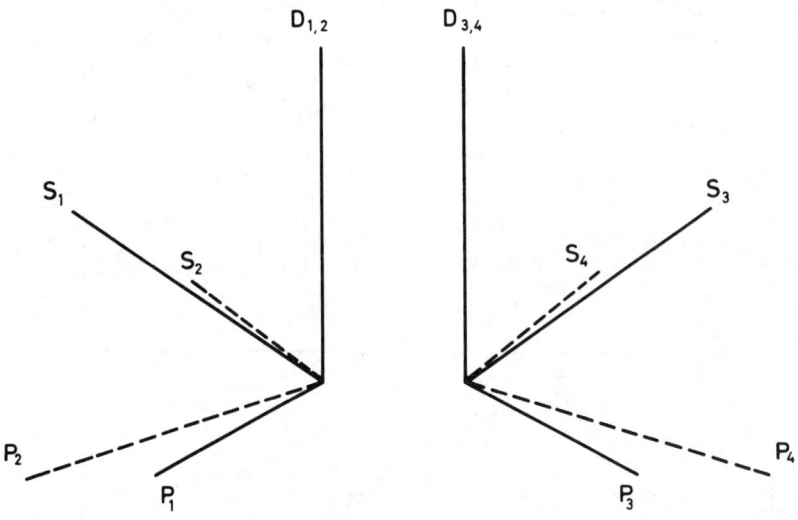

Figure 15   The four solutions at $M_{\pi\pi} = 1.18$ GeV.

solutions exist whenever Im $z_i \approx 0$. For example, for $M_{\pi\pi} \gtrsim 1.26$ GeV we have to use continuity of the zeros to decide which is solution 1 and which is solution 2. We also see that solutions 3 and 4 will need to be considered for $M_{\pi\pi} > 1.4$ GeV as Im z_2 is becoming small.

The actual solutions need not have exactly complex conjugate zeros, since, for example, the predicted absorption in the production process may differ for the solutions and so lead to a different extrapolated $\pi\pi$ cross-section, $d\sigma_{\pi\pi}/d\Omega$. Absorption decreases rapidly with increasing $M_{\pi\pi}$ and in the region above 1 GeV we do, in fact, have solutions with approximately complex conjugate zeros. In Fig. 16 we also show the positions of the zeros obtained from the phase shifts in the elastic region. The two solutions there are not, in general, due to complex conjugate zeros, but arise because the production amplitude analysis leads to different extrapolated $d\sigma_{\pi\pi}/d\Omega$. For example, at $M_{\pi\pi} \approx 600$ MeV there are two additional solutions with Im $z_1 > 0$ but with partial waves that do not satisfy unitarity.

It is illuminating to draw the contours of Re z_i on the Mandelstam plot. They are shown in Fig. 17. The continuation of the contour z_1 towards the Mandelstam triangle has been

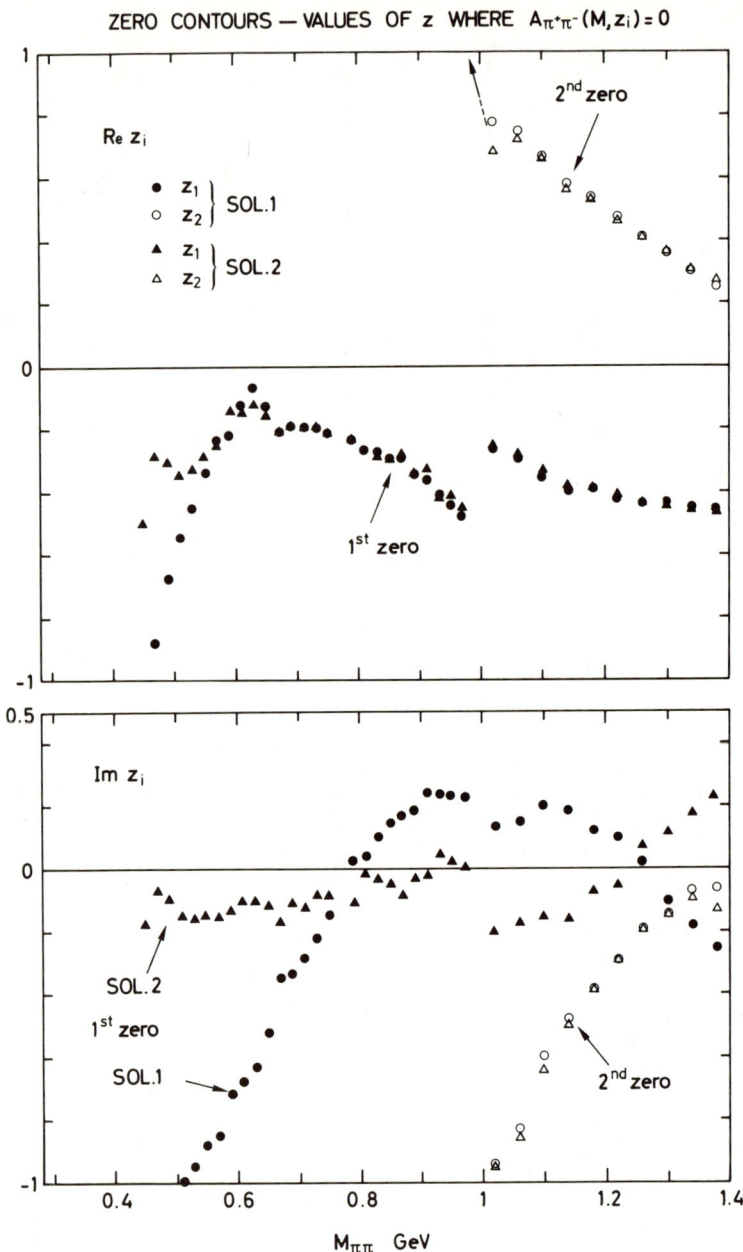

Figure 16 The positions of the zeros, $z = z_i$, of the $\pi^+\pi^-$ amplitude as calculated from the two partial wave solutions listed in Tables II and III. z is the cosine of the $\pi\pi$ c.m. scattering angle.

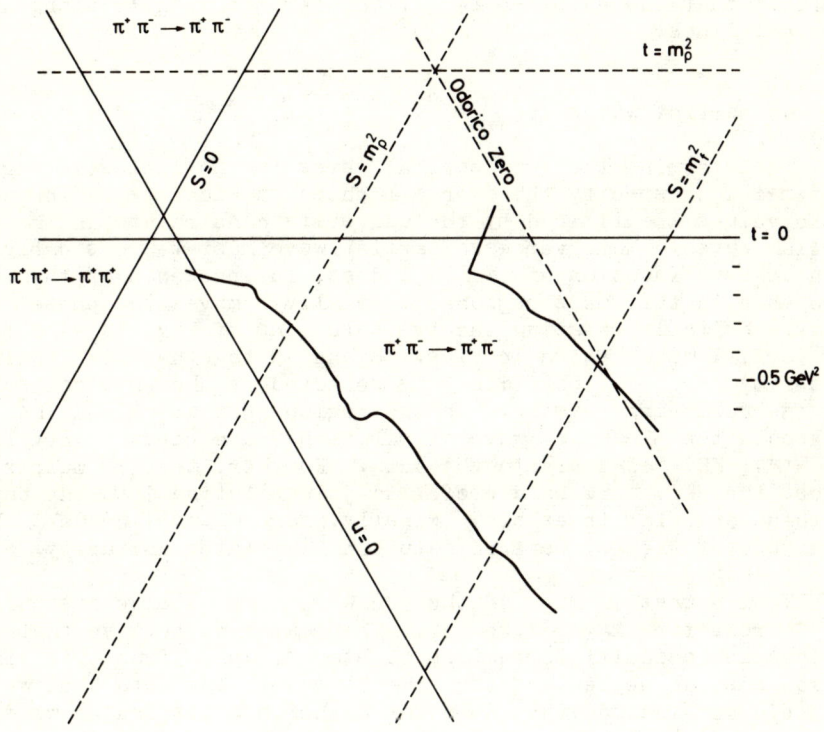

Figure 17 The zero contours in the Mandelstam plot

associated[33] with the on-shell appearance of the Adler zero. The contour z_2 is reasonably consistent with the proposal by Odorico[8] that the double-pole killing zeros propagate along straight lines.

c) $\pi\pi$ partial waves :

To determine the $\pi\pi$ partial waves in the inelastic region it remains to specify the over-all phase at each $M_{\pi\pi}$. The possible values are limited by the unitarity constraints on the three partial waves. Moreover each partial wave must be reasonably continuous as a function of $M_{\pi\pi}$. Indeed, the presence of the f resonance in this mass region essentially removes the phase ambiguity. Suitably rotating the two solutions of Fig. 14 we obtain the partial waves shown in Figs. 18 and 19 together with their unitarity circles. Solution 1 is selected as the physical solution for the following reasons. In the region just above the $K\bar{K}$ threshold the $I=0$ S wave of solution 2 contributes very little to $\sigma(\pi\pi \to K\bar{K})$ contrary to the data. Further, the M matrix fits across the $K\bar{K}$ threshold prefer to join solution 1 to the physical solution below threshold. Finally, for $M_{\pi\pi} > 1.26$ GeV the magnitude of the S wave of solution 2 violates unitarity.

Notice that in Fig. 18 the P wave lies outside its unitarity circle for $M_{\pi\pi} \gtrsim 1.26$ GeV. The reason we believe that the picture is basically correct as it stands, apart from this violation, is the neglect of the $\pi\pi$ F wave. The data that we used did not include the $J=5$ or higher moments and so we were unable to determine the $L=3$ partial wave. On the other hand we investigated the stability of the analysis to the inclusion of elastic F waves with $\delta_F \lesssim 5°$. We find that the D wave is essentially unchanged and that the S wave is only slightly altered. The major change is in the P wave which decreases and rotates in the clockwise direction (for example, for $\delta_F = 4°$ at $M_{\pi\pi} = 1.38$ GeV, we find $\Delta\delta_P = -10°$ and $\Delta r_P/r_P = -0.12$).

To determine the parameters of the f resonance we fit the resonance form, Eq. (17) with

$$D_2(y) = 9 + 3y + y^2,$$

to the values of $|r_D|$ in the region $1.14 < M_{\pi\pi} < 1.38$ GeV. Since $|r_D|$ is well determined and independent of the over-all phase this procedure should be reliable. We find

$$M_f = 1271 \pm 2 \text{ MeV}, \quad \Gamma_f = 182 \pm 4 \text{ MeV}$$

$$x_f = 0.81 \pm 0.01, \quad r_f = 0.70 \pm 0.08 f.$$

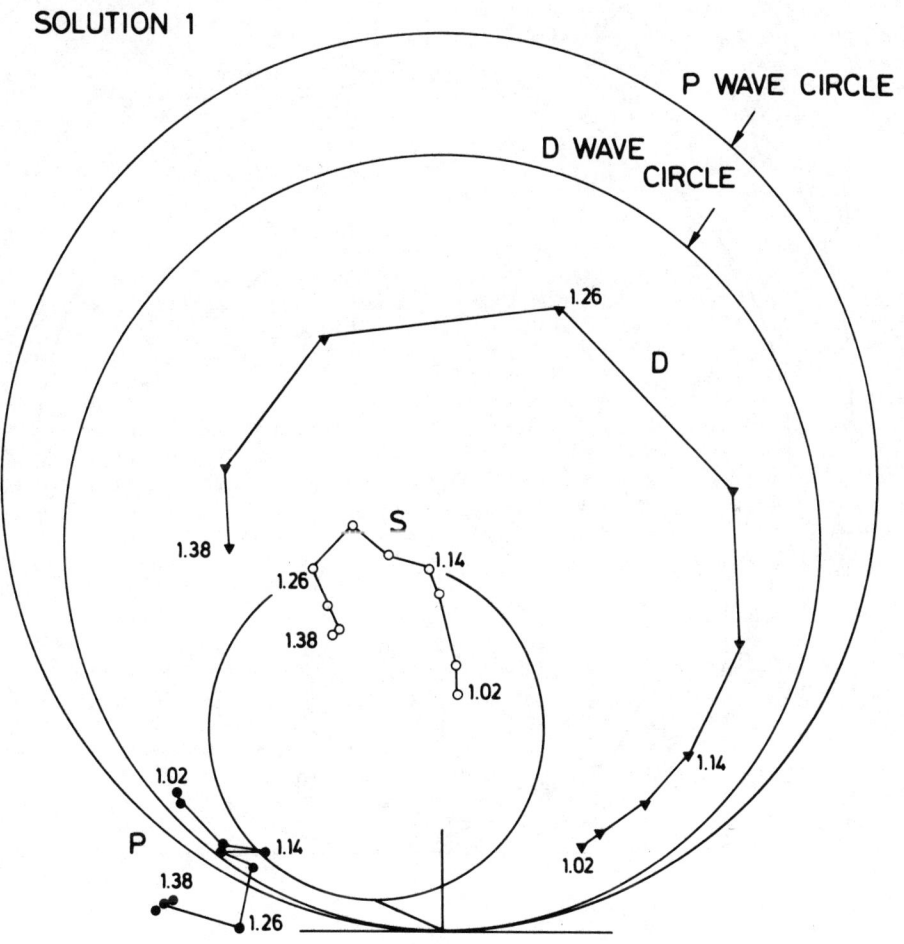

Figure 18 The physical solution, solution 1, for the $\pi\pi$ partial wave amplitudes above 1 GeV. The I = 0 S wave, P wave and I = 0 D wave unitarity circles are in the ratio $\frac{2}{3} : \sqrt{3} : \frac{2}{3}\sqrt{5}$.

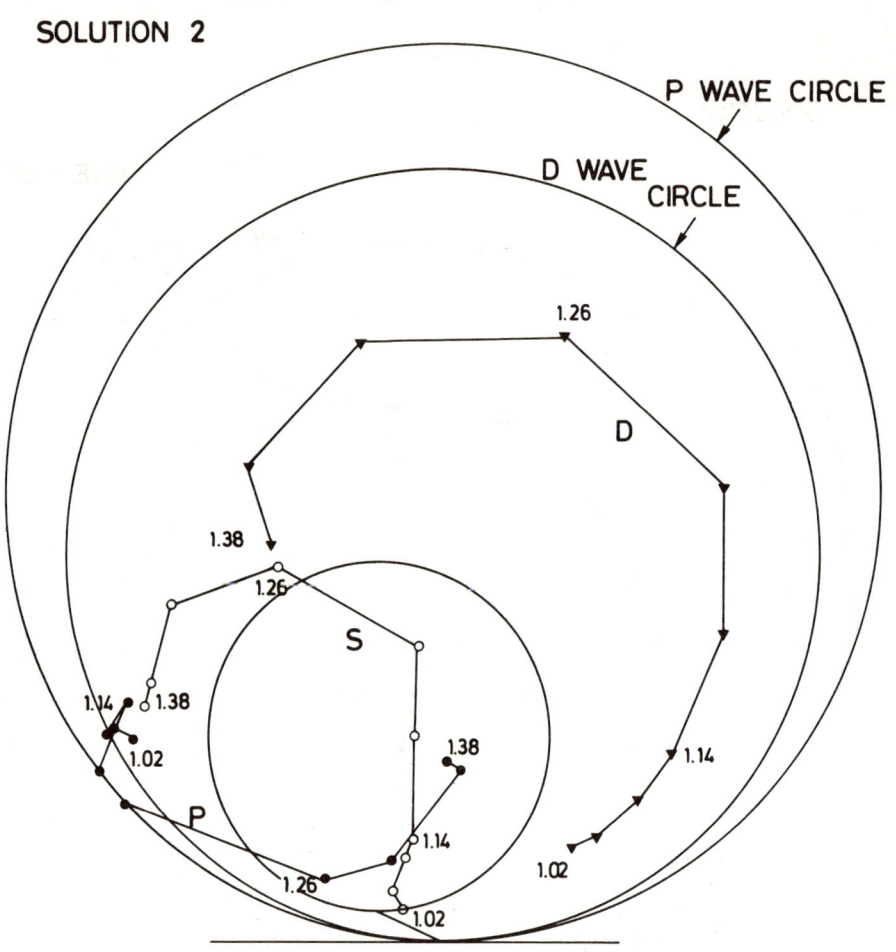

Figure 19 The ππ partial waves for the unphysical solution, solution 2.

Figure 18 also indicates the presence of a resonant $I=0$ S wave under the f, with a mass and width of roughly 1240 and 200 MeV respectively. To confirm these parameters we wish to include F waves and to extend the analysis to higher $M_{\pi\pi}$.

d) Phase coherent analysis :

As in the elastic region we have performed a phase shift analysis with a simplified form of parametrization, cf. Eqs. (20). We neglect A_2 exchange and assume that the absorptive background C is real relative to π exchange. We include a common slope factor, $\exp[b(t-\mu^2)]$, in all amplitudes. In addition to the parameters C and b, we have the magnitudes and relative phases of the S, P, D partial waves. In each 40 MeV mass bin these seven parameters give a good fit to the s channel moments in the region $0 < -t < 0.1$ GeV^2. The results for C, b and the $I=0$ S wave are shown in Fig. 11, where the over-all phase has been fixed by requiring the P wave to be elastic. The partial waves are very similar to the solution 1 results of Fig. 18. Moreover, we also find a solution almost identical to solution 2.

A surprising result[34,35] is the rapid decrease of the strength of absorption, C, with increasing $M_{\pi\pi}$. This could be anticipated from the data by inspection of the positions of the crossover zeros in the s channel $<Y_1^0>$ moments. For example, comparing the moments shown at $M_{\pi\pi} = 710$ MeV (Fig. 6) and at $M_{\pi\pi} = 1140$ MeV (Fig. 13) we see that at the higher mass the zeros occur at smaller $|t|$. The Williams' model, with $C=1$, is known[36,15] to give a good description of the small t data in the ρ region, but is unsatisfactory in the f region.

8. $K\bar{K}$ THRESHOLD

The data and the phase shift results indicate a dramatic effect in the $I=0$ S wave in the region of the $K\bar{K}$ threshold. Moreover, the effect occurs in a small range of $M_{\pi\pi}$. Up to 980 MeV and beyond 1 GeV the partial wave amplitudes do not change rapidly, and yet, in between, the $I=0$ S wave amplitude has altered drastically. Clearly to investigate this effect properly we require the moments for $\pi^-p \to \pi^-\pi^+n$, together with those for $\pi^-p \to K^-K^+n$ (and $K_1^0K_1^0n$), in smaller $M_{\pi\pi}$ bins. However, even a study of the existing data and phase shifts is illuminating.

First we performed an S, P, D wave M matrix fit directly on the $\pi^-p \to \pi^-\pi^+n$ data in the region $920 < M_{\pi\pi} < 1080$ MeV. That is we parametrized the production amplitudes as a function of $M_{\pi\pi}$, as well as of t. We did fits with and without effective range terms in the M matrix. As an alternative approach we also fitted the $I=0$ S wave phase shifts (of Figs. 7, 18, 19) to an M matrix over the same $M_{\pi\pi}$ region. We discuss the results of the second method first. The preferred fit was the one which joined the phase shifts of solution 1 below the $K\bar{K}$ threshold to those of solution 1 above threshold. An example of such a fit is shown in Fig. 20, corresponding to an $I = 0$ two channel $(\pi\pi, K\bar{K})$ M matrix

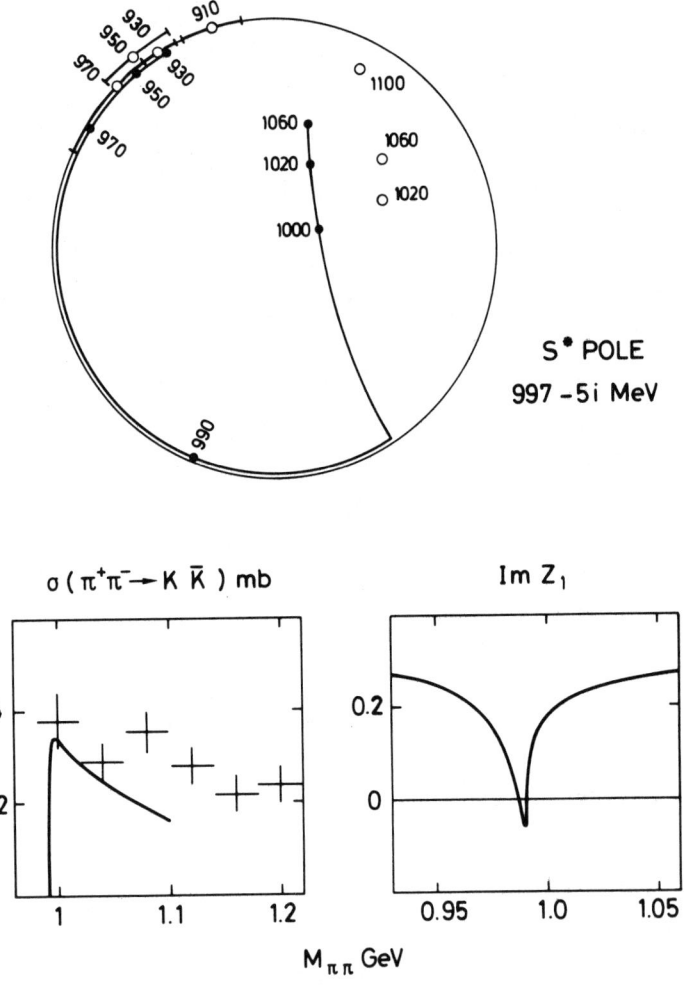

Figure 20 The $I = 0$ S wave unitarity circle. The open circles are solution 1 of the energy independent analysis. The black dots come from the sample M matrix fit. The last curve is the behaviour of Im z_1 (cf. Fig. 16) calculated using the M matrix and assuming that $\delta_S^2 = -20°$, $\delta_P = 156°$, $\delta_D^0 = 6°$.

$$M = \begin{pmatrix} 0.27 & 0.83 \\ 0.83 & -0.09 \end{pmatrix} f^{-1} \quad (25)$$

The resulting S wave amplitude has a pole on the second Riemann sheet of the complex energy plane at

$$S^*(pole) = 997 - i5 \text{ MeV}. \quad (26)$$

The existence of such a pole was suggested by Hoang[37] and subsequently confirmed by Protopopescu et al.[24] whose favoured solution gives $S^* = 997 - i27$ MeV. In Fig. 20 we compare the S wave contribution to $\sigma(\pi^+\pi^- \to K\bar{K})$, calculated from Eq. (25) to the data of Ref. 24. The narrower the peak in this cross-section, the larger the S^* coupling to the $K\bar{K}$ channel[9].

It is interesting to compare the above results with our first M matrix fit directly to the $\pi^-\pi^+$n data. There we find a wider S^* structure. The reason is that, although the fit basically follows phase shift solution 1, it jumps to solution 2 for a range of $M_{\pi\pi}$ in the immediate vicinity of the $K\bar{K}$ threshold. This is likely to happen in any energy dependent fit to such a narrow structure, and is well illustrated in Fig. 7 where the solution of Protopopescu et al.[24] goes from the region of our solution 1 to solution 2 just below the $K\bar{K}$ threshold. The situation is more confused as the energy independent analysis of Ref. 5 showed some indication of preferring a switch from solution 1 to solution 2 just below the $K\bar{K}$ threshold.

The rapidly changing S partial wave in the region of the $K\bar{K}$ threshold produces a sharp structure in the zero contours. For example, in Fig. 20 we show the behaviour of Im z_1 calculated using the parameters of Eq. (25) and reasonable constant values for the other phase shifts. In the energy dependent fit Im z_1 is found to dip to zero less sharply. Comparison with the energy independent results of Fig. 16 again emphasizes the need for data in smaller $M_{\pi\pi}$ bins in this mass region.

Since we estimate such a small width for the S^* it is clear that the parameters of Eqs. (25) and (26) are not reliably determined from the data in 20-40 MeV mass bins. However, the point we wish to make is that the S^* structure appears to be narrower than hitherto thought.

ACKNOWLEDGEMENTS

We thank C. Michael, D. Morgan and W. Ochs for interesting discussions.

REFERENCES

1. C. Goebel, Phys.Rev.Letters $\underline{1}$, 337 (1958).
2. G.F. Chew and F.E. Low, Phys.Rev.Letters $\underline{113}$, 1640 (1959).
3. See, for example, the review articles by :
   P.E. Schlein, Proc. of the International School of Subnuclear Physics, Erice (1970) ;
   J.L. Petersen, Physics Reports $\underline{2C}$, 157 (1971) ;
   D. Morgan, Proc. of the VII Finnish Summer School, Loma-Koli (1972).
4. CERN-Munich Collaboration : G. Grayer, B. Hyams, C. Jones, P. Schlein, P. Weilhammer, W. Blum, H. Dietl, W. Koch, E. Lorenz, G. Lütjens, W. Männer, J. Meissburger, W. Ochs and U. Stierlin, to be published.
5. G. Grayer et al., Experimental Meson Spectroscopy (AIP), Proc. of III Philadelphia Conference (1972), p. 5.
6. G. Grayer et al., Phys.Letters $\underline{35B}$, 610 (1971).
7. J.T. Carroll et al., Phys.Rev.Letters $\underline{28}$, 318 (1972).
8. R. Odorico, Phys.Letters $\underline{38B}$, 411 (1972).
9. S.M. Flatté et al., Phys.Letters $\underline{38B}$, 232 (1972).
10. G.L. Kane and M. Ross, Phys.Rev. $\underline{177}$, 2353 (1969) ;
    C.D. Froggatt and D. Morgan, Phys.Rev. $\underline{187}$, 2044 (1969).
11. G.L. Kane, Experimental Meson Spectroscopy (Columbia University Press, 1970), p. 1.
12. A. Kotanski and K. Zalewski, Nuclear Phys. $\underline{B4}$, 559 (1968) ; $\underline{B20}$, 236 (1970) E ; $\underline{B22}$, 317 (1970).
13. G. Grayer et al., Nuclear Phys. $\underline{B50}$, 29 (1972).
14. W. Männer, Contribution to International Conference on $\pi\pi$ Scattering and Associated Topics, Tallahassee (March 1973).
15. P. Estabrooks and A.D. Martin, Phys.Letters $\underline{41B}$, 350 (1972).
16. See, for example :
    M. Ross, F.S. Henyey and G.L. Kane, Nuclear Phys. $\underline{B23}$, 269 (1970).
17. C. Michael, Springer Tracts of Modern Physics $\underline{55}$, 174 (1970).
18. P.K. Williams, Phys.Rev. $\underline{D1}$, 1312 (1970).
19. C.D. Froggatt and D. Morgan, Phys.Letters $\underline{40B}$, 655 (1972).
20. J.D. Kimel and E. Reya, Phys.Letters $\underline{42B}$, 249 (1972).
21. W. Hoogland, Contribution to International Conference on $\pi\pi$ Scattering and Associated Topics, Tallahassee (March 1973).
22. P. Baillon et al., Phys.Letters $\underline{38B}$, 555 (1972).
23. J.P. Baton, G. Laurens and J. Reignier, Phys.Letters $\underline{33B}$, 525 and 528 (1970).
24. S.D. Protopopescu et al., Phys.Rev. $\underline{D7}$, 1279 (1973).
25. See, for example :
    G.C. Fox, Experimental Meson Spectroscopy (AIP), Proc. of III Philadelphia Conference (1972).
26. W.D. Apel et al., Phys.Letters $\underline{41B}$, 542 (1972).
27. E.I. Shibata, D.H. Frisch and M.A. Wahlig, Phys.Rev.Letters $\underline{25}$, 1227 (1970) ;
    A. Skuja et al., LBL Preprint 1020 (March 1973).
28. A. Barbaro-Galtieri, Advances in Particle Physics (Wiley, 1968), Vol. 2, p. 212.

29. W. Beusch, Experimental Meson Spectroscopy (Columbia University Press, 1970), p. 185.
30. A. Gersten, Nuclear Phys. B12, 537 (1969).
31. E. Barrelet, Nuovo Cimento 8A, 331 (1972).
32. C. Schmid, Proc. of 1971 Amsterdam International Conference on Elementary Particles (North Holland, 1972), p. 275.
33. M. Pennington and C. Schmid, Phys.Rev. D7, 2213 (1973).
34. W. Ochs and F. Wagner, to be published in Nuclear Phys.
35. A.D. Martin and P. Estabrooks, to be published in Proc. of VIII Rencontre de Moriond (1973).
36. P. Baillon et al., Phys.Letters 35B, 453 (1971).
37. F.T. Hoang, Nuovo Cimento 61A, 325 (1969).

HIGH ENERGY π-π SCATTERING

Prepared by

W. D. Walker
Department of Physics, Duke University
Durham, North Carolina 27706

ABSTRACT

A study has been done of $\pi^- - \pi^+$ and $\pi^- - \pi^-$ scattering using the reactions $\pi^- + p \rightarrow \pi^- + \pi^+ + n$ and $\pi^- + p \rightarrow \pi^- + \pi^- + \Delta^{++}$. The elastic scattering cross section shows a steady decrease from 120 mb (for $\pi^- - \pi^+$) in the region of the ρ^0 to 1 - 2 mb in the 2 - 4 GeV/c^2 mass range. Estimates of the total cross section are also made using the optical theorem. Both the $\pi^- - \pi^+$ and $\pi^- - \pi^-$ cross sections seem to be about 15 mb in the high mass region.

The study of high energy π-π scattering is about as interesting as the study of high energy π-p or p-p scattering. The things that one learns are about the same. One learns something about the size and transparency of the particles colliding. One other thing that comes as a by-product of these studies is the possibility of determining, by using the optical theorem, the total π-π cross section, which is a quantity of considerable interest. The data that is reported here is the result of the study of $\pi^- p \rightarrow \pi^- + \pi^+ + n$ at 25 GeV/c and a series of experiments on $\pi^- p \rightarrow \pi^- + \pi^- + \Delta^{++}$ which were done at bombarding energies of 5, 7, 7.5, 13, 16, 20, 25 GeV/c. Those events were collected in the course of experiments done at Illinois, Wisconsin, Harvard, and SLAC.[1] The numbers of events that we have are listed in Table I.

Table I. Data tabulation

P_{in}	No. Events $(t_{p\Delta}) \leq .5$	All $\Delta^{++} + \pi^-\pi^-$	Total No. $(3\pi)^- + p$
5	3345	5645	28 K
7 - 7.5	1945	2826	16 K
13 - 16	592	808	10 K
20 - 25	794	1046	

The first data to be discussed is that obtained in the course of the Wisconsin experiment of W. Robertson et al.[2] This work was on the time honored process $\pi^- + p \rightarrow \pi^- + \pi^+ + n$. The experiment was done at a bombarding momentum of 25 GeV/c. The π-π cross sections were determined using the method of Durr and Pilkuhn.[3] By restricting the analysis to events with small Δ^2, it is possible to ignore the spin dependent form factor corrections.

Our mass spectra are shown in Figures 1 and 2. In Figure 1 one can see the ρ^0 and f^0 clearly in the data, but not the g^0. The π^+-π^- dipion mass spectrum is rather flat out to a mass of 4 GeV/c^2. This fact alone tells one that the π-π elastic cross section must fall fairly rapidly as a function of dipion mass, in that there are phase space weighting factors that tends to increase like $m^{*3}_{\pi-\pi}$.

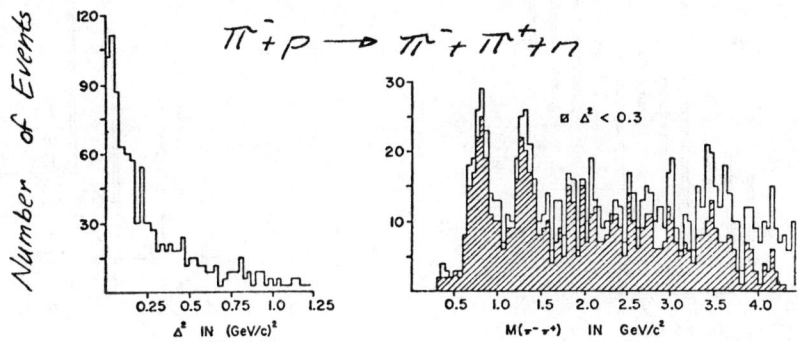

Fig. 1. $\pi^+\pi^-$ Mass spectrum with momentum Transfer Spectrum.

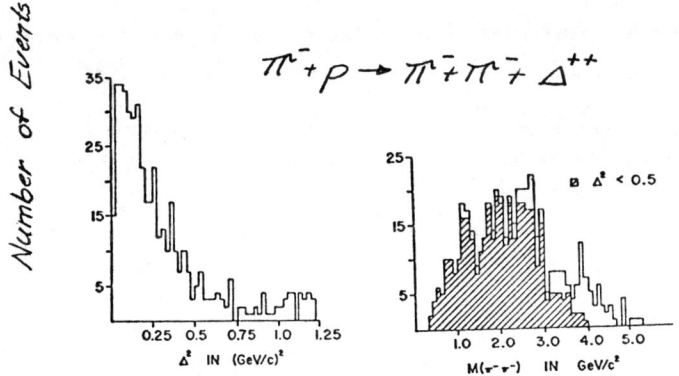

Fig. 2. $\pi^-\pi^-$ Mass spectrum from 25 GeV/c experiment.

The angular distributions are shown in Figure 3. These angular distributions seem typical of diffraction scattering. It seems likely that the imaginary part of the π-π scattering amplitude dominates as more and more partial waves are involved. Thus it is likely that one can obtain a reasonable estimate of the total π-π cross section by using the optical theorem. The results of the estimates of the total cross section are shown in Figure 4.

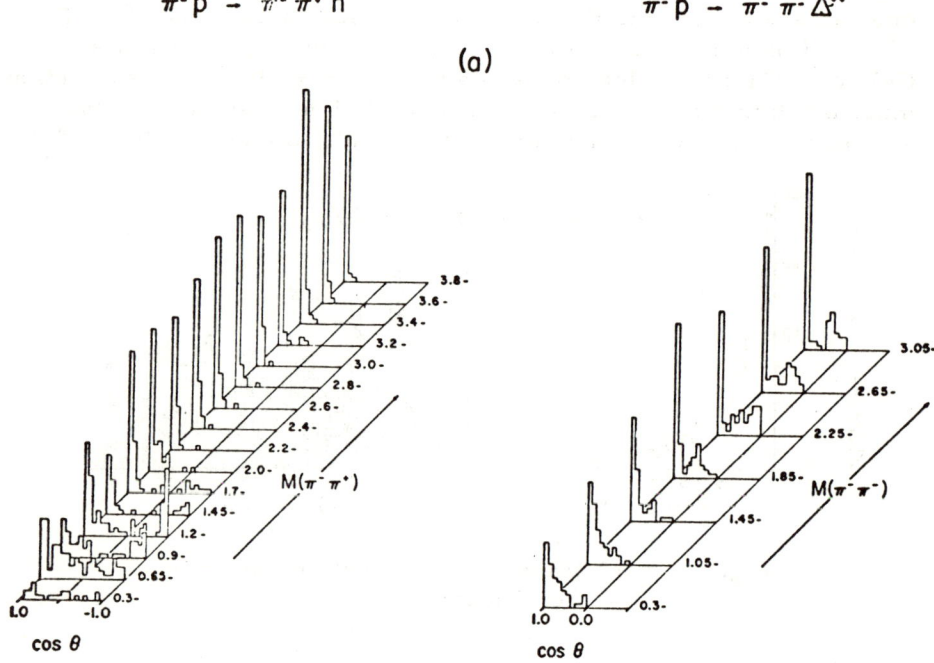

Fig. 3a. π-π Angular distribution from the 25 GeV/c experiment.

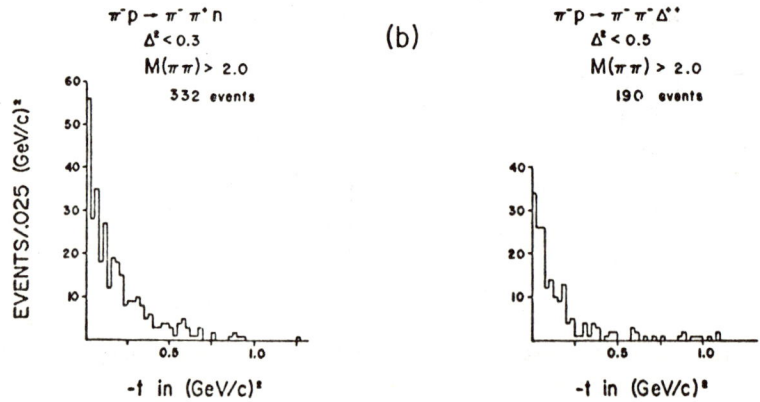

Fig. 3b. Momentum transfer spectrum for the high mass event.

The results are interesting in that they show a decrease in the π^+-π^- cross section from 120 mb in the region of the ρ to 1 - 2 mb in the high mass region. The π^--π^- cross section shows very similar behavior to the π^+-π^- in the high mass region.

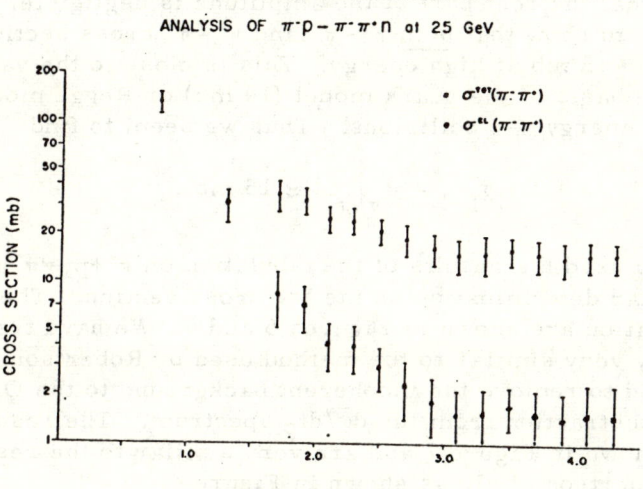

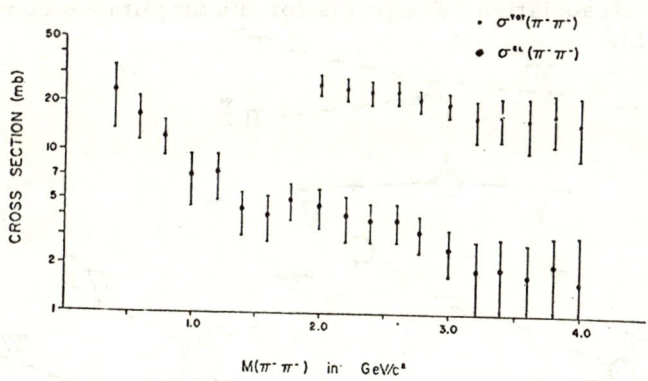

Fig. 4. π-π Elastic cross sections and optical theorem estimates of the total cross section.

The form of the optical theorem we used is

$$\left(\frac{d\sigma}{dt}\right)_{t=0} = \frac{1}{16\pi} \sigma_T^2 .$$

We assume that the real part of the amplitude is negligible. The results seem to show that both π^+-π^- and π^--π^- cross sections tend to a value of ~ 15 mb at high energy. This is close to the value expected on the basis of the quark model (14 mb) or Regge model (16 mb) for high energy π-π collisions. Thus we seem to find

$$\sigma_{I=2} = \sigma_{\pi^+\pi^-} \approx 15 \text{ mb} .$$

We have also used the results of the tabulation on $\pi^-+p \rightarrow \pi^-\pi^-+\Delta^{++}$ to do a similar determination on the I=2 cross section. The results of this tabulation are shown in Figures 5 and 6. We have treated the data in a way very similar to the method used by Robertson et al.[2] We have tried to remove the incoherent background to the O.P.E. process by subtraction from the $d\sigma/dt_p$ spectrum. The results obtained are shown in Figure 7 and are very similar to the results obtained by Robertson et al. as shown in Figure 4.

Next we will discuss the difficulty with competing amplitudes that are coherent with the usual OPE amplitude.[4] The complications come from the presence of competing amplitudes. The simplest amplitudes which are likely to be present are those which contribute to diffractive dissociation. Diagrams for the amplitudes considered are shown below.

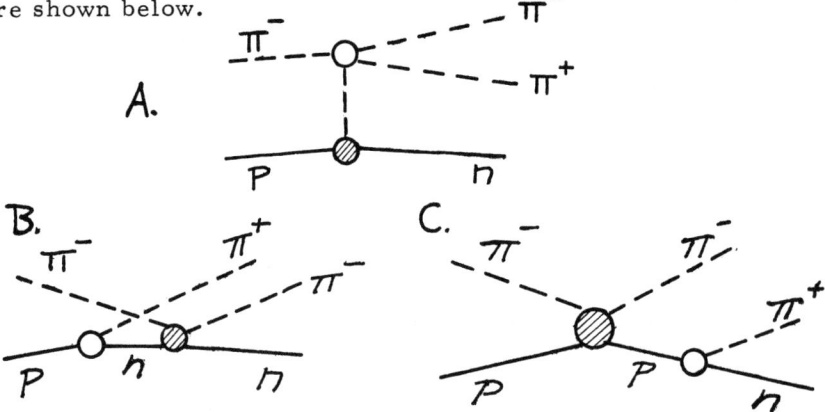

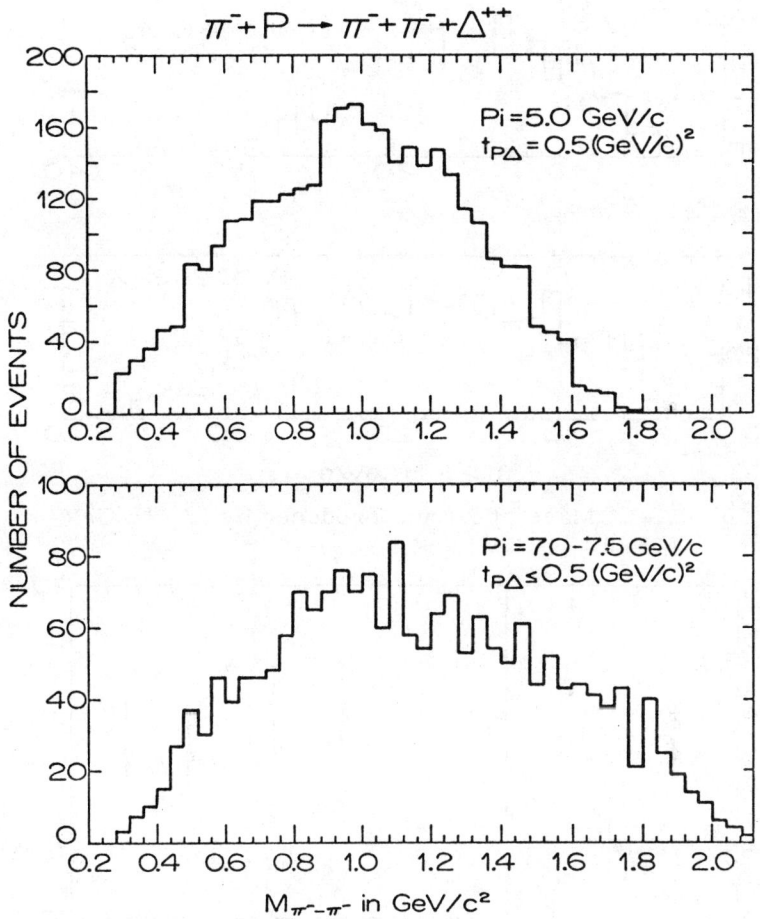

Fig. 5. $\pi^- - \pi^-$ Mass spectrum of events produced by 5 and 7 - 7.5 GeV/c π^-.

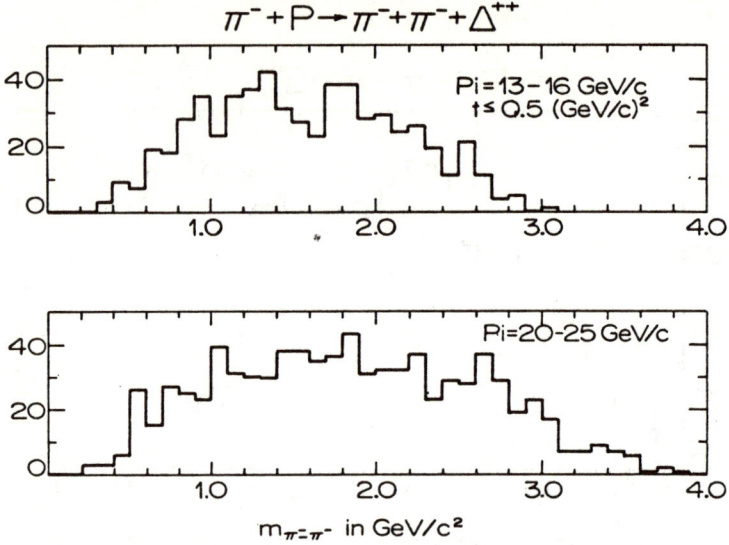

Fig. 6. $\pi^- - \pi^-$ Mass spectrum produced by 13 - 16 GeV/c and 20 - 25 GeV/c π^-.

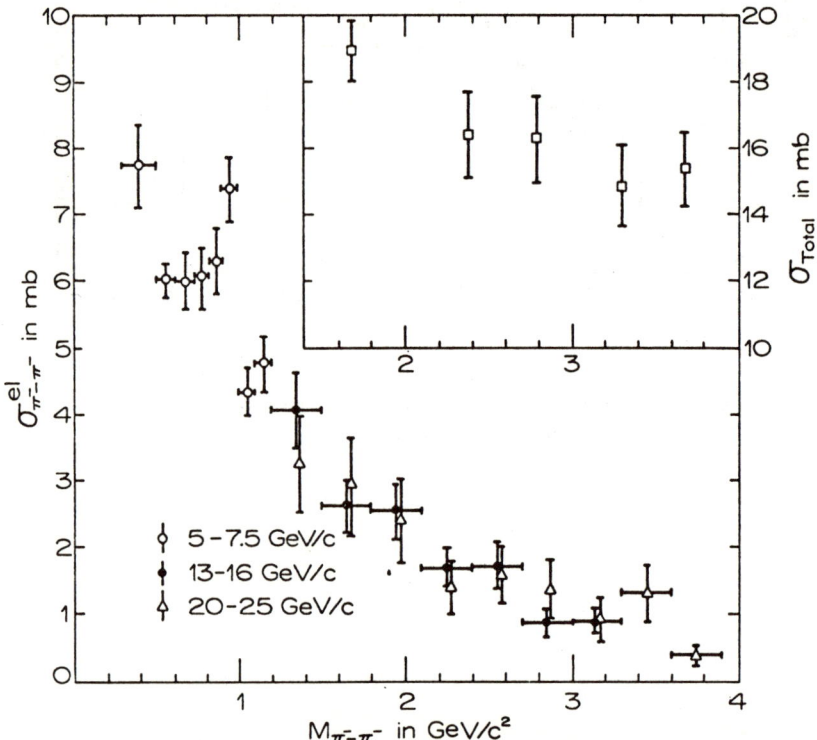

Fig. 7. Elastic and total cross section deduced from the mass spectra of Figures 5 and 6.

Amplitudes B and C which have opposite signs and consequently tend to cancel each other are given by the following expressions:[4]

$$\langle \lambda' | T | \lambda \rangle = \sqrt{2} g \, (U'_\lambda | \gamma_5 \, \not{K}'' | U_\lambda) \times \left[\frac{T_{\pi^- p}}{S_{\pi^+ n} - M^2} + \frac{T_{\pi^- n}}{\mu^2 - 2\varepsilon'' M} \right]$$

where

$$T_{\pi-N} = i \frac{M^*_{\pi-N} K_{\pi-N}}{M} e^{1/2 \beta t_{\pi-\pi}} \sigma_{\pi-N}$$

$\varepsilon'' = $ Energy of the π^+ in the lab.

For our 25 GeV/c $\pi^- p \to \pi^- + \pi^+ + N$ data we went through on an event by event basis and calculated the three amplitudes. One has to do something about damping the pion emission since

$$\langle \gamma_5 \not{K}'' \rangle \sim |K''|$$

which we damped with an $e^{-1/2 \beta K''^2_\pi}$ term. Even after doing this we find a sizable contribution to the amplitude (10 - 20%) at small values of $t_{\pi-\pi}$.

We have calculated the effects of these amplitudes previously in the mass region of the ρ and f^0 and found them to be quite small.[4] Since the $\pi-\pi$ cross sections tend to become quite small in the high mass region, these other two amplitudes tend to give appreciable effects (10 - 20% at small $t_{\pi-\pi}$). It is certainly worthwhile to do a more careful calculation of these amplitudes as the experiments improve in statistical accuracy. The effect of these competing amplitudes is to decrease the cross section in the 2 - 3 BeV/c² range and to raise the cross section in the 3 - 4 BeV/c² range. It seems that the amplitudes must be considered when determining a small $\pi-\pi$ cross section. Essentially one can think of the nucleon undergoing the Yukawa process of emitting a virtual pion. In the intermediate state there is a nucleon and a virtual pion present. The incident particle can scatter off of either particle. If the $\pi-\pi$ cross section is large, then the OPE diagram will dominate. For small values of $\sigma_{\pi-\pi}$ and $t_{\pi-\pi}$ these amplitudes tend to become rather comparable in size.

In conclusion it appears that the $\pi^- - \pi^-$ cross section and the $\pi^- - \pi^+$ cross section are essentially equal for high mass values. It would be interesting to see with what precision this equality holds.

A way of doing this would be to look at the process $\pi^- + \pi^+ \to \pi^\circ + \pi^\circ$ at high mass value. The amplitude for this process is $\sqrt{2/3}[A_2 - A_0]$. A measurement of the charge exchange cross section would give at least an upper limit on the difference of the $I = 0$ and $I = 2$ cross section.[5] The angular distribution likewise would give unique information concerning the difference in these two scattering amplitudes. The π-π cross section is apt to be very small if the present results are any indication. This process would be affected by similar coherent processes as referred to above.

I should thank Dr. W. Robertson and Mr. S. Dhar for their generous help in preparing this talk. I also wish to thank my colleagues at Illinois, Harvard and SLAC for the use of their data.

REFERENCES

1. G. Ascoli et al., Phys. Rev. Letters **26**, 929 (1971). A portion of the events from this compilation are used. No CERN events are included.
2. W. Robertson, W. D. Walker and J. L. Davis, Phys. Rev. D (1973).
3. H. P. Dürr and H. Pilkuhn, Nuovo Cimento **40**, 899 (1965).
4. B. Y. Oh, Ph.D. thesis, University of Wisconsin, 1968.
   B. Y. Oh, A. F. Garfinkel, R. Morse, W. D. Walker, J. D. Prentice, E. C. West and T. S. Yoon, Phys. Rev. D**1**, 2494 (1970).
5. For a calculation of what the π°-π° cross section at high energy might look like see G. Cohen-Tannoudji, R. Lacaye, F. S. Henyey, D. Richards, W. J. Zakrzewski and G. L. Kane, Nuc. Phys. B **45**, 109 (1972).

Chapter 2. π-π Amplitudes and Inelastic Effects

ZEROS IN $\pi\pi$ SCATTERING

M. R. Pennington[*]
Lawrence Berkeley Laboratory, University of California
Berkeley, California 94720

ABSTRACT

We discuss how zeros in $\pi\pi$ scattering relate various dynamical aspects of the scattering amplitude. In particular, we consider the way in which many differing roles, such as the Legendre zeros of resonances, double-pole-killing zeros, the high energy crossover zero and the Adler zero, can be played by the same zero contour. We emphasize that the study of the scattering amplitude is essentially one of looking at zeros travelling across the Mandelstam plane.

INTRODUCTION TO ZEROS IN $\pi\pi$ SCATTERING

The purpose of this talk is to advertise a particularly useful tool for studying the scattering amplitude and that is the investigation of the zeros of the amplitude. We will illustrate, in a fairly straightforward way, the close relationship between the zero structure of the amplitude and various dynamical mechanisms. We shall, of course, only consider zeros in $\pi\pi$ scattering, but many of the properties we discuss are features of more general scattering processes.[1,24]

Before we can begin this discussion of what role zeros play we must introduce zeros into the Mandelstam plane. This we do in the simplest possible way. Whilst we cannot determine the scattering amplitude, $F(s,t)$, everywhere, we know that in certain limited energy regions we can make a reasonable approximation to it. For example, at high energies, near the forward or backward directions, we can supposedly represent $F(s,t)$ by a Regge expansion, involving only the rightmost Regge singularities. Similarly, in the low energy region, in the neighborhood of a narrow resonance, we can reasonably well describe the amplitude by a simple Breit-Wigner form. So for $s \in (m^2 - m\Gamma, m^2 + m\Gamma)$, where m, Γ are the resonance mass and total width respectively, we have:

$$F(s,t) = (2\ell + 1) \frac{m\Gamma x}{m^2 - s - im\Gamma} P_\ell\left(1 + \frac{2t}{s - 4\mu^2}\right) \qquad (1)$$

+ background

[*]Work supported by the U. S. Atomic Energy Commission.

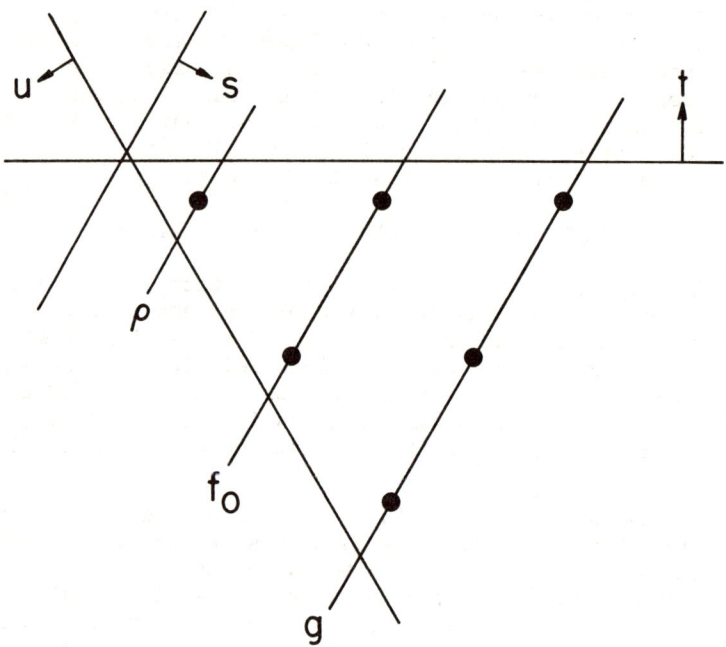

Fig. 1. The three most important low energy $\pi\pi$ resonances with spin are shown in the s-channel of the Mandelstam plane. The black dots mark the Legendre zeros of these resonances.

(ℓ is the spin of the resonance and $x = \Gamma_{e\ell}/\Gamma$). Such a simple form is a good description of the amplitude at least for narrow resonances, e.g., for the ρ, f_0, and g. We deliberately omit the ϵ or σ and S^* resonances, as neither is well-described by Eq. (1).

Let us notice a very important feature of the resonant part of Eq. (1), and that is the Legendre polynomial $P_\ell(\cos \theta_s)$. This function has ℓ zeros inside the physical region, i.e., for $\cos \theta_s \in [-1,+1]$, where ℓ is the spin of the resonance. Indeed we can regard these zeros as dynamical objects which give the spin to the resonance. We see that even though we are looking at the amplitude close to a pole of the S-matrix the zero structure is equally important in determining the amplitude. If the background in Eq. (1) is negligible, the whole amplitude will have ℓ zeros in the neighborhood of the spin ℓ resonance. In practice, of course, the background is not negligible and the zeros are shifted from their simple Legendre zero positions. However, as we shall see, unitarity generally constrains these zeros not to move far from where $P_\ell(\cos \theta) = 0$.

Let us look where these zeros occur in the Mandelstam plane, which is where all the action takes place (see Fig. 1). The ρ has one zero, the f_0 two, the g three and so on. In Fig. 1 we have marked, for simplicity, the position of the Legendre zeros of these resonances, as if the background were negligible. So we now have zeros sitting in the Mandelstam plane, what do they do?

Well, the usefulness of zeros rests in the remark that the zeros of an analytic function of two variables are not isolated, as shown in Fig. 1, but are continuous. Thus we have zero contours (to be defined below), which traverse the Mandelstam plane relating various dynamical aspects of the scattering amplitude. As we shall see, not only are the Legendre zeros of different resonances related to each other, but are also related to high energy Regge zeros. In this way zero contours provide a natural vehicle for duality. We shall also discuss how zero contours enable us to relate low energy resonance dynamics to chiral dynamics.

Before we can connect up the isolated zeros plotted in Fig. 1, we have to define what we mean by a zero contour and this is what we do next.

ZERO CONTOUR DEFINED

Consider the scattering amplitude, $F(s,t)$, in the s-channel with s real (since this is where we have its value experimentally), then $F(s,t) = 0$ for complex values of $t = t_0(s)$. We then refer to the projection of this complex curve on the real t axis as a zero contour, i.e.,

$$t = \operatorname{Re} t_0(s) \qquad (2)$$

is a zero contour. In some cases it will be more useful to treat the amplitude as a function of s and $z = \cos\theta_s = 1 + 2t/(s-4\mu^2)$ rather than s and t. Then the same zero contour will be at $z = \text{Re } z_0(s)$ where $z_0 = 1 + 2t_0/(s - 4\mu^2)$.

Since experimentally we look not only at real s but also real z, we next ask what does a zero at $z = \text{Re } z_0 + i \text{ Im } z_0$ mean in terms of measurable quantities, in particular for $|\text{Re } z_0| < 1$. Experimentally we measure $d\sigma/d\Omega$ or $|F|^2$, so let us look at that. Well, under certain conditions, we will see a minimum in the angular distribution at $z = \text{Re } z_0(s)$, so that the zero contours defined above are just the paths of dips in the angular distribution. Now we know what $\text{Re } z_0$ means, what about $\text{Im } z_0$? This also has a well-defined interpretation. $\text{Im } z_0$ is related to the magnitude of $|F|$ at $z = \text{Re } z_0$. This is easy to see, since the smaller $|\text{Im } z_0|$ is, the closer the zero is to a real zero and so the closer $|F|$ at the minimum is to zero. Now it is only nearby zeros, (with $|\text{Re } z_0| < 1$), which have small imaginary parts, that produce marked dips in the differential cross sections. However, these are not the only zeros which play a role in understanding the scattering amplitude as we shall see, so we will, in fact, need our definition of zero contours given by Eq. (2) which is more general than that of just paths of minima of $d\sigma/d\Omega$.

ZEROS NEAR RESONANCES

Before we actually plot zero contours in the Mandelstam plane of Fig. 1, we discuss briefly what we might expect these contours to do when they approach resonance structures. We consider two examples of physical interest, for the purposes of illustration:
(i) a narrow p-wave resonance with a smooth s-wave background,
(ii) a narrow s-wave structure with a smooth p-wave background.
(i) For definiteness, consider $\pi^-\pi^0$ elastic scattering in the s-channel. Then we ask: where does $F^{s1}(s,z) + F^{s2}(s,z) = 0$ in the region of the ρ resonance? In this region we have just s and p waves and so consider*

---

* In terms of partial waves it is clear why our definition of zero contours of Eq. (2) considers the amplitude for real s and complex t or z. Experimentally we know the partial waves only for real s, but we can use the partial wave expansion to continue the amplitude to complex z. More general definitions of zero contours with both complex s and t are not useful phenomenologically although formally well defined in model amplitudes.[2]

$$3f_1^1(s)z + f_0^2(s) = 0 \ . \tag{3}$$

In the ρ region we can represent $f_1^1(s)$ by a unitary Breit-Wigner form and use a phase shift representation for $f_0^2(s)$, so that

$$\text{Re } z_0(s) = -\frac{1}{3}\frac{(m^2-s)}{m\Gamma}\frac{1}{2}\sin 2\delta_0^2(s) - \frac{1}{3}\sin^2 \delta_0^2(s). \tag{4}$$

We know that if the background were negligible the amplitude vanishes at $z = 0$, so we first ask how far is the zero shifted at $s = m^2$ by a nonzero background? From Eq. (4) we have

$$\text{Re } z_0(m^2) = -\frac{1}{3}\sin^2 \delta_0^2(m^2) \tag{5}$$

which implies

$$0 \geq \text{Re } z_0(m^2) \geq -1/3 \ . \tag{6}$$

We see that even if the background were as large as possible, unitarity constrains the zero contour to cross $s = m^2$ not far from $z = 0$. Indeed in this particular case the background is exotic, so $\sin^2 \delta_0^2$ is small and $\text{Re } z_0(m^2) \simeq -1/20$.

Next we consider the energy dependence of the zero away from $s = m^2$, e.g., $s\epsilon(m^2 - m\Gamma, m^2 + m\Gamma)$. If δ_0^2 is slowly varying, the zero contour is close to linear [Eq. (4)] as depicted in Fig. 2a. If we know nothing about δ_0^2, other than it is between 0 and $-\pi/2$, the zero contour is constrained by unitarity to pass between the hatched boundaries of Fig. 2a.

For $\pi^+\pi^-$ elastic scattering the background in the ρ region has both $I = 0$ and $I = 2$ s-waves. Nonetheless unitarity still requires

$$0 \geq \text{Re } z_0(m^2) \geq -1/3$$

so that the zero cannot be shifted too far away from $z = 0$, in this case too.

In general for resonances with spin $\ell \geq 1$, zero contours will cross the resonance positions smoothly and close to where the appropriate Legendre polynomial vanishes. However, when we have an s-wave resonance this need not be the case and this is what we next consider.

(ii) The situation of a smooth p-wave background with a rapidly varying s-wave occurs between 900 and 1000 MeV in $\pi^+\pi^-$ elastic

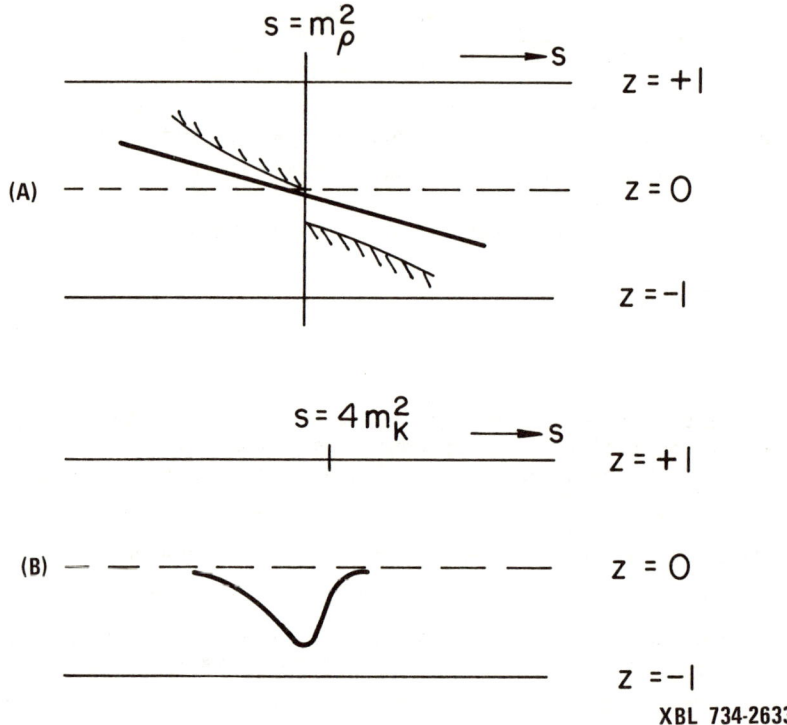

XBL 734-2633

Fig. 2. (a) The solid line marks the zero contour in $\pi^-\pi^0$ scattering at it crosses the ρ resonance, with δ_0^2 assumed to be a slowly varying function of energy for $s \simeq m_\rho^2$. With $\delta_0^2 \in (-\frac{\pi}{2}, 0)$, but otherwise unknown, the zero is constrained by unitarity to pass between the hatched boundaries.
(b) An example of how a zero contour behaves when crossing a narrow s-wave structure, with a smooth p-wave background, as occurs in $\pi^+\pi^-$ elastic scattering near $K\bar{K}$ threshold.

scattering. Then the zero contour varies rapidly as shown in Fig. 2b, using $\delta_1^1 \sim 150°$ and δ_0^0 varying from $80°$-$200°$. We should therefore expect to see such a wiggle in the zero contour near $K\bar{K}$ threshold where such a rapidly varying δ_0^0 has been found.[3]

We are now almost in a position to plot zero contours on Fig. 1. However, a remark is called for about the number of zeros increasing with increasing energy. Near threshold the amplitude is just s-wave. As the p-wave begins to grow, a zero enters the physical region from outside, either through the backward or forward direction. It steadily moves towards the center of the physical region when the p-wave resonates. Then as the d-wave begins to grow, a second zero enters the physical region. If the d-wave is $I = 0$ this zero, together with the Legendre zero of the ρ, will become the two Legendre zeros of the f_0. As higher and higher partial waves grow, more and more zeros enter the physical region producing higher and higher spin resonances. Whether the zero contour enters the physical region through the forward or backward direction depends on the relative sizes and signs of the contributing phase shifts in that energy region. However, the ℓth zero's entrance is mainly controlled by the ℓth and $(\ell-1)$th partial waves. As the energy increases we should therefore expect to see increasing numbers of zeros entering the physical region.

ZERO CONTOURS FROM THE DATA

Let us look at the actual pattern of zeros[4] for the two reactions $\pi^+\pi^-$ and $\pi^-\pi^0$ elastic scattering as determined from the s-, p-, and d-wave phase shifts of Protopopescu et al.[3] up to 1100 MeV (Fig. 3).

$\pi^+\pi^-$ We see the first zero enters the physical region through the backward direction and moves towards the center of the physical region, crosses $s = m_\rho^2$ close to $z = 0$, then wiggles near $K\bar{K}$ threshold as expected because of the rapid variation of δ_0^0. The zero then lines up ready to become one of the Legendre zeros of the f_0 resonance. As the $I = 0$ d-wave grows a zero approaches the physical region and enters through the forward direction* and moves ready to become the other Legendre zero of the f_0. Note that since the s and t channels are identical, the first zero looks as though it may well pass through the Mandelstam triangle as suggested by the dotted line — the possible significance of this will be discussed later.

---

* We cannot determine the zero contour far outside the physical region as a finite number of partial waves provides a poor representation there.

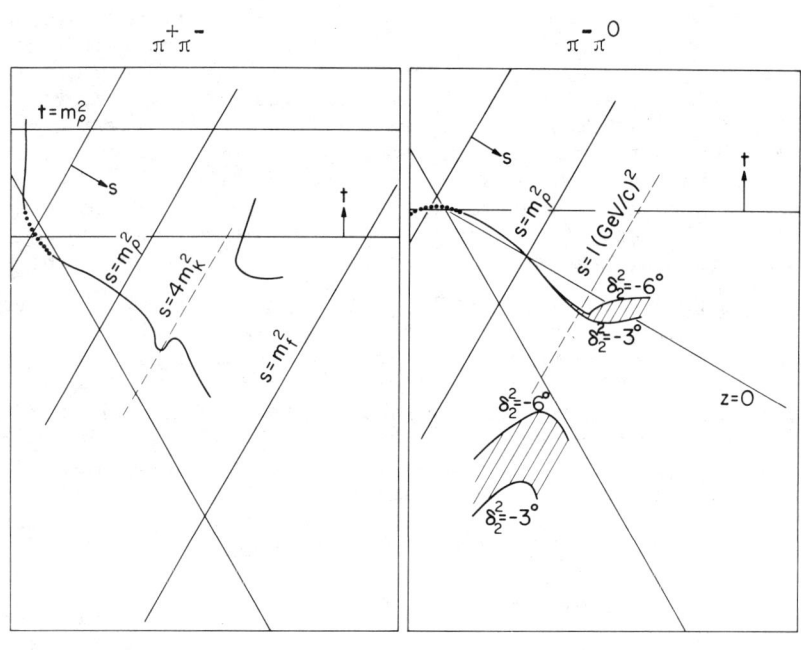

Fig. 3. The zero contours for $\pi^+\pi^-$ and $\pi^-\pi^0$ elastic scattering as determined in Ref. 4 from the s-, p-, and d-wave phase shifts of Protopopescu et al. and Baton et al. The dotted line is shown to illustrate a possible continuation of the first zero contour through the Mandelstam triangle. Since in $\pi^-\pi^0$ elastic scattering the $I = 2$ d-wave is not well determined we have a range of positions of the zero contours, marked by the shaded regions. These were obtained by varying δ_2^2 at 1 GeV from $-3°$ to $-6°$.

97

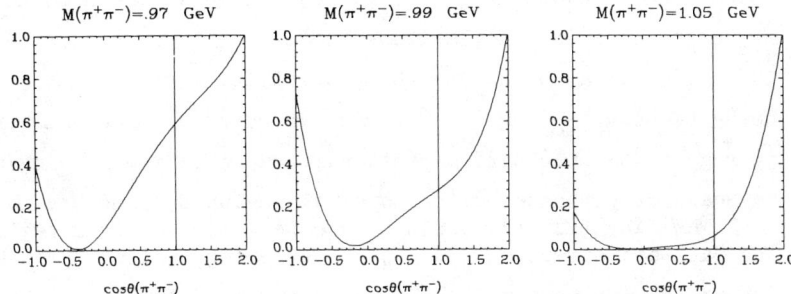

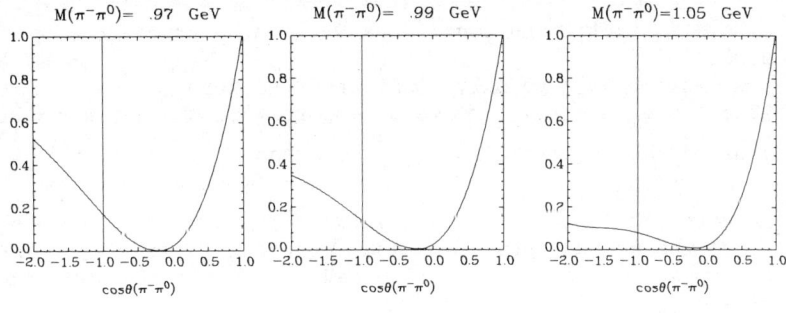

Fig. 4. The angular distributions (in arbitrary units) for $\pi^+\pi^- \to \pi^+\pi^-$ and $\pi^-\pi^0 \to \pi^-\pi^0$ scattering at 970, 990, and 1050 MeV.

In $\pi^-\pi^0$ elastic scattering the s and u channels are the same. The first zero leaves the region of the Mandelstam triangle enters the physical region through the forward direction crosses the line $z = 0$, this time very close to $s = m_\rho^2$ since the s-wave background is exotic (as in Fig. 2a). Then as the zero contour moves on there is some latitude because δ_2^2 is rather uncertain. The shaded area represents the possibility for δ_2^2 at 1 GeV being between $-3°$ and $-6°$. As can be seen, a second zero approaches the backward direction, because of the growing $|f_2^2|$, but has not yet entered the physical region by 1100 MeV.

The fact that the first zero enters the physical region through the backward direction in $\pi^+\pi^-$ scattering, whilst through the forward direction for $\pi^-\pi^0$ scattering is determined by $\mathrm{sgn}(2\delta_0^0 + \delta_0^2) = \mathrm{sgn}(\delta_1^1)$ and $\mathrm{sgn}(\delta_0^2) = -\mathrm{sgn}(\delta_1^1)$, respectively. Similarly, the second zero approaches the forward direction for $\pi^+\pi^-$ scattering as $\delta_2^0 > 0$, whilst for $\pi^-\pi^0$ scattering the d-wave is exotic and negative and the zero approaches the backward direction.

So far we have only considered the zero contours, i.e., the lines $z = \mathrm{Re}\, z_0(s)$, and we do not know if the zeros produce marked dips in the angular distributions or not. This depends on how large $\mathrm{Im}\, z_0(s)$ is.

In Fig. 4 we show the angular distributions[4] in the region where the second zero nears and enters the physical region, i.e., for $\sqrt{s}$ between 970 and 1050 MeV. As can be seen for both $\pi^+\pi^-$ and $\pi^-\pi^0$ elastic scattering, the first zero--the Legendre zero of the ρ--is always a very nearby zero producing a marked dip in the angular distribution (not just in the energy region shown). However, the second zero in $\pi^+\pi^-$ scattering has a large imaginary part and produces no marked dip in the differential cross-section until it is right inside the physical region, when its imaginary part decreases rapidly ready to become the Legendre zero of the f_0. In $\pi^-\pi^0$ scattering the second zero is not nearby even at 1.05 GeV.

WHERE DO THE ZERO COME FROM?

As mentioned above, when the energy increases, higher partial waves grow and an increasing number of zeros enter the physical region, resulting in higher and higher spin resonances where do these zeros come from? It appears that all the zeros except the first, the Legendre zero of the ρ, may have a common origin, which we next discuss, leaving consideration of the first zero till later.

Consider $\pi^+\pi^-$ elastic scattering in the s and t channels and note that the u-channel is exotic having $I_u = 2$. Imagine we already have the first zero inside the physical region and we have the ρ resonance (Fig. 5). Now we look in the unphysical region near $s = m_\rho^2$, $t = m_\rho^2$. There the amplitude is approximately given by

$$F(s,t) \simeq \frac{a}{m^2 - s} + \frac{a}{m^2 - t} \tag{7}$$

where we can consider m^2 to be complex to include the width. Rewriting Eq. (7) as

$$F(s,t) = \frac{a(2m^2 - s - t)}{(m^2 - s)(m^2 - t)} \tag{8}$$

we see F has a double pole at $s = m^2$, $t = m^2$, but with a zero intersecting this double pole at $s + t = 2m^2$, i.e., F = 0 for $u = 4\mu^2 - 2m^2$. The occurrence of such a zero which "kills" the double pole in Eq. (8) is completely general. Let us forget, for the moment, the zero contours we saw above from the data and let us pretend that the fixed u zero, we have just seen occurs at the intersection of the ρ poles in the s and t channels, continues to traverse the Mandelstam plane at fixed u.[5] This zero contour then crosses $s = m_f^2$ remarkably close to $z = +1/\sqrt{3}$ (i.e., where $P_2(z) = 0$) and then crosses $s = m_g^2$ close to $z = 0$ (where $P_3(z) = 0$). Recall we already have the ρ's Legendre zero in the physical region, so we now have two zeros, which takes care of the f_0, but to have a spin 3 g resonance we need another zero to enter the physical region between $s = m_f^2$ and $s = m_g^2$. Where does this come from? Well, we now have double poles in the unphysical region where the ρ and f_0 poles in the s and t channels intersect. Again we must have a double pole killing zero, this time at $\text{Re } u = 4\mu^2 - m_\rho^2 - m_f^2$ which crosses $s = m_g^2$ very close to where $P_3(z) = 0$. As each resonance occurs we have more and more double poles in the unphysical region, which are killed by more and more zeros, which enter the physical region producing higher and higher spin resonances and so the cycle continues. This structure of fixed u zeros is epitomized by the single term Veneziano amplitude.[6]

This amplitude, you recall, for $\pi^+\pi^-$ elastic scattering involves the product of poles in the s and t channels

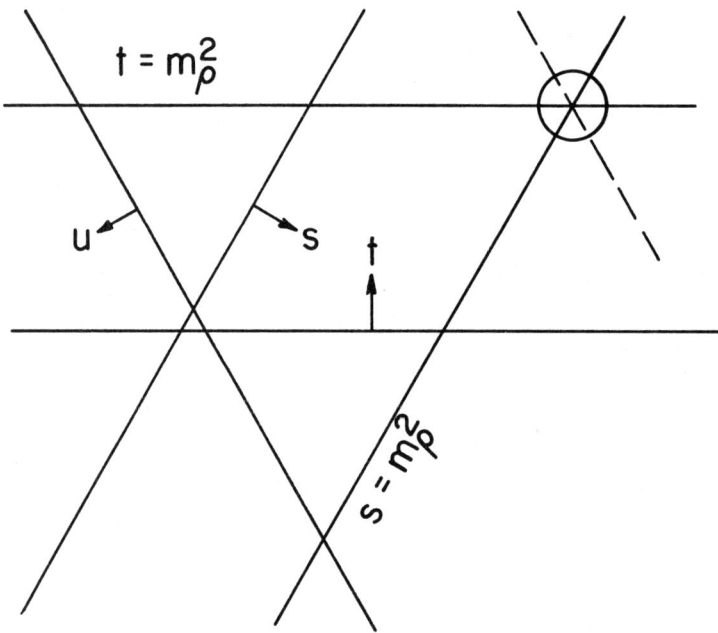

Fig. 5. The Mandelstam plane for $\pi^+\pi^-$ elastic scattering showing the ρ poles in the s and t channels. The circle, centered on the double pole in the unphysical region, indicates where Eqs. (7,8) may be a reasonable approximation. The dashed line marks the fixed u zero which "kills" the double pole.

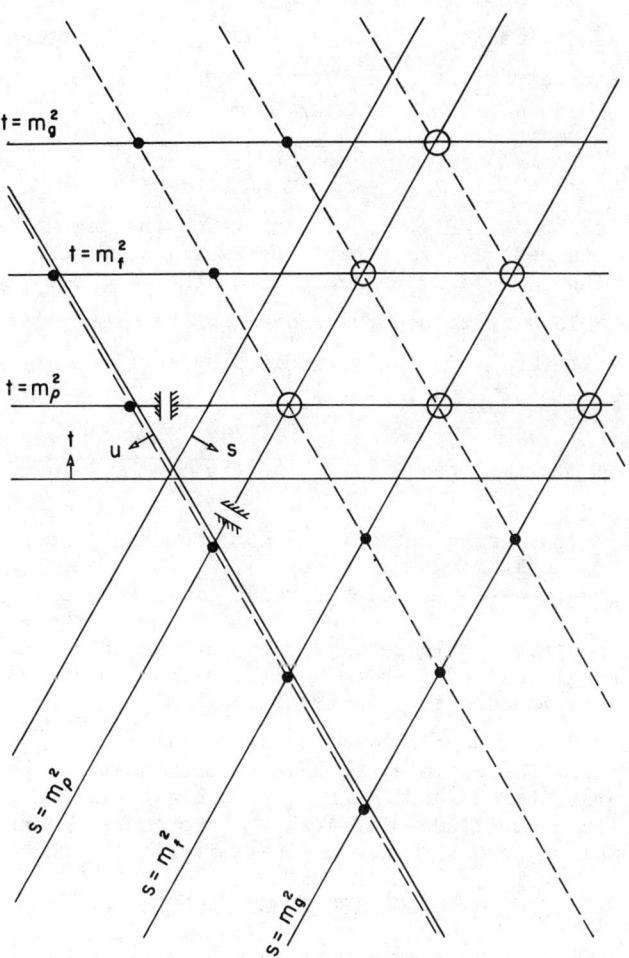

Fig. 6. The zeros (dashed lines) and poles (solid lines) of the Lovelace-Veneziano model, Eq. (10). The open circles mark the double poles and the black dots the Legendre zeros of resonances. The hatched band at $s = m_\rho^2$ and at $t = m_\rho^2$ indicates where unitarity would constrain the Legendre zero of the ρ to be.

$$V_{st} \sim \Gamma(1 - \alpha_s)\,\Gamma(1 - \alpha_t)$$

where $\alpha_x = \alpha_0 + \alpha' x$. To kill the forbidden double poles the above form is divided by[6] $\Gamma(2 - \alpha_s - \alpha_t)$. Let us look at the zero and resonance structure of this amplitude

$$V_{st} = \frac{\Gamma(1 - \alpha_s)\,\Gamma(1 - \alpha_t)}{\Gamma(2 - \alpha_s - \alpha_t)} \quad . \tag{9}$$

This is shown in Fig. 6. Ignoring for the moment the zero close to $u = 0$, we have a zero structure exactly as described before. However, the ρ has no Legendre zero, the f_0 one, the g two, and so on. This amplitude therefore has the wrong spin structure for $\pi\pi$ scattering. This was remedied by Lovelace[7] by multiplying Eq. (9) by $(1 - \alpha_s - \alpha_t)$ to give

$$F_{LV}(s,t) = \frac{\Gamma(1 - \alpha_s)\,\Gamma(1 - \alpha_t)}{\Gamma(1 - \alpha_s - \alpha_t)} \quad . \tag{10}$$

This introduces an extra zero, not a double-pole-killing zero, at $u = 4\mu^2 - \dfrac{(1 - 2\alpha_0)}{\alpha'}$ which Lovelace fixes at $u = 2\mu^2$. As can be seen in Fig. 6, this zero crosses through the Mandelstam triangle (the significance of which will be discussed later) and gives the correct spin structure to the ρ, f_0, g etc. However, this zero is outside the physical region and clearly violates unitarity, since at the ρ, for example, we have already seen it should pass through the marked bands shown in Fig. 6. This violation occurs because the s-wave amplitude with its ϵ resonance, degenerate in mass with the ρ, grossly exceeds the unitarity bound.

ODORICO ZEROS INTRODUCED

Although the Lovelace-Veneziano model[7] violates unitarity it provides a beautiful example of how an amplitude, which is essentially real, is determined by knowing its poles and its zeros. However, the physical world is supposedly just the unitarization of some dual model and perhaps some features of the simple Veneziano model, for example, are preserved by this unitarization. Certainly the real parts of the pole positions are unchanged, perhaps then the real parts of the zero positions are still approximately at fixed u. Indeed this has been suggested by Odorico.[5] However, his hypothesis of straight line zeros does not just apply to amplitudes represented by single Veneziano terms, but also to amplitudes with resonances in all three channels, when the single term

Virasoro model[8] is an explicit example having straight line zeros (SLZ). Using this SLZ hypothesis we can build up the structure of zeros and poles in the Mandelstam plane for the "true" amplitude without recourse to a specific (nonunitary) model. Of course, to make any sense of this prescription Odorico avoids the most blatant violation of unitarity in $\pi^+\pi^-$ elastic scattering by shifting the first zero, which Lovelace fixes at $u = 2\mu^2$, to $u = 4\mu^2 - \frac{1}{2} m_\rho^2$. This zero, which you recall is not a double-pole-killing zero, then becomes the Legendre zero of the rho with an s-wave background consistent with unitarity. It is rather remarkable, as we shall mention again later, that this fixed u zero of Odorico's now also crosses both the f_0 and g resonances very close to where $P_2(z) = 0$ and $P_3(z) = 0$, respectively.

Let us look at Odorico's SLZ hypothesis in the two cases of $\pi^+\pi^-$ and $\pi^-\pi^0$ elastic scattering[4] (see Fig. 7) and compare it with the data of Fig. 3.

$\pi^+\pi^-$ The first zero certainly follows an approximate straight line from $s = m_\rho^2$ to $s = m_f^2$, just as suggested by Odorico, aside from the wiggle at $K\bar{K}$ threshold. This is not surprising since no such coupled channel effect is built into the Veneziano model. However, closer to $\pi\pi$ threshold the data suggests that this zero leaves through the backward direction, while Odorico's zero leaves through the forward direction (corresponding to $\delta_0^0 < 0$) and passes far from the Mandelstam triangle.

The second zero also does not follow a very straight line, however, it is very likely that it is in fact related to the double-pole-killing zero in the unphysical region near $s = m_\rho^2$, $t = m_\rho^2$.

$\pi^-\pi^0$ Here the pattern of Odorico zeros is more complicated, and here too the first zero does not follow Odorico's lines very closely. Nonetheless, it is very likely that the second zero is once again related to the double-pole-killing zero. However, Odorico's lines suggest that, in the physical region, this second zero becomes the Legendre zero of the g at $z = 0$, while the rho's Legendre zero becomes that of the g at $z = -\sqrt{3/5}$. From the data it seems that these zero contours do not cross, but that the double-pole-killing zero becomes the g's Legendre zero in the backward hemisphere (more on this later).

COMMENTS ON STRAIGHT LINE ZEROS

It is appropriate here to make a few general remarks about Odorico zeros. It is very likely that the increasing number of zeros which enter the physical region from outside are the double-pole-killing zeros in the unphysical region (as also emphasized by A. D. Martin and W. J. Ochs in these proceedings). This

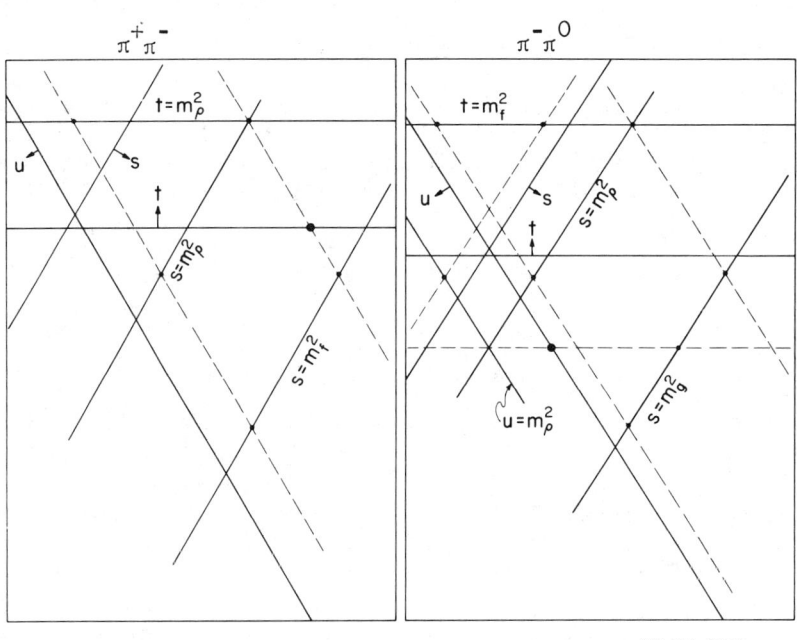

Fig. 7. Structure of Odorico zeros for $\pi^+\pi^-$ and $\pi^-\pi^0$ elastic scattering. The dashed lines denote the zeros.

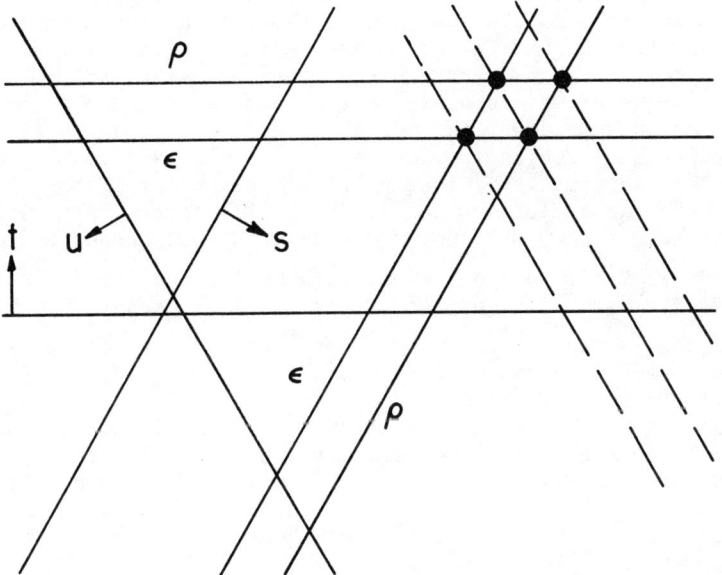

Fig. 8. The Mandelstam plane for $\pi^+\pi^-$ elastic scattering showing the ϵ and ρ poles in the s and t channels. (There is no significance to the ϵ resonance being drawn at lower mass than the ρ: they are interchangeable). The black dots mark the four double poles in the unphysical region which must be killed by three different zeros (dashed lines) if only straight line zeros are allowed.

provides us with a simple way of keeping track of the zeros and poles of the scattering amplitude, at least in a somewhat idealized world, e.g., one with the mass spectrum of the Lovelace-Veneziano model. The idea of straight line zeros is a further idealization which need not occur in reality. However, it may well be that if we look at the Mandelstam plane as a whole and not in any local region, particularly near threshold, that straight line zeros are a reasonable first, or perhaps zeroth, approximation to reality.

To see simply that the straight line zero hypothesis refers to an idealized world with Veneziano model mass spectrum, let us consider a simple example. Let us return to Fig. 5 and introduce the ϵ or σ resonance. This introduces new double poles in the unphysical region. In the Lovelace-Veneziano model, all poles are real and the ϵ and ρ poles are degenerate in mass. Then a single real fixed u zero can kill the $\rho\rho$, $\epsilon\epsilon$, and $\epsilon\rho$ double poles. However, it is clear that in nature, if the ϵ pole exists, it is not degenerate with the ρ. Then the complex double poles in the unphysical region are not degenerate and each double pole must be killed separately. Clearly, if we only allow fixed u zeros, we must have three distinct zeros, which will enter the physical region before 1200 MeV and produce a spin four resonance in the f_0 region (Fig. 8). Of course, no such zeros occur here. Such a difficulty with straight line zeros not only occurs here but everywhere the ϵ and S^*, or any other "daughter" resonance, in one channel cross resonances in the other.

If we abandon the SLZ hypothesis then one zero can be very clever and wiggle around to kill the four double poles in Fig. 8. Then there is certainly no need for this zero to enter the physical region along a fixed u line. Alternatively, it is possible that there are a number of zeros in the unphysical region killing the double poles, but with only one of them coming near to and entering the physical region.

HIGH ENERGY ZEROS

We next briefly discuss what happens to zeros near the forward or backward directions at high energies. It has been noted by Schmid,[9] Harari,[10] Cohen-Tannoudji et al.[11] and many others that if we look at the near forward Legendre zeros of the ρ, f_0, and g of Fig. 1, then they are at the same t-value: $t \sim -0.3 (\text{GeV}/c)^2$. Similarly, their near backward Legendre zeros lie at fixed $u \sim -0.3 (\text{GeV}/c)^2$. Perhaps such a fixed t or u zero persists at higher energies, as we may expect from duality considerations.

If the amplitude considered involves isospin one in the t-channel, then we expect to see a fixed t zero at high energy. This is the well-known crossover zero which arises from the vanishing of the imaginary part of the ρ exchange amplitude. In a model of just Regge poles and strong exchange degeneracy this vanishing occurs at $t \sim -0.6(\text{GeV}/c)^2$, whereas absorption models with no strong exchange degeneracy place this zero at $t \sim -0.2(\text{GeV}/c)^2$.

In the absorption model of Cohen-Tannoudji et al.[11] which includes duality, it is conjectured that in $\pi^+\pi^-$ scattering (see Figs. 3,7) that a zero leaves the region of the Mandelstam triangle becomes the rho's Legendre zero and then moves on at fixed t becoming the near forward Legendre zeros of the f_0 and g and then the cross over zero. Such a fixed t zero is to be contrasted with Odorico's fixed u zero. We see the data does not locally indicate such a behavior. However, it appears that instead of one zero producing this fixed t effect it is mocked by different zeros in different energy regions, and so exhibiting "average" but not "local" duality.

Similarly backward $\pi^-\pi^0$ scattering is controlled by ρ exchange. We might then expect a zero to leave the Mandelstam triangle become the rho's Legendre zero and move off at fixed u along the line of Odorico's zero. Again such a high energy zero is built up not by one zero but by different zeros in different energy regions. In both forward $\pi^+\pi^-$ and backward $\pi^-\pi^0$ scattering the high energy zero pattern may be distorted by the presence of $I = 0,2$ and $I = 2$ exchange amplitudes, respectively, in addition to ρ exchange, so let us isolate the amplitude with just isospin one in the t-channel to see what happens. The zero contours for this amplitude are shown in Fig. 9 from the data of Carroll et al.[12] as determined by Eguchi et al.[13] * Here the approximately fixed t zero of Cohen-Tannoudji et al. and others appears quite distinctly in this amplitude, which involves just ρ exchange in the t channel. Indeed, the zero contour (Fig. 9) is approximately at $t = -0.3(\text{GeV}/c)^2$. This dip in the differential cross section is usually thought of in terms of the vanishing of the imaginary part of the amplitude at a real value of t, rather than as a zero of the whole amplitude at a complex value of t. Tryon[14] has, in fact, determined the imaginary part of the $\rho\pi\pi$ Regge residue function $r_\rho(t)$, from the same data[12] by performing a physical region crossing sum rule calculation. Tryon finds r_ρ

---

* Eguchi et al.[13] also determine the zero contours for $\pi^+\pi^-$ and $\pi^-\pi^0$ elastic scattering from the data of Carroll et al. Apart from near $K\bar{K}$ threshold, their results are absolutely identical to those found by Pennington and Protopopescu,[4] using the data of Protopopescu et al. (and shown in Fig. 3) despite the comments made by Eguchi et al. implying the contrary.

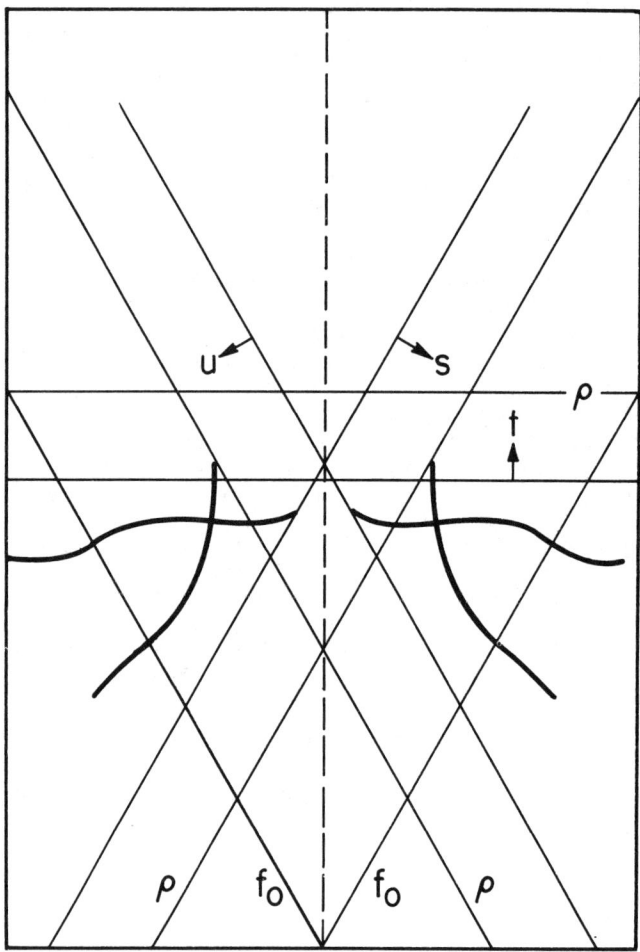

Fig. 9. The zero contours for the $\pi\pi$ amplitude with isospin one in the t-channel from Eguchi et al. (Ref. 13) using the data of Carroll et al. (Ref. 12). This amplitude vanishes on the line $s = u$ (the vertical dashed line) by Bose symmetry.

vanishes at $t = -0.52(GeV/c)^2$ which looks rather like the zero implied by strong exchange degeneracy in a pure pole model. However, in Ref. 15 it is reported that Basdevant and Schomblond[16] suggest that this zero is nearer -0.2 than $-0.6(GeV/c)^2$. We will have to wait to see the details, before we can draw any conclusions.

THE LEGENDRE ZERO OF THE RHO AND CURRENT ALGEBRA?

We next discuss the most important zero in very low energy $\pi\pi$ scattering and that is the ρ's Legendre zero. We have seen that this zero contour appears to pass close to threshold and near to the Mandelstam triangle. It is in this region that current algebra has some predictions to make and so let us see if they are relevant in any way.

Within and near the Mandelstam triangle the amplitude is controlled by just the s- and p-waves $f_0^0(s)$, $f_0^2(s)$, and $f_1^1(s)$. If the exotic channel is weak (in a sense defined below), these partial waves will be quasilinear given by

$$f_0^0(s) \simeq \tfrac{3}{2} a_1^1 (2\mu^2 + s - \tfrac{5}{2} s_A)$$
$$f_0^2(s) \simeq \tfrac{3}{4} a_1^1 (4\mu^2 - s - 2s_A) \qquad (11)$$
$$f_1^1(s) \simeq \tfrac{1}{4} a_1^1 (s - 4\mu^2)$$

on using crossing symmetry, so that

$$\frac{a_0^0}{a_0^2} \simeq \tfrac{5}{2} - \frac{6\mu^2}{s_A} . \qquad (12)$$

The forms of Eq. (11) only apply for values of s in, or close to, the gap in the cut s-plane. For values of s_A in this same range, the parameter s_A of Eqs. (11, 12) is the value of s at which the s-wave of the Chew-Mandelstam invariant amplitude $A(s,t) = \tfrac{1}{3}(F^{s0} - F^{s2})$ vanishes. We see that the amplitude, in the neighborhood of the Mandelstam triangle, is determined essentially by two parameters: a_1^1 to set the scale and the dimensionless ratio a_0^0/a_0^2. It is for these two parameters that current algebra makes predictions, which we will shortly discuss.

Let us remark that by a weak exotic channel we mean that the contribution of the $I = 2$ absorptive part, both in the direct channel and in the crossed channel, to the Froissart-Gribov integrals for the p- and d-wave amplitudes below threshold, is very much less than the corresponding contributions of the $I = 0$ and 1 absorptive parts. These conditions are justified by the nonexistence of exotic resonances and hence of leading exotic Regge trajectories.

This weakness of the $I = 2$ channel has implied that the s and p waves of Eq. (11) are quasilinear below threshold with derivatives determined by the p-wave scattering length, a_1^1. This scattering length can also be simply calculated using the weakness of the exotic channel. From an unsubtracted fixed s dispersion relation for $F^{s1}(s,t)$, we have

$$a_1^1 = \frac{4}{3\pi} \int_{4\mu^2}^{\infty} \frac{dt}{t^2} A^{s1}(t, 4\mu^2) \qquad (13)$$

where $A^{s1}(t,s)$ is the t-channel absorptive part with $I_s = 1$. Using the crossing matrix we have

$$a_1^1 = \frac{4}{3\pi} \int_{4\mu^2}^{\infty} \frac{dt}{t^2} [A^{t1}(t, 4\mu^2) - A^{t2}(t, 4\mu^2) + A^{s2}(t, 4\mu^2)]$$

$$\simeq \frac{4}{3\pi} \int_{4\mu^2}^{\infty} \frac{dt}{t^2} A^{t1}(t, 4\mu^2) \qquad (14)$$

by the weakness of the $I = 2$ absorptive part in both the direct and crossed channels. To a good approximation the integrand of Eq. (14) is dominated by the ρ resonance, so we have

$$a_1^1 \simeq a_1^1(\rho) = \frac{4\Gamma_\rho}{m_\rho^3}\left(1 + \frac{10\mu^2}{m_\rho^2 - 4\mu^2}\right) \qquad (15)$$

using the narrow resonance approximation. Numerically this gives*

---

* This simple way of estimating a_1^1, which neglects all contributions except that of the rho, gives a value for a_1^1 which is 10-15% smaller than a more exact calculation[15,17] using the data and high energy estimates in Eq. (13). So our simple calculation is rather good.

$a_1^1(\rho) = 0.032$ in pion mass units. We see, from Eq. (15), how the ρ resonance essentially determines the scale of the near threshold $\pi\pi$ amplitude,[17] using the weakness of the $I = 2$ amplitude [see Eq. (11)].

We now discuss what current algebra has to say about the $\pi\pi$ amplitude in the neighborhood of the Mandelstam triangle. Current algebra and PCAC make definite predictions about the 4π amplitude when one or two pions have zero 4-momentum, that is for the amplitude at certain kinematic points on the planes $s + t + u = 3\mu^2$ or $2\mu^2$. To compare these predictions with experiment we have to know how to extrapolate to the plane $s + t + u = 4\mu^2$. Assuming this extrapolation is smooth and simple, two definite predictions from PCAC become relevant.

(1) The Adler-Weisberger relation expresses the derivative of the $I_s = 1$ amplitude at $s = 0$, $t = u = \mu^2$, in terms of the pion decay constant f_π (experimentally $f_\pi = 93$ MeV). The derivative of the $I = 1$ amplitude near threshold is just the p-wave scattering length, so that the Adler-Weisberger relation, together with smoothness, predicts[18]

$$a_1^1 = \frac{1}{24\pi f_\pi^2} \quad (= 0.030 \text{ in pion mass units}). \quad (16)$$

Equating the algebraic conditions Eq. (15) and Eq. (16) we obtain

$$\frac{1}{f_\pi^2} = \frac{96\pi \Gamma_\rho}{m_\rho^2}\left(1 + \frac{10\mu^2}{m_\rho^2 - 4\mu^2}\right). \quad (17)$$

With $\mu = 0$ this condition is the well-known KSRF relation.[19] Since this condition is empirically well satisfied PCAC predicts the correct "scale" for low energy $\pi\pi$ dynamics. The second prediction of PCAC, we consider, concerns zeros of the amplitude:

(2) The Adler self-consistency condition[20] states that the $\pi\pi \to \pi\pi$ amplitude vanishes at the point

$$s = t = u = \mu^2. \quad (18)$$

Weinberg has given an explicit extrapolation of this condition to the plane $s + t + u = 4\mu^2$:

$$A(s = \mu^2, t, u) = 0 \quad \text{for all } t, u \in (0, 4\mu^2) \quad (19)$$

where

$$A(s,t) = \frac{1}{3}(F^{s0} - F^{s2}) = \frac{1}{2}(F^{t1} + F^{t2}). \quad (20)$$

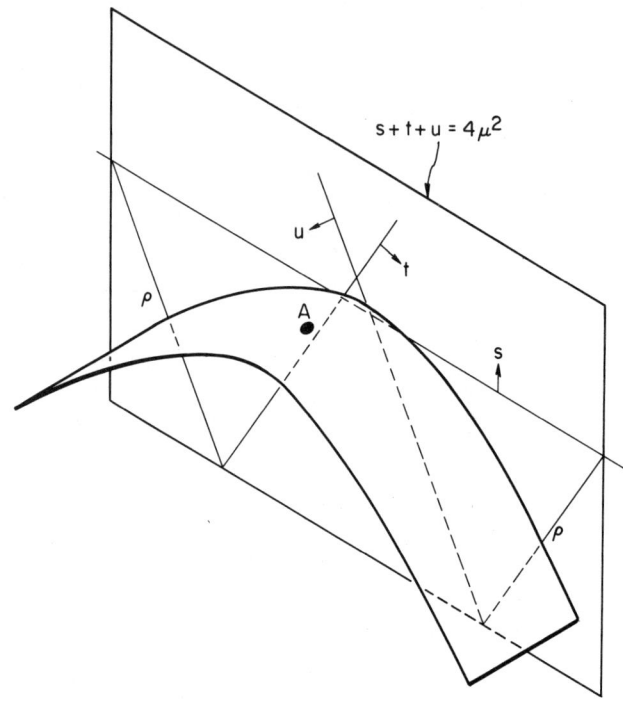

Fig. 10. A simple three dimensional illustration of how a surface of zeros for the invariant amplitude $A(s,t)$ can relate the Adler zero at the point $s = t = u = \mu^2$ (marked by A) to the Legendre zero of the ρ. This zero extrapolates to the plane $s + t + u = 4\mu^2$ like Weinberg's amplitude, inside the Mandelstam triangle. The amplitude $A(s,t)$ is the $\pi^+\pi^- \to \pi^0\pi^0$ amplitude in the s-channel and the $\pi^-\pi^0$ elastic scattering amplitude in the t and u channels. On the physical plane this figure is to be compared with Fig. 3. The three dimensions are specified by the orthogonal coordinates $s, t - u$ and $s + t + u$. The latter variable is normal to the plane $s + t + u = 4\mu^2$ and represents varying one pion mass at a time from 140 MeV to zero.

This condition, Eq. (19), implies that the s-wave projection of A vanishes at $s_A = \mu^2$, which we recall from Eq. (12) predicts $a_0^0/a_0^2 = -7/2$. We refer the reader to Weinberg's paper[18] for a discussion of the assumptions involved in this particular extrapolation. This same extrapolation of the Adler zero implies that the $\pi^+\pi^-$ elastic amplitude vanishes on the line $u = 2\mu^2$ (in the Mandelstam triangle), which is where Lovelace imposes the zero in his single term Veneziano model [see Eq. (10) and Fig. 6] for all values of s.[7] Different on mass-shell extrapolations give different values for s_A, and hence for a_0^0/a_0^2, but most[21] give s_A not far from μ^2.

From the data shown in Fig. 3 we have seen that the zero contour, which is the Legendre zero of the ρ, may very well pass through or near the Mandelstam triangle, as suggested by the dotted curves. In Fig. 10 we show a simple extrapolation of the Adler zero on-mass-shell consistent with Weinberg's model for the Chew-Mandelstam invariant amplitude, Eq. (20), inside the Mandelstam triangle and consistent with unitarity in the region of the ρ resonance. (Note no zero at fixed s, t, or u can be consistent with both of these.) Whilst this simple zero contour agrees with the data beyond 500 MeV, we have to look in a different way to see that there is a zero closer to the Mandelstam triangle. Let us look at Re $A(s,t)$ along the line $t = 0$ as a function of $(s - u)$ as found from the data. From Fig. 11, it is clear that a zero of the amplitude really must occur near the Mandelstam triangle, though, of course, not necessarily exactly where Weinberg's model places it. In fact, a zero will pass through the Mandelstam triangle for $1 > a_0^0/a_0^2 > -\infty$, and just below it (in the sense of Fig. 10) if the ratio a_0^0/a_0^2 is large and positive. From Eqs. (11) and (12), it is clear that an exact knowledge of s_A, an as yet poorly known parameter, is required before we can know the s-wave scattering lengths in any way but crudely.

By relating the on-mass-shell appearance of the Adler zero to the Legendre zero of the ρ we understand immediately why if we take a model like that of Weinberg's or any other having s-waves with near threshold zeros and unitarize it, we obtain a resonating p-wave.[22] The near-threshold zeros of the s-wave imply a zero in the invariant amplitude which unitarity forces to curve down into the t and u channel physical regions (Fig. 10), and when this zero contour crosses the line $z_t = 0$ we have to have a p-wave resonance,[23,2] at least if the exotic amplitude is weak. However, the actual parameters of this p-wave resonance are not quite so simply understood.

Having discussed most of the zeros of present physical interest all that remains is to summarize.

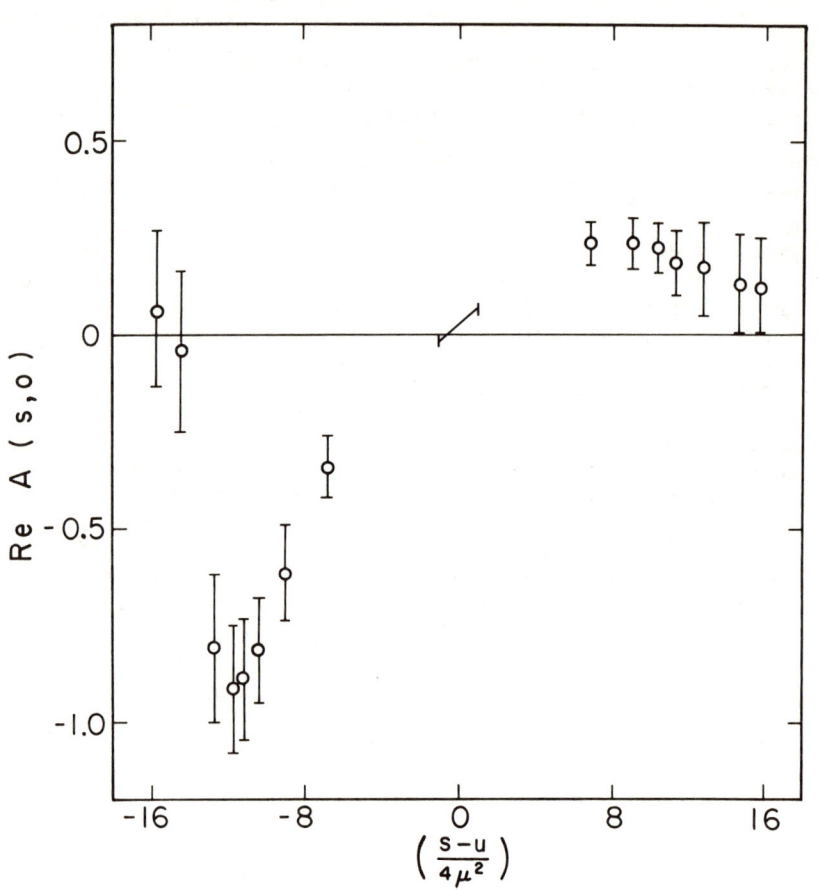

Fig. 11. A plot of the Re $A(s, t = 0)$ against $(s - u)/4\mu^2$ from the data of Protopopescu et al. and Baton et al., illustrating that there must be a zero of this amplitude between $(s - u)/4\mu^2 = -5$ and $+5$. The solid line is Weinberg's prediction for this amplitude inside the Mandelstam triangle. The amplitude $A(s,t)$ describes $\pi^+\pi^- \to \pi^0\pi^0$ scattering in the s-channel. For similar plots for other $\pi\pi$ amplitudes, see D. Morgan (Ref. 1).

CONCLUSIONS

Zeros provide us with a particularly beautiful, yet simple, way of relating various dynamical aspects of the scattering amplitude. We have discussed how double-pole-killing zeros become the Legendre zeros of resonances and how these zeros appear in the high energy Regge region. Finally we have discussed how the near-threshold zeros of the $\pi\pi$ s-wave amplitudes are related to the existence of a resonating p-wave and how this Legendre zero of the ρ may be thought of as the on-mass-shell manifestation of the Adler zero appearing in the physical scattering amplitude. In this way,[24] the investigation of zeros may help us to understand how unitarity, analyticity, and crossing build the dynamics of the scattering amplitude. Indeed, we can think of the study of the scattering amplitude, both theoretically and experimentally, as essentially one of looking at zeros travelling across the Mandelstam plane.

ACKNOWLEDGMENTS

It is a very great pleasure to thank my many colleagues for numerous helpful discussions about zeros. In particular I wish to thank Dr. Christoph Schmid, Dr. Serban Protopopescu, and Dr. Paul Pond.

REFERENCES

1. E. Barrelet, Nuovo Cimento $\underline{8A}$, 331 (1972); D. Morgan, Rutherford Laboratory report RPP/T/27 (1972), "Questions in $\pi\pi$ scattering and related problems."
2. A. Arneodo, F. Guerin, and J. T. Donohue, University of Nice preprint NTH 73/2 (March 1973), "Zero Contours and ρ-Dominance in Low Energy $\pi\pi$ Scattering."
3. S. D. Protopopescu et al., Phys. Rev. $\underline{D7}$, 1279 (1973); G. Grayer et al. in "Experimental Meson Spectroscopy-1972" edited by K. W. Lai and A. H. Rosenfeld (A.I.P., New York, 1972).
4. M. R. Pennington and S. D. Protopopescu, Phys. Letters $\underline{40B}$, 105 (1972).
5. R. Odorico, Nucl. Phys. $\underline{B37}$, 509 (1972); Phys. Letters $\underline{38B}$, 37 and 411 (1972); in "Experimental Meson Spectroscopy-1972" edited by K. W. Lai and A. H. Rosenfeld (A.I.P., New York, 1972).
6. G. Veneziano, Nuovo Cimento $\underline{57A}$, 190 (1968).
7. C. Lovelace, Phys. Letters $\underline{28B}$, 264 (1969).
8. M. A. Virasoro, Phys. Rev. $\underline{177}$, 2309 (1969).
9. C. Schmid, Phys. Rev. Letters $\underline{20}$, 689 (1968).
10. H. Harari, "Hadronic Interactions of Electrons and Photons," p. 529, edited by J. Cumming and H. Osborn (Academic Press, London, 1971).
11. G. Cohen-Tannoudji et al., Nucl. Phys. $\underline{B45}$, 109 (1972).
12. J. T. Carroll et al., Phys. Rev. Letters $\underline{28}$, 318 (1972).

13. T. Eguchi, M. Fukugita, and T. Shimada, Phys. Letters $\underline{48B}$, 56 (1973).
14. E. P. Tryon, Hunter College preprint (July 1972), "The $\rho\pi\pi$ Regge residue from a new class of sum rules."
15. J. L. Basdevant, C. D. Froggatt, and J. L. Petersen, University of Glasgow preprint (March 1973), "Phenomenology of $\pi\pi$ Scattering."
16. J. L. Basdevant and C. Schomblond (in preparation).
17. M. R. Pennington and S. D. Protopopescu, Phys. Rev. $\underline{D7}$, 1429 (1973).
18. S. Weinberg, Phys. Rev. Letters $\underline{17}$, 616 (1966).
19. K. Kawarabayashi and M. Suzuki, Phys. Rev. Letters $\underline{16}$, 255 (1966); Riazuddin and Fayyazuddin, Phys. Rev. $\underline{147}$, 1071 (1966).
20. S. L. Adler, Phys. Rev. $\underline{137B}$, 1022 (1965).
21. See, for example, R. G. Levers, Nuovo Cimento $\underline{60A}$, 575 (1969); I. Bars, Phys. Rev. $\underline{D2}$, 1630 (1970); D. Morgan and G. Shaw, Nucl. Phys. $\underline{B43}$, 365 (1972).
22. See, for example, J. Dilley, Nucl. Phys. $\underline{B25}$, 227 (1971); L. S. Brown and R. L. Goble, Phys. Rev. Letters $\underline{20}$, 346 (1968) and Phys. Rev. $\underline{D4}$, 723 (1971); M. R. Pennington and P. Pond, Nuovo Cimento $\underline{3A}$, 548 (1971); O. Piguet and G. Wanders, Nucl. Phys. $\underline{B46}$, 295 (1972).
23. M. R. Pennington, thesis submitted to the University of London, 1971 (unpublished); M. R. Pennington and C. Schmid, Phys. Rev. (to be published, April 1973).
24. C. Chiu, R. J. Eden and C. I. Tan, Phys. Rev. $\underline{170}$, 1490 and 1516 (1968) should also be seen.

COUPLED CHANNEL ANALYSIS IN THE $K\bar{K}$ THRESHOLD REGION

G. Grayer[*], B. Hyams, C. Jones, P. Schlein[**] and P. Weilhammer
CERN, Geneva, Switzerland

and

W. Blum, H. Dietl, W. Koch, E. Lorenz, G. Lütjens,
W. Männer, J. Meissburger, W. Ochs and U. Stierlin
Max-Planck-Institut für Physik und Astrophysik,
Munich, Germany

ABSTRACT

Results of an energy-dependent phase-shift analysis for $\pi\pi$ scattering between 900 and 1120 MeV mass are presented using data of the reactions $\pi^{\pm}p \to \pi^{\pm}\pi^{+}n$ and $\pi^{-}p \to K^{+}K^{-}n$. The I = 0 S-wave becomes highly inelastic at $K\bar{K}$ threshold and accounts for the structure in the $\pi\pi$ system. A zero effective range approximation for the S-wave leads to a pole of the S-matrix which is located on the second Riemann sheet at $(1012 \pm 6) - i(16 \pm 5)$ MeV.

1. INTRODUCTION

Most of the present knowledge of $\pi\pi$ scattering comes from the study of the reaction $\pi N \to \pi\pi N$, where information on $\pi\pi$ scattering is extracted by pole extrapolation. One of the outstanding problems in $\pi\pi$ scattering is the structure in the cross-section and decay angular distribution at around 1 GeV $\pi\pi$ mass. Furthermore, experiments studying the reaction $\pi^{-}p \to K\bar{K}n$ [1,2] showed a pronounced threshold enhancement which was found to be dominated by an I = 0 S-wave state produced by one-pion exchange. A first attempt to determine the elasticity of the I = 0 $\pi\pi$ S-wave gave an upper limit of $\eta_{0}^{0} = 0.45 \pm {}^{0.15}_{0.43}$ [3] within 30 MeV above $K\bar{K}$ threshold. A further analysis of the threshold anomaly was performed by Protopopescu et al.[4] using data of the reaction $\pi^{+}p \to \pi^{+}\pi^{-}\Delta^{++}$, $K^{+}K^{-}\Delta^{++}$ at 7.1 GeV/c. New information from high statistics experiments, especially on the channel $\pi^{-}p \to K^{+}K^{-}n$, allows a more detailed study of the $\pi\pi$ phase shifts in the region of the $K\bar{K}$ threshold.

In the following we describe an energy-dependent phase-shift analysis between 900 and 1120 MeV invariant mass. The input data come from the following samples of reactions (in brackets, the number of events for the total samples):

$\pi^{-}p \to \pi^{+}\pi^{-}n$ 39 000   (290 000) events, p_{lab} = 17.2 GeV/c   (1)

$\pi^{+}p \to \pi^{+}\pi^{+}n$ 3 800   (17 500) events, p_{lab} = 12.5 GeV/c   (2)

$\pi^{-}p \to K^{+}K^{-}n$ 3 300   (21 000) events, p_{lab} = 9.8 GeV/c .   (3)

---

[*] Now at M.P.I., Munich, Germany.

[**] Permanent address: University of California, Los Angeles, USA.

All three reactions were measured with the CERN-Munich wire spark chamber spectrometer under similar conditions.

In order to extract the information for on-shell $\pi\pi$ scattering, all cross-sections and moments were extrapolated to the pion pole. To simplify the analysis the following assumptions were made:

a) only partial waves up to $\ell = 2$ were assumed to contribute;

b) the mass difference between $K^{\pm}$ and K^0 is neglected, therefore the mass range between 988 and 1000 MeV is excluded from the fit;

c) the channel $\pi\pi \to K\bar{K}$ is assumed to be the only inelastic channel which contributes to the $\pi\pi$ scattering within the range of the analysis.

2. THE REACTION $\pi^-p \to \pi^+\pi^-n$

Details of the experiments $\pi^-p \to \pi^+\pi^-n$ at 17.2 GeV/c are published elsewhere[5]. Figures 1 and 2 show the on-shell cross-section for $\pi^+\pi^- \to \pi^+\pi^-$ scattering and the extrapolated $\langle Y_\ell^0 \rangle$ moments up to $\ell = 4$ in the mass range between 900 and 1300 MeV. The extrapolation procedure is described in another paper[6]. The cross-section is constrained such that the P-wave reaches the unitarity limit at the ρ-peak. (Using our cross-section calibration we obtain a value for the ρ-peak which is 90 ± 5% of the unitarity limit depending on the extrapolation function. The experimental error on σ is 5%.)

The moments $\langle Y_5^0 \rangle$ and $\langle Y_6^0 \rangle$ are compatible with zero within the range of this analysis.

3. THE REACTION $\pi^+p \to \pi^+\pi^+n$

The reaction $\pi^+p \to \pi^+\pi^+n$ contains only $I = 2$ $\pi\pi$ states; thus this experiment provides a separate measurement of the $S(I = 2)$ and $D(I = 2)$ waves. Details of this experiment at 12.5 GeV/c incident momentum were presented by Hoogland[7] at this conference. The relevant results for the mass spectrum and the phases for the S- and D-waves are displayed in Figs. 3 and 4. The phases were calculated under the assumption that up to 1.5 GeV $\pi\pi$ mass the inelasticity is negligible.

4. THE REACTION $\pi^-p \to K^+K^-n$

Previous experiments of this reaction suffer from low statistics or high background contribution, therefore in this section we discuss in detail the results of our new experiment at 9.8 GeV/c incident momentum.

In this experiment the $\pi^+\pi^-n$ final state was suppressed by a large Čerenkov counter (n = 1.00145, pion threshold = 2.7 GeV/c, efficiency > 99% for pions above 5 GeV/c momentum). The final states $K\bar{K}n$ and $p\bar{p}n$ were separated on the basis of a kinematic fit. Figure 5 shows the distribution of the missing mass squared using only the K^+K^- hypothesis. A clear separation of both states is observed.

The acceptance for K^+K^- pairs of our apparatus decreases for increasing four-momentum transfer and mass. For a K^+K^- mass between 1.0 and 1.12 GeV and a $|t| < 1$ $(GeV/c)^2$ the acceptance as a function of the K^+K^- decay angular distribution is always greater than zero. For $|t| < 0.15$ $(GeV/c)^2$ this condition is fulfilled up to 1.4 GeV mass. The correction for the acceptance loss was calculated by rotating the observed events around the beam axis for a vertex randomly distributed along the target. Figure 6 shows the raw data and the corrected mass spectrum for $|t| < 0.15$ $(GeV/c)^2$. Decay-in-flight losses and some topology-dependent losses (caused by δ-rays, neutrons triggering the anticoincidence counters, and secondary interactions in the target) are included in the correction factor.

The data contain about 8% background from $p\bar{p}$ events; its approximate shape is indicated in Fig. 6. Above 1.1 GeV mass the small inefficiency of the Čerenkov counter causes a background contribution of $\pi^+\pi^-$ pairs. We estimate this contribution to be between 2% and 6% of the raw data. This background is expected to peak between 1.3 and 1.4 GeV K^+K^- mass. The contribution of events with an extra π^0 is estimated to be less than 2%.

The most striking feature of the mass spectrum is the steep rise at threshold. The spectrum rises within 20 MeV and then stays nearly constant up to 1.2 GeV. Below $|t| < 0.15$ $(GeV/c)^2$, ϕ production is negligible, while at higher momentum transfer, the contribution is significant.

Figure 7 displays the corrected differential cross-section $d\sigma/dt$ for the mass region between 0.988 and 1.12 GeV. The region between 1.015 and 1.025 GeV around the ϕ resonance was excluded. The expected distribution for one-pion exchange (Born term amplitude modified by the Dürr-Pilkuhn form factor for the $pn\pi$ vertex) is shown normalized to the second data point. For low momentum transfer we observe agreement.

The on-shell cross-section $\pi^+\pi^- \to K^+K^-$ was extracted using the Chew-Low method modified by the Dürr-Pilkuhn form factor. Details are given in Ref. 3. The results of evasive and non-evasive extrapolations using the data between t_{min} and $|t| = 0.4$ $(GeV/c)^2$ are displayed in Fig. 8. Within the statistical error we observe agreement between the cross-sections obtained by the different extrapolations. The cross-section rises directly to the S-wave unitarity limit and stays close to the limit up to 1200 MeV. No contribution of the ϕ resonance is observed.

The contribution of the $p\bar{p}$ background to the extrapolated cross-section is expected to be very small because the $p\bar{p}n$ events give rise to a very flat background in $d\sigma/dt (\sim e^{-3 \cdot |t|})$ when interpreted as K^+K^-.

Figure 9 shows the moments $\langle Y_\ell^0 \rangle$ up to $\ell = 4$ as a function of mass for $|t| \leq 0.15$ $(GeV/c)^2$. Also shown for larger mass bins are the moments at the pion pole obtained by a linear extrapolation. In the threshold region the moment $\langle Y_1^0 \rangle$ is significant, showing that at least a P-wave must be present. In the low mass region the moments with m ≠ 0 are all compatible with zero. It should be mentioned that the structure of the higher moments is affected by the $p\bar{p}$ background, and thus our conclusions on the P- and D-waves must be considered tentative.

5. THE ENERGY-DEPENDENT ANALYSIS

The partial wave amplitudes in terms of the isospin decomposition for the elastic and inelastic channels are

$$\pi^+\pi^- \to \pi^+\pi^- \quad \begin{aligned} T_S &= \tfrac{2}{3} T_0^0 + \tfrac{1}{3} T_0^2 \\ T_P &= T_1^1 \\ T_D &= \tfrac{2}{3} T_2^0 + \tfrac{1}{3} T_2^2 \end{aligned} \quad (4)$$

$$\pi^+\pi^- \to K^+K^- \quad \begin{aligned} T_S' &= \tfrac{1}{\sqrt{3}} T_0^{0\prime} \\ T_P' &= \tfrac{1}{\sqrt{2}} T_1^{1\prime} \\ T_D' &= \tfrac{1}{\sqrt{3}} T_2^{0\prime} \end{aligned} \quad (5)$$

The superscript denotes the isospin, the subscript the angular momentum.

The cross-section and moments $\langle Y_\ell^0 \rangle$ are:

$$\sigma(\pi^+\pi^- \to \pi^+\pi^-) = 4\pi \lambdabar^2 \sum_{\ell=0}^{\ell=2} (2\ell+1) \cdot |T_\ell|^2 , \quad (6)$$

where

$$\lambdabar = \frac{\hbar c}{q} ; \quad q = \sqrt{(m^2/4) - m_\pi^2} ; \quad \hbar c = 0.624 \text{ GeV mb}^{1/2}$$

with m = invariant mass and m_π = pion mass.

$$\begin{aligned} \langle Y_1^0 \rangle &= 2\sqrt{\tfrac{3}{4\pi}} \left\{ \text{Re}(T_S T_P^*) + 2\,\text{Re}(T_P T_D^*) \right\} \cdot \alpha \\ \langle Y_2^0 \rangle &= 2\sqrt{\tfrac{5}{4\pi}} \left\{ \tfrac{3}{5} |T_P|^2 + \text{Re}(T_S T_D^*) + \tfrac{5}{7} |T_D|^2 \right\} \cdot \alpha \\ \langle Y_3^0 \rangle &= \tfrac{9}{\sqrt{7\pi}} \left\{ \text{Re}(T_P T_D^*) \right\} \cdot \alpha \\ \langle Y_4^0 \rangle &= \tfrac{30}{7\sqrt{4\pi}} \left\{ |T_D|^2 \right\} \cdot \alpha \end{aligned} \quad (7)$$

$$\text{with } \alpha = \frac{4\pi \lambdabar^2}{\sigma} .$$

Replacing the amplitudes for the elastic scattering by the inelastic ones, one obtains the equations for the cross-section and moments for the process $\pi^+\pi^- \to K^+K^-$.

6. THE PARAMETRIZATION OF THE PARTIAL WAVES

Assuming the channel $\pi\pi \to K\bar{K}$ to be the only inelastic channel, the whole scattering process for the I = 0 S-wave is described by the 2 × 2 T-matrix:

$$\widetilde{T}_0^0 = \begin{pmatrix} T_{11} & T_{12} \\ T_{12} & T_{22} \end{pmatrix} \quad (8)$$

$T_{11} (= T_0^0)$ describes the $\pi\pi \to \pi\pi$ scattering process

$T_{12} (= T_0^{0\prime})$ describes the $\pi\pi \to K\bar{K}$ scattering process

T_{22} describes the $K\bar{K} \to K\bar{K}$ scattering process.

A convenient way of describing the threshold effects is to express the T-matrix in terms of the reaction matrix K [8]:

$$\widetilde{T} = \frac{\widetilde{K}}{\widetilde{1} - i\,\widetilde{q}\widetilde{K}} \quad (9)$$

where

$$\widetilde{1} = \delta_{ik}, \quad \widetilde{q} = \delta_{ik} \cdot \sqrt{(m^2/4) - m_i^2}$$

K is a real and symmetric matrix. For the energy dependence of the K-matrix we used the zero effective range approximation of Dalitz[9]:

$$\widetilde{K} = \widetilde{q}^{1/2}\, \widetilde{K}'\, \widetilde{q}^{1/2} \quad (10)$$

where

$$\widetilde{q}^{1/2} = \delta_{ik}\sqrt{q_i}, \quad \widetilde{K}' = \text{scattering length matrix (real and symmetric).}$$

This leads to the following expression for the T-matrix:

$$\widetilde{T}_0^0 = \frac{1}{D} \begin{pmatrix} \sqrt{q_1} & 0 \\ 0 & \sqrt{q_2} \end{pmatrix} \begin{pmatrix} K'_{11} - iq_2(K'_{11}K'_{22} - K'^2_{12}) & K'_{12} \\ K'_{12} & K'_{22} - iq_1(K'_{11}K'_{22} - K'^2_{12}) \end{pmatrix} \begin{pmatrix} \sqrt{q_1} & 0 \\ 0 & \sqrt{q_2} \end{pmatrix} \quad (11)$$

with

$$D = \text{Det}\,(\widetilde{1} - i\widetilde{q}\widetilde{K}') = 1 - i(q_1 K'_{11} + q_2 K'_{22}) - q_1 q_2 (K'_{11}K'_{22} - K'^2_{12}) \quad (12)$$

Below threshold: $q_2 = i\sqrt{m_K^2 - m^2/4}$.

The energy dependence of the $I = 0$ S-wave process is then described by the three real constants of the scattering length matrix, namely K'_{11}, K'_{12}, K'_{22}.

For the other partial waves we chose the representation for the elastic and inelastic amplitudes in terms of phases and elasticities:

$$\pi\pi \to \pi\pi \qquad T_\ell^I = \frac{\eta_\ell^I e^{2i\delta_\ell^I} - 1}{2i}$$

$$\pi\pi \to K\bar{K} \qquad T_\ell^{I'} = \frac{\eta_\ell^{I'} e^{2i\delta_\ell^{I'}}}{2i} \qquad (13)$$

with $(\eta_\ell^I)^2 + (\eta_\ell^{I'})^2 = 1$ by unitarity.

The energy dependences of the $I = 1$ P-wave and $I = 0$ D-wave are parametrized arbitrarily in terms of polynomials where the coefficients are determined by the fit:

$$(I = 1) \text{ P-wave:} \begin{cases} \delta_1^1 = a_1 + a_2 m^2 + a_3/(m^4 - 0.5) \\ \eta_1^1 \begin{cases} = 1 \text{ below threshold} \\ = 1/[1 + (b_1 + b_2 m^2)^2] \text{ above threshold} \end{cases} \\ \delta_1^{1'} = c_1 + c_2 m^2 \end{cases} \qquad (14)$$

$$(I = 0) \text{ D-wave} \begin{cases} \delta_2^0 = d_1 + d_2 m^2 \\ \eta_2^0 \begin{cases} = 1 \text{ below threshold} \\ = 1/[1 + (e_1 + e_2 m^2)^2] \text{ above threshold} \end{cases} \\ \delta_2^{0'} = f_1 + f_2 m^2 \end{cases} \qquad (15)$$

The information on the $I = 2$ S- and D-waves was taken from the study of the separate channel $\pi^+ p \to \pi^+\pi^+ n$. The energy dependence of the phases can be sufficiently described by:

$$\delta_0^2 = -0.07 - 0.30 \cdot m$$

$$\eta_0^2 = 1$$

$$\delta_2^2 = 0.019 - 0.034 \cdot m$$

$$\eta_2^2 = 1$$

(m in GeV, δ in radians).

Altogether the complete energy-dependent process is described by 16 parameters. The parameters were determined by a χ^2 minimization using the program MINUIT [9]. As input data we used the cross-sections and moments at four mass points below threshold and three mass points above threshold (910, 930, 950, 970, 1020, 1060, and 1100 MeV), thus giving rise to 50 data points. The minimization gave a χ^2 of 39.2 for 33 degrees of freedom *). As cross-section for the reaction $\pi^+\pi^- \to K^+K^-$ we used the results of the non-evasive extrapolation rebinned in 40 MeV steps.

7. RESULTS OF THE ENERGY-DEPENDENT ANALYSIS

Table 1 lists the parameters of Eqs. (11) and (13-15) as determined by the χ^2 minimization. The errors are calculated with the subroutine MINOS of the minimization program for 1 standard deviation.

Table 1. Results for the parameters of a χ^2 minimization

Scattering length matrix elements	$K'_{11} = -0.13 \pm 0.28$	GeV^{-1}
	$K'_{12} = 4.05 \pm 0.30$	GeV^{-1}
	$K'_{22} = -2.96 \pm 0.50$	GeV^{-1}
δ^1_1 parameters	$a_1 = 3.03 \pm 0.05$	rad
	$a_2 = -0.193 \pm 0.068$	rad GeV^{-2}
	$a_3 = -0.068 \pm 0.011$	rad GeV^4
η^1_1 parameters	$b_1 = -0.850 \pm 0.41$	
	$b_2 = 0.81 \pm 0.64$	GeV^{-2}
$\delta^1_1{}'$ parameters	$c_1 = -2.00 \pm 0.44$	rad
	$c_2 = 3.38 \pm 0.42$	rad GeV^{-2}
δ^0_2 parameters	$d_1 = -0.39 \pm 0.04$	rad
	$d_2 = 0.57 \pm 0.05$	rad GeV^{-2}
η^0_2 parameters	$e_1 = -0.016 \pm 0.012$	
	$e_2 = 0.068 \pm 0.035$	GeV^{-2}
$\delta^0_2{}'$ parameters	$f_1 = 0.10 \pm 0.11$	rad
	$f_2 = -0.66 \pm 0.18$	rad GeV^{-2}

---

*) The χ^2 is slightly lower than the results presented at the conference because a different parametrization of the P-wave phase is used. Also we include the latest results of the phases of the I = 2 S- and D-waves.

The results for the energy dependence of the cross-sections and moments are indicated by solid curves on Figs. 1, 2, 8, and 9.

8. THE I = 0 S-WAVE

The I = 0 S-wave accounts for most of the structure observed in the cross-section and moments because both its phase and elasticity undergo a rapid variation. Figure 10 shows the Argand diagram for the elastic $\pi\pi$ S-wave and Fig. 11 displays separately the phase and elasticity as a function of mass. The phase has already passed through 90° by 900 MeV and then rapidly increases to 180° at the $K\bar{K}$ threshold. Above threshold the S-wave becomes extremely inelastic. The elasticity reaches a minimum value of about 0.16 around 1020 MeV. Below threshold the phase joins the so-called "down" solution and agrees with the phase-shift analysis of Protopopescu et al., while above threshold a difference both for the phase and elasticity is observed.

The channel $\pi^+\pi^- \to K^+K^-$ near threshold is entirely produced in the S-wave state. The P- and D-waves contribute less than 10% to the total cross-section below 1120 MeV.

It was suggested by many authors (see Ref. 1) that the whole effect can be explained by a virtual bound state below threshold or a resonance close above threshold. In the concept of particles as poles in the complex energy plane the position of this pole can be found at Re D = Im D = 0, where D is the denominator of the complete T-matrix [Eq. (11)]. We found a pole on the second Rieman sheet at $(1012 \pm 6) - i(17 \pm 5)$ MeV.

9. THE P-WAVE AND I = 0 D-WAVE

Both the phases of the P- and D-waves show a slow variation as a function of mass. The P-wave phase rises from 140° at 900 MeV to 154° at the $K\bar{K}$ threshold and then stays nearly constant at this value. The D-wave rises from 4° at 900 MeV to 11° at 1120 MeV. The inelasticities for both the P- and D-waves are very small, typically 1%.

* * *

REFERENCES

1 W. Beusch, review paper, *in* Proc. Int. Conf. on Experimental Meson Spectroscopy (Columbia University Press, N.Y. and London, 1970), p. 185 (and further references given therein).

2 B.D. Hyams, W. Koch, D.C. Potter, L. von Lindern, E. Lorenz, G. Lütjens, U. Stierlin and P. Weilhammer, Nuclear Phys. B22, 189-204 (1970).

3 B.D. Hyams, W. Koch, D.C. Potter, L. von Lindern, E. Lorenz, G. Lütjens, P. Schlein, U. Stierlin, P. Weilhammer, W. Beusch, W. Wenzel, D. Johnson, V. Stenger and P. Wohlmut, Proc. Int. Conf. on Experimental Meson Spectroscopy (Columbia University Press, N.Y. and London, 1970), p. 41.

4   S.D. Protopopescu, M. Alston-Garnjost, A. Barbaro-Galtieri,
    S.M. Flatté, J.H. Friedman, T.A. Lasinski, G.R. Lynch,
    M.S. Rabin and F.T. Solmitz, Phys. Rev. $\underline{D7}$, 1279 (1973).

5   G. Grayer, B. Hyams, C. Jones, P. Schlein, W. Blum, H. Dietl,
    W. Koch, E. Lorenz, G. Lütjens, W. Männer, J. Meissburger,
    W. Ochs, U. Stierlin and P. Weilhammer, contributed talk to
    the 4th Int. Conf. on High-Energy Collisions, Oxford, 1972.

6   G. Grayer, B. Hyams, C. Jones, P. Schlein, W. Blum, H. Dietl,
    W. Koch, E. Lorenz, G. Lütjens, W. Männer, J. Meissburger,
    W. Ochs, U. Stierlin and P. Weilhammer, Proc. Int. Conf. on
    Experimental Meson Spectroscopy, Philadelphia, 1972
    (American Institute of Physics), p. 5.

7   G. Grayer, B. Hyams, C. Jones, P. Weilhammer, W. Blum, H. Dietl,
    W. Koch, E. Lorenz, G. Lütjens, W. Männer, J. Meissburger,
    W. Ochs, U. Stierlin and W. Hoogland, contributed paper to
    this conference.

8   R.H. Dalitz, Strange Particles and Strong Interactions (Oxford
    University Press, N.Y., 1961).

9   MINUIT: a minimization program by F. James and M. Roos, CERN
    Computer Library.

10 P. Baillon, R.U. Carnegie, E.E. Kluge, D.W.G.S. Leith,
    H.L. Lynch, B. Ratcliff, B. Richter, H.H. Williams and
    S.H. Williams, Phys. Letters $\underline{38}$ B, 555 (1972).

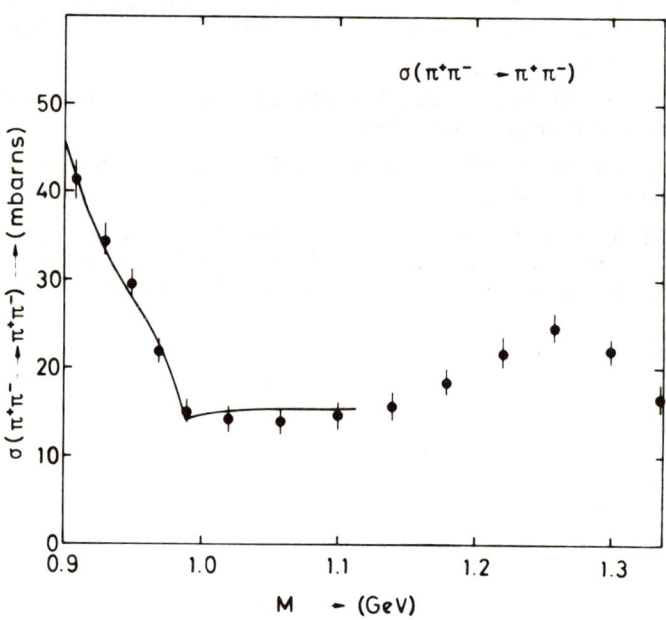

Fig. 1 On-shell cross-section $\pi^+\pi^- \to \pi^+\pi^-$. Error bars: extrapolated values from $\pi^-p \to \pi^+\pi^-n$ at 17.2 GeV.
Solid curve: result of energy-dependent fit.

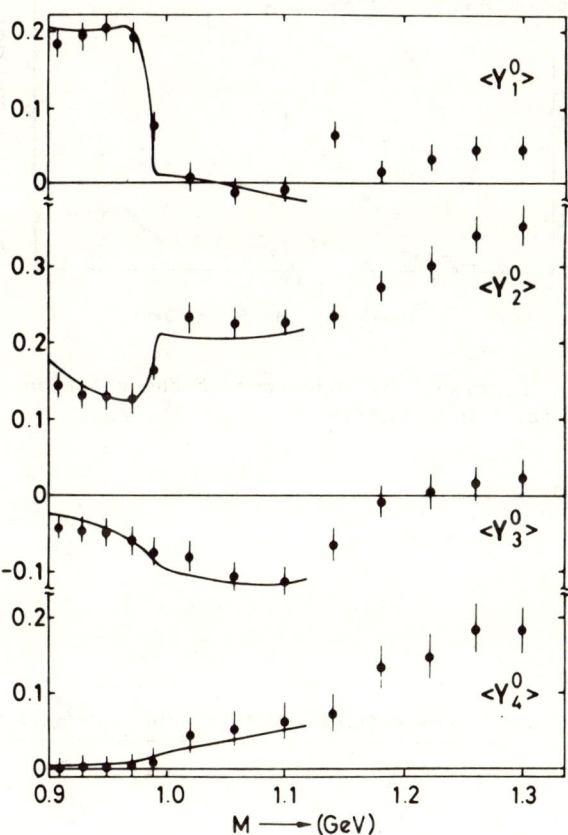

Fig. 2   On-shell moments $\langle Y_\ell^0 \rangle$ for $\pi^+\pi^- \to \pi^+\pi^-$ scattering. Error bars: extrapolated moments from $\pi^- p \to \pi^+\pi^- n$ at 17.2 GeV. Solid curves: results of energy-dependent fit.

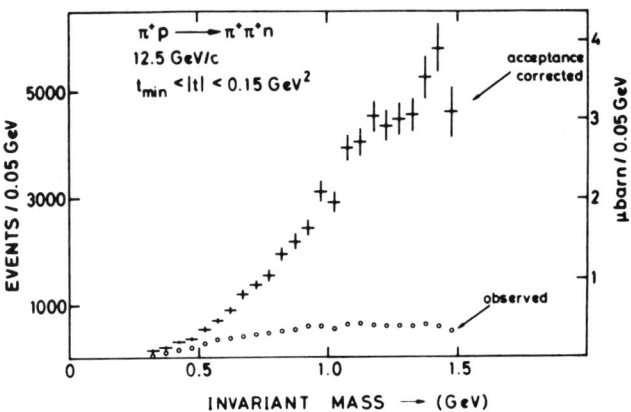

Fig. 3 $\pi^+\pi^+$ invariant mass spectrum of the reaction $\pi^+p \to \pi^+\pi^+n$ for $|t| < 0.15$ GeV².

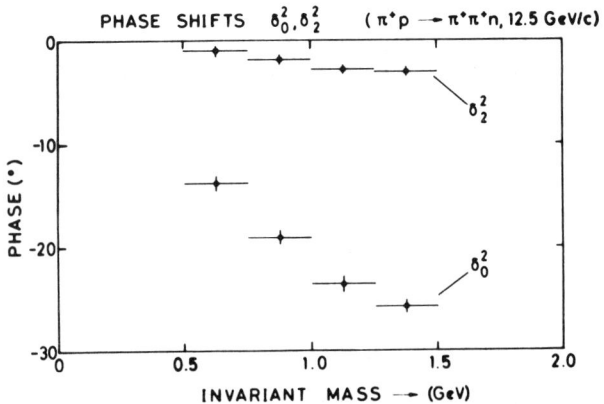

Fig. 4 $\pi\pi$ phase shifts for I = 2 S- and D-waves.

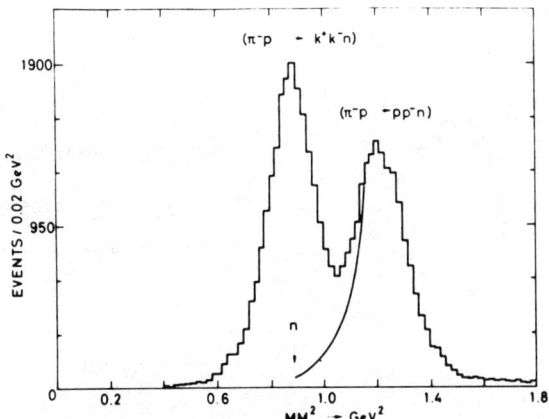

Fig. 5 (Missing mass)^2 distribution for the reaction $\pi^- p \to K^+ K^- n$, $p\bar{p}n$ at 9.8 GeV/c, for the K^+K^- hypothesis.

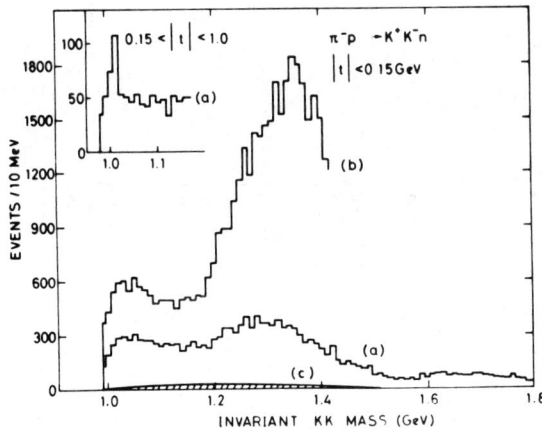

Fig. 6 K^+K^- invariant mass spectrum of the reaction $\pi^- p \to K^+K^- n$ for $|t| < 0.15$ GeV^2 at 9.8 GeV. Curve a) = raw data, curve b) = corrected mass spectrum, and curve c) = approximate background distribution from $\pi^- p \to p\bar{p}n$ in the sample of raw data.

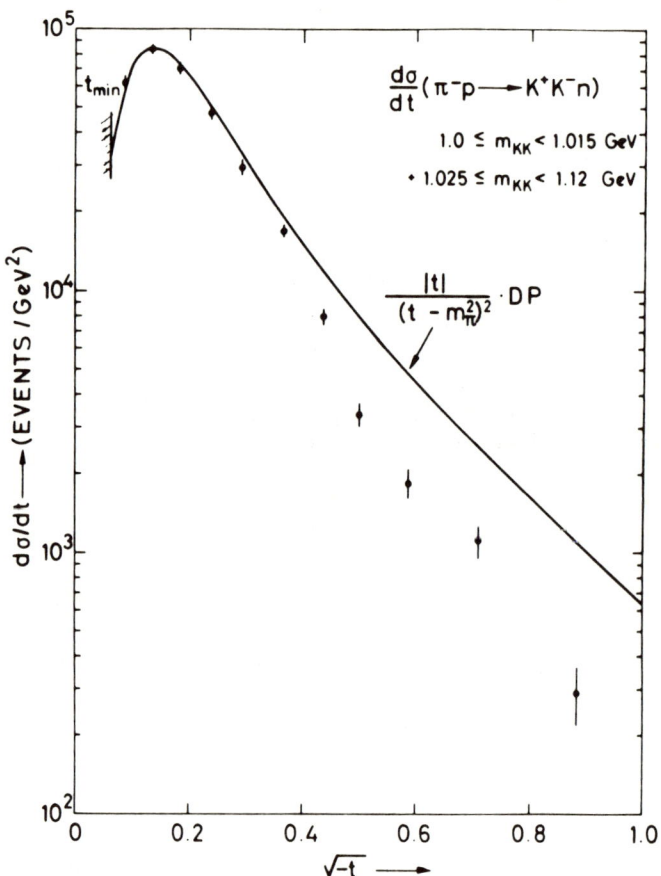

Fig. 7 The differential cross-section $d\sigma/dt$ for the reactions $\pi^- p \to K^+ K^- n$ for $K^+ K^-$ events in the threshold region. The horizontal scale is expanded as $\sqrt{-t}$. Error bars: corrected data. Solid curve: expected distribution for one-pion exchange (Born term amplitude modified by the Dürr-Pilkuhn form factor). The curve is normalized to the second data point.

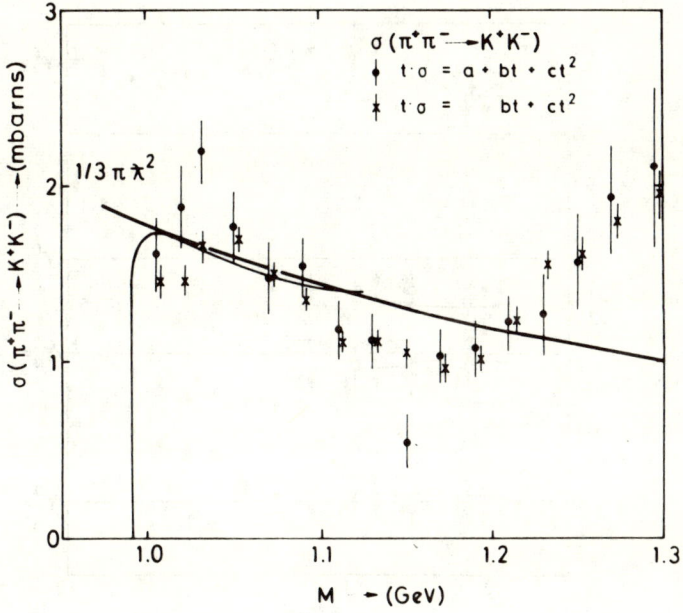

Fig. 8 The cross-section $\pi^+\pi^- \to K^+K^-$ between threshold and 1300 MeV. Error bars: results of pole extrapolation from $\pi^-p \to K^+K^-n$ at 9.8 GeV/c. Solid curve: result of energy-dependent fit. Also shown is the unitarity limit for the S-wave cross-section.

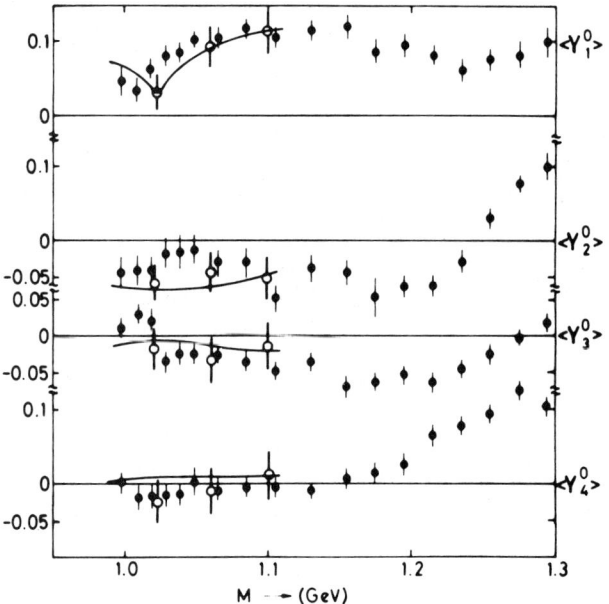

Fig. 9 t-channel off-shell moments $\langle Y_\ell^0 \rangle$ in the K^+K^- rest frame for $|t| < 0.15$ GeV^2.
φ: extrapolated moments for $\pi^+\pi^- \to K^+K^-$ scattering. Solid curve: result for an energy-dependent analysis.

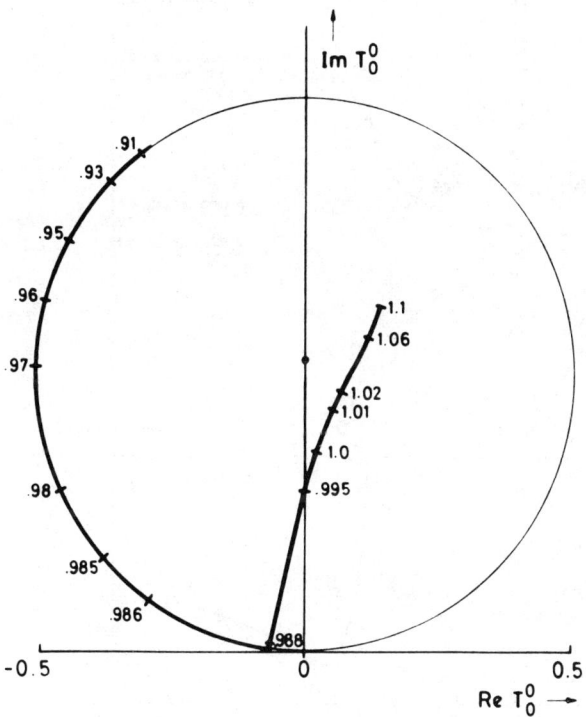

Fig. 10   Argand diagram for the $\pi^+\pi^- \to \pi^+\pi^-$ I = 0 S-wave.

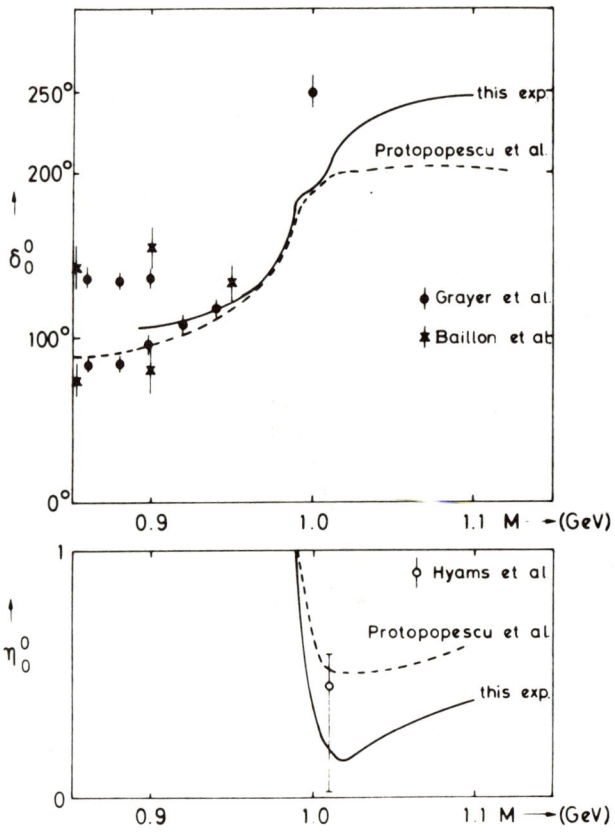

Fig. 11 Phase and elasticity for the I = 0 S-wave between 900 and 1120 MeV. Data points from Refs. 3, 4, 6, and 10.

π-π COUPLED CHANNELS*

P. K. Williams
The Florida State University, Tallahassee, Fla. 32306

ABSTRACT

A few observations are made relating to: various inelastic channels which couple to the π-π system; coupled channel analysis especially near the $K\bar{K}$ threshold; the π-exchange production mechanism and absorption. Finally, a method for measuring the π-π total cross section from the inclusive reaction $\pi^+ p \to \Delta^{++}$ + anything is discussed.

CHANNELS COUPLED TO π-π

The π-π scattering process is reasonably well understood only when it is purely elastic. To date, not a great deal is known about the various inelastic channels which couple to the π-π system. The π-π total cross section in fact can only be guessed very crudely by adding up observed contributions and allowing for the missed or unobserved contributions, or by using the optical theorem with the π-π elastic differential cross section.[1] These estimates are further complicated by the experimental fact that one of the incident pions is virtual. To extrapolate all the results to the pion mass shell would require much more data than are presently available. However, the efforts to estimate $\sigma_{total}(\pi\pi)$ indirectly are indeed heroic, and we shall return to this question in the latter part of this paper.

Of the inelastic channels which couple to the π-π system, the most notable is the $K\bar{K}$ channel. The importance of this channel for understanding π-π physics has been realized only during the last two years.[2] Indeed, it required the observation of the S* effect near the $K\bar{K}$ threshold in π-π scattering to emphasize the necessity of coupled-channel unitarity and analyticity to understand the variations of the π-π s-wave phase shift.

The single most important elementary observation to remember about the $K\bar{K}$ system is the G parity rule

$$G_{K\bar{K}} = (-1)^{I+J},$$

so that e.g. $\pi^+\pi^- \to K^+K^-$ only through $J^P = 0^+, 2^+$ --- with I = 0 or through $J^P = 1^-, 3^-$ --- with I = 1; only the former sequence is allowed for the $K^0_1 K^0_1$ final state. The G parity rule is also useful for identifying quantum numbers for three-particle final states with $K\bar{K}$ subsystems.

---

*In part supported by the U.S. Atomic Energy Commission and by the National Science Foundation.

Secondly, one should remember that $K\bar{K}$ systems produced by different t-channel mechanisms can in principle interfere in the physical region. So, for example, the $K\bar{K}$ system can have interfering contributions from decays of both A_2^0 and f^0 even though A_2 and f have different isospin, G parity and different production mechanisms in πN collisions.[5] (This is analogous to ρ-ω interference in the $\pi\pi$ system). Another example is $\delta(962)$ or $\pi_N(1016)$ decays to $K\bar{K}$ which could interfere with $S^* \to K\bar{K}$. Of course, these interferences go away at the pion pole, but their presence in the physical region indicates that one must do the pole extrapolation carefully.

Aside from the $K\bar{K}$ channel, the other two-body channels which might be of importance are $\eta\eta$, $\pi\omega$, $\pi\phi$, $\eta\rho$, πA_1, πA_2, $\bar{K}K^*$, $\omega\omega$, etc, and at high enough energy of course the $N\bar{N}$ channel opens and has been observed.[3] Of these, only the $\pi\omega$ channel is known to be of importance in understanding the onset of p-wave inelasticity[2], although a small $\bar{K}K^*$ branching ratio of the f' has been reported.[4] The p-wave inelasticity below about 1100 MeV should be almost entirely explained in terms of the $\pi\omega$ channel and the known coupling constant $g_{\rho\omega\pi}$ (the mechanism is identical to that for ω decay) so no new parameters may need to be introduced (e.g. additional background terms) in the description of the p-wave coupled channel system. This observation may be of help in the already complicated $\pi\pi$ phase shift analyses in the 800-1100 MeV region, in that it saves parameters.

With the exception of $\eta\eta$, $\pi\omega$ and $\omega\omega$, the two-body channels listed above can end up as final states of three pseudo-scalar mesons. Since this is the next least complicated configuration we will now consider it. For $\pi\pi \to$ 3-pseudo-scalars we have the additional J^P selection rules that $J^P = 0^+$ is forbidden and 0^+ subsystems of 2-pseudo-scalars are forbidden. In addition, for $\pi\pi \to K\bar{K}\pi$, the G-parity of the $K\bar{K}$ system must be negative so that only $K\bar{K}$ with $J^P(K\bar{K}) = 1^-, 3^- \text{---}$ and $I = 0$ or $J^P(K\bar{K}) = 2^+, 4^+ \text{---}$ and $I = 1$ are allowed.

Thus, for example, in the process

$$\pi^+ + \pi^- \to K^+ + K^- + \pi^0,$$

the two-body state $A_2^0 \pi^0$ is possible, but we must have $I = 0, 2$ and $J^P = \text{even}^+$ so that the orbital angular momentum between A_2^0 and π^0 must be odd. Consequently, $f' \to A_2^0 \pi^0$ with a p-wave decay, not to mention that $f \to K\bar{K}\pi$ is allowed and can go through an intermediate A_2^0 virtual state, via the $fA_2\pi$ coupling constant. This should be looked for. Similarly, $\phi\pi$ with $I = 1$, $J^P = \text{odd}^-$ and $\ell(\phi\pi) = \text{odd}$ is possible, and $g \to \phi\pi$ ($\ell = 3$), $\rho' \to \phi\pi$ ($\ell = 1$) could be observed. In the charged mode ($\pi^\pm \pi^0 \to K^+ K^- \pi^\pm$, for example) both g and ρ' can decay to $A_2 \pi$ through an orbital d-wave. Finally, the two-body state $\bar{K}K^*(+ K\bar{K}^*)$ certainly competes with $A_2\pi$ and $\phi\pi$, but $f' \to \bar{K}K^*$ with $\ell = 2$, so the p-wave decay to $A_2\pi$ should be competitive even though there is more phase space available for the $\bar{K}K^*$ mode. On the other hand, $g \to \bar{K}K^*$ ($\ell = 3$), $\rho' \to \bar{K}K^*$ ($\ell = 1$) might not compete with $\phi\pi$ in the neutral mode due to phase space, but in the charged mode $g \to A_2\pi$ may compete with both $\bar{K}K^*$ and $\phi\pi$ due to the favorable barrier factor.

The only other final state of three pseudo-scalars is $\eta\pi\pi$ which couple to $A_2\pi$ with $(-1)^I = (-1)^J = -(-)^{\ell(A_2\pi)}$; or to $f\eta$ with $I = 0$, J^P = even$^+$ and $\ell(f\eta)$ = odd. So $f' \to \pi\pi\eta$ possibly through a virtual f in a p-wave decay, but probably the $A_2\pi$ mode which is also p-wave would win out.

Thus, due to the interplay of phase space and angular momentum barriers, the observed branching ratios of resonances formed in $\pi\pi$ scattering to various quasi-two-body states may be quite different than, for example, one might expect[6]. In addition, the unravelling of these branching ratios from the data will require rather sophisticated partial wave analysis of the 2 → 3-body reaction. Cashmore[7] has addressed himself to this question. It appears likely that information on the branching ratios can be obtained. Such analyses may not be terribly good for establishing the existence of new resonances, unless they have very small coupling to $\pi\pi$ and $K\bar{K}$ channels.

COUPLED CHANNEL ANALYSIS

Now we will turn attention to coupled channel analysis. This subject has been reviewed recently by Morgan[8], Williams[9], and by Petersen[10], with special emphasis on the $\pi\pi$-$K\bar{K}$ coupled channel problem and the S* effect. Consequently, there will be little that is really new in this section. Basically one writes the partial wave amplitude matrix in terms of the K matrix, or M matrix as follows[8-11]

$$[a_\ell] = K[1-iK]^{-1} = k^{\ell+\frac{1}{2}}[M-ik^{2\ell+1}]^{-1} k^{\ell+\frac{1}{2}},$$

where k is a diagonal matrix of channel momenta. Near the threshold of a particular inelastic channel, the scattering in that channel can be described by a complex scattering length $A = a+ib$. For the two-channel s-wave case, the matrices above can be expressed in terms of a, b and $\alpha \equiv K_{11}$ as follows

$$[a_0] = \begin{bmatrix} \dfrac{T_B + ik_2 b}{1-ik_2 A} S_B & \dfrac{\sqrt{k_2 b}}{1-ik_2 A} S_B^{\frac{1}{2}} \\ \dfrac{\sqrt{k_2 b}}{1-ik_2 A} S_B^{\frac{1}{2}} & \dfrac{k_2 A}{1-ik_2 A} \end{bmatrix}$$

where $T_B = (S_B - 1)/2i = \alpha(1-i\alpha)^{-1}$, k_2 is the channel momentum,

$$[K] = \begin{bmatrix} \alpha & [(1+\alpha^2)k_2 b]^{\frac{1}{2}} \\ [(1+\alpha^2)k_2 b]^{\frac{1}{2}} & k_2(a+\alpha b) \end{bmatrix}$$

$$[M] = \dfrac{1}{\alpha a - b} \begin{bmatrix} k_1(a+\alpha b) & [(1+\alpha^2)k_1 b]^{\frac{1}{2}} \\ [(1+\alpha^2)k_1 b]^{\frac{1}{2}} & \alpha \end{bmatrix}$$

The advantage of M is that it is analytic at $k_2^2 \simeq 0$ so can be expanded as $M_{ij} = M_{ij}^0 + R_{ij} k_2^2$, which corresponds to expanding a, b, α above. The usual effective range expansion of the complex scattering length is $A \rightarrow [(a+ib)^{-1} - \frac{1}{2}rk_2^2]^{-1}$, whereas the dependence of α on the center of mass energy may be almost anything (i.e. continuation of the behavior far below threshold). Finally, we write $[a_0]_{11}$ in terms of the reduced K-matrix[11] element α_{11}^R as $[a_0]_{11} = \alpha_{11}^R (1-i\alpha_{11}^R)^{-1}$, where (with $r = 0$),

$$\alpha_{11}^R = \alpha + \dfrac{ik_2 b(1+\alpha^2)}{1-ik_2(a+\alpha b)}$$

where it is to be noted that, below threshold $k_2 \rightarrow i|k_2|$, so that $\alpha_R \rightarrow$ real, corresponding to pure elastic scattering. This parameterization of the two-channel problem in terms of the "background" phase shift $\delta_B = \tan^{-1}(\alpha)$ and the complex scattering length A may be useful for visualizing the physics of the problem. The use of an M matrix, expanded near threshold, is more general, and recent analyses[2] have properly used it. However, it is interesting to note that the effect range case ($r \neq 0$) yeilds a natural connection to a resonance pole formulation, for a pole near threshold, either above or below. We write for α_{22}^R:

$$\alpha_{22}^R = k_2 \left[\frac{1}{a+ib} - \tfrac{1}{2} r k_2^2 \right]^{-1} = \frac{x}{\varepsilon - i}$$

where $x = \Gamma_2/\Gamma_1 = k_2 \gamma_2/\Gamma_1$ and $\varepsilon = 2(W_0-W)/\Gamma_1$ using obvious notation, so $\gamma_2 = 2/\mu r$, $\Gamma_1 = \gamma_2 b/(a^2+b^2)$ and $W_0 = W_t + \tfrac{1}{2}\gamma_2 a/(a^2+b^2)$, where μ is the reduced mass and W_t the threshold energy of channel 2. Consequently,

$$[a_0]_{22} = \frac{\Gamma_2/2}{\Delta - i\Gamma/2}$$

$$[a_0]_{11} = T_B + \frac{i}{\varepsilon + i} \cdot \frac{\Gamma_2/2}{\Delta - i\Gamma/2} S_B$$

where $\Delta = W_0 - W$ and $\Gamma = \Gamma_1 + \Gamma_2$. Defining $S_B = S_B'(\varepsilon+i)/(\varepsilon-i)$:

$$[a_0]_{11} = T_B' + \frac{\Gamma_1/2}{\Delta - i\Gamma/2} S_B'$$

These two forms for $[a_0]_{11}$ correspond respectively to

$$\alpha_{11}^R = \alpha + \frac{(1+\alpha^2)}{(\varepsilon+\alpha) - i\,(1+\varepsilon^2)/x}$$

$$= \alpha' + \frac{(1+\alpha'^2)}{(\varepsilon-\alpha') - ix}$$

where $\alpha' = \tan \delta_B' = \tan[\ln(S_B')/2i]$. The interpretation of a virtual bound state for a pole below threshold would correspond to α being slowly varying (no pole in the K-matrix), whereas the resonance interpretation would imply that α' is slowly varying in the vicinity of the phenomenon. Parameterizing the energy dependence of α (or α') would require at least two parameters (e.g. as tan (c+dW)); so including the original three parameters a, b and r, there are altogether at least five parameters which may be necessary to distinguish between the resonance and virtual bound state interpretations. Finally, note that, if α is slowly varying, the appropriate form for $[a_0]_{11}$ exhibits a complex singularity structure with an additional pole on an underlying sheet which has ImW > 0 at the pole (see ref. 2). It also exhibits the asymmetric shape characteristic of the virtual bound state interpretation for a pole below threshold.

THE PRODUCTION MECHANISM

We now turn to consideration of the production mechanism. Recent developments in analysis of the high statistics data of the CERN-MUNICH[12] experiment as well as that of SLAC[13] have taught us as much about the production mechanism as about the π-π phase shifts. Future experiments (on polarized targets,[14] for example) will yeild information primarily on the production mechanism.

Consequently, it is apparent that the whole field is moving from being primarily spectroscopy oriented to being primarily dynamics oriented. This is a very pleasant development for theorists especially, as some pressing questions remain unanswered, as indicated in the talks by Martin[15], Kane[16], Matthews[17] and Field[18]. Of course, as we learn more about the dyanmics, the pendulum of attention will swing back to spectroscopy, and so on, presumably for ever and ever!

The recent studies[12-19] of details of the production mechanism have taught us that absorption is important for modifying the basic exchange mechanism, and that the basic non-strange charge exchanges carry at least the quantum numbers of the π and A$_2$ and the neutral exchanges carry in addition ω quantum numbers. At small t, π-exchange does dominate due to the proximity of the pion pole, but other t-channel quantum numbers, such as those induced by absorption and A$_2$ exchange, are not completely negligible. Any complete analysis must therefore take these things into account. The ability to use the dipion decay angular distribution as an analyser of t-channel natural and unnatural parity contributions has proved a formidable tool for study of exchange mechanisms. Progress has been made on models as a result. The data quality is good enough that one can begin to study details of the absorption mechanism and phase relationships between various helicity amplitudes! As a result, simple models may not be good enough to describe the data, and we may have to revise even some of our oldest pheomenological assumptions in constructing adequate models for the production mechanism.

The possibility of studying details of absorption is indeed an exciting one. The recent work of Wagner and Ochs[20] and Estabrooks and Martin[21] on the so-called OPE-δ[22] or "Poor Man's Absorption" (PMA)[23] model may point to a revision of previous ideas on absorption, namely the assumption that absorption does not depend appreciably on the dipion mass. The OPE-δ(PMA) model represents an attempt to get the gist of absorption modifications correct by smoothing out the wrinkles of the pure OPE helicity amplitudes both in J and in t, consistent with the minimal requirements of angular momentum conservation. This can be done without resorting to the usual partial wave expansion associated with absorptive corrections simply by evaluating part of each helicity amplitude at the pion pole. The resulting amplitudes for production of a dipion with spin ℓ, helicity μ are of the form (large s, small t)

$$H^{(\ell)}_{\lambda\lambda'\mu} = \gamma_\ell(m_{\pi\pi})\,(-t')^{n/2}\,F_n(t)\,M^{(\ell)}_{\lambda\lambda'\mu}(s, t = m_\pi^2)/(m_\pi^2 - t)$$

where γ_ℓ is essentially the π-π scattering amplitude, n is the net helicity flip, $(-t')^{n/2}$ is required by angular momentum, $F_n(t)$ is a collimating factor introduced to approximate the smooth long-range effects of absorption and $M_{\lambda\lambda'\mu}(\ell)(s,t)$ is a polynomial in t appearing in the Born term, here evaluated at the pole. In particular, one of the (n = 0) amplitudes for production of a transverse dipion with nucleon helicity flip is of the form $F_0[t/(t-m_\pi^2) - C]$ where C = 1 in the OPE-δ(PMA), which corresponds to evaluating t in the numerator at the pole. (The absorptive correction to A_2 also would contribute effectively to C). The usual absorptive modification gives C ~ 1, almost independent of ππ mass. The recent work alluded to above[20,21] uses C as a parameter and finds C decreases from ~ 1 at $\hat{m}_{\pi\pi} \lesssim 1$ GeV/c² to about ~ 0.4 above the f-mass - a substantial variation! This would appear to indicate not only that OPE-δ(PMA) is wrong but also that absorption modification might be in real trouble.

It is possible that to escape this difficulty would require resorting to the full flexibility of Reggeized absorption, and in addition invoking a substantial amount of A_2 exchange which should couple progressively weakly to high $m_{\pi\pi}$ states (like ~ $m_{\pi\pi}^{-1}$) relative to pion-exchange as predicted by Hoyer, Roberts and Roy[24]. Estabrooks and Martin[15], and Kane[16] have suggested something of this nature, which in Reggeized absorption models would also probably require final state (dipion-nucleon) cross section which decreases with increasing $m_{\pi\pi}$. The calculations should be done.

On the other hand, OPE-δ(PMA) is a specific model for the pion exchange part of the production mechanism. If C is taken to be a fitted parameter, and if it is found to decrease from ~ 1, then there must be additional contributions to the production mechanism. If A_2 exchange is important, it can be added to OPE-δ(PMA). The collimating factors F_n are then adjusted to fit the data, retaining C = 1.[19] But if A_2 weakens with increasing $m_{\pi\pi}$, then F_n may have to depend on $m_{\pi\pi}$ (it is essentially $(F_1/F_0)C$ that is measured sensitively[10]), but any substantial variation of F_n with $m_{\pi\pi}$ would violate another assumption of the model. It remains to be seen if OPE-δ(PMA) can retain its usefulness by adding A_2 exchange, but it is fairly certain that OPE-δ(PMA) by itself is not sufficient to explain the data at $-t \gtrsim .15$ or $m_{\pi\pi} \gtrsim 1.2$.

MEASURING THE π-π TOTAL CROSS SECTION

Finally, we return to the problem of measuring the π-π total cross section, a problem referred to in the beginning. We would like to suggest here the possibility of doing it through the reaction

$$\pi^+ + p \to \Delta^{++} + \text{anything}.$$

The extraction of the π-exchange contribution and the extrapolation to the pion pole are hindered by the width of the Δ^{++}, but are helped by the fact that the exchanged pion is a "fully-fledged" pion, in the terminology of Fox[25], and by the fact that the Δ^{++} decays, the analysis of which facts makes it possible to purify the π-exchange part somewhat before extrapolating to the pion pole. We restrict ourselves to small $-t_{N\Delta} \lesssim 0.2$ to take full advantage of the fully-fledged pion[25], and analyse the Δ^{++} decay, the angular distribution of which is given by

$$W_\Delta(\theta,\phi) = \frac{3}{4\pi}\left[\rho_{33}\sin^2\theta + \rho_{11}(\tfrac{1}{3}+\cos^2\theta) - \frac{2}{\sqrt{3}}\mathrm{Re}\rho_{3-1}\sin^2\theta\cos 2\phi \right.$$
$$\left. - \frac{2}{\sqrt{3}}\mathrm{Re}\rho_{31}\sin 2\theta\cos\phi\right]$$

In the t-channel (Gottfried-Jackson) frame, pure pion exchange contributes only to ρ_{11}. We **now** take advantage of an observation by Fox[25] that M1-ρ-exchange <u>and all</u> natural parity exchanges in a quark model <u>and</u> pion-absorption-generated natural parity exchange all give approximately the Stodolsky-Sakurai distribution (SSD) for decay of the Δ^{++}, namely that for which $\rho_{33} = \rho_{3-1}\sqrt{3} = 3/8$. Assuming approximate incoherence between all natural and unnatural parity exchange contributions to the ρ_{ij}, and assuming $\mathrm{Re}\rho_{31}\simeq 0$ (experimentally and in all models, true), the Δ^{++} decay is

$$W_\Delta^{\text{inclusive}} \simeq \frac{3}{4\pi}\left[\frac{3r+1}{8}(\tfrac{1}{3}+\cos^2\theta) + \frac{3(1-r)}{8}\sin^2\theta(1 - \tfrac{2}{3}\cos 2\phi)\right]$$

where $r = r(s, t, m_{\pi\pi}^2)$ is the fraction of the inclusive reaction which is pure pion exchange, and $r \to 1$ at $t = m_\pi^2$.

The quantity

$$\left[r\frac{d\sigma^{\text{inclusive}}}{dt\,dm_{\pi\pi}^2}(t-m_\pi^2)^2\right]_{t=m_\pi^2} = K\sigma_{\pi\pi}(\text{total}),$$

$$\left[(1-r)\frac{d\sigma^{\text{inclusive}}}{dt\,dm_{\pi\pi}^2}(t-m_\pi^2)^2\right]_{t=m_\pi^2} = 0$$

where K is the usual kinematic factor of proportionality.[2] The quantity $rd\sigma^{\text{inclusive}}$ for $t<0$ **should** show even G parity meson signals (ρ, f ---) and $(1-r)d\sigma^{\text{inclusive}}$ should show mostly the G = - signals (ω, A_2 ---). The point here is that the use of r might work much better than simple extrapolations of $d\sigma^{\text{inclusive}}$ to the pion pole, even better than one has any a priori right to expect, simply because so many <u>non-pion</u> exchanges contribute only the SSD to W_Δ.

In connection with the suggestion to use this method, it should be cautioned[26] that (i) there are substantial backgrounds e.g. from the dissociation $p \to \Delta^{++}\pi^-$, which are not elliminated by the small-$t_{N\Delta}$ cut and which must then be extrapolated away. If these contribute substantially to ρ_{31}, the method must be modified; (ii) the experiment should not be done at too high an energy because the resolution in missing mass for small $t_{N\Delta}$ gets progressively worse. However, notwithstanding the need for caution, we feel the method can give improved estimates of the $\pi\pi$ total cross section. This information could be of crucial importance to π-π phase shift analysis in the 1-2 GeV/c^2 region of $m_{\pi\pi}$.

ACKNOWLEDGEMENTS

The Anthor acknowledges helpful conversations with Professor W. D. Walker, Professor J. D. Kimel and Dr. E. Reya.

REFERENCES

1. W. D. Walker, report to this conference.
2. S. D. Protopoescu, et.al., Phys. Rev. (to be published); Experimental Meson Spectroscopy - 1972, Edited by A. Rosenfeld and K. W. Lai, American Institute of Physics, New York, 1972, p. 17; E. Lorenz, report to this conference.
3. G. Grayer, et.al., Phys. Letters 42B, 249 (1972).
4. Particle Data Group, Rev. Mod. Phys. 43, 561 (1971).
5. N. N. Biswas, et.al., Phys. Rev. D5, 1564 (1972); R. Diebold, report to this conference.
6. See, for example, F. J. Gilman, Experimental Meson Spectroscopy, Edited by A. Rosenfeld and K. W. Lai, American Institute of Physics, New York, 1972, p. 460.
7. R. J. Cashmore, report to this conference.
8. D. Morgan, RHEL preprint RPP/T/27.
9. P. K. Williams, Phys, Rev. D6, 3178 (1972).
10. J. L. Petersen, Physics Reports 2C, 155 (1971).
11. R. Dalitz, Ann. Rev. Nucl. Sci. 13, 339 (1963).
12. G. Grayer, et.al., Experimental Meson Spectroscopy - 1972, Edited by A. Rosenfeld and K. W. Lai, American Institute of Physics, 1972, p. 5; P. Estabrooks and A. D. Martin, Phys. Letters 41B, 350 (1972).
13. P. Baillon et.al., Phys. Letters 35B, 453 (1971).
14. E. Lorenz, report to this conference; R. Carnegie, report to this conference.
15. A. D. Martin, report to this conference.
16. G. L. Kane, report to this conference.
17. J. A. J. Matthews, report to this conference.
18. R. Field, report to this conference.
19. J. D. Kimel and E. Reya, report to this conference.
20. W. Ochs and F. Wagner, "On the Strength of Absorption in Single Pion Production Reactions," München (M.P.I.) preprint.
21. A. D. Martin and P. Estabrooks CERN.TH. 1647. See also Ref. 15.
22. P. K. Williams, Phys. Rev. D1, 1312 (1970); L. Chan and P. K. Williams, Phys. Rev. 188, 2455 (1969).
23. G. C. Fox, Phenomenology in Particle Physics 1971, Edited by C. B. Chiu and G. C. Fox (California Institute of Technology, Pasadena, Cal., 1971), p. 703.
24. P. Hoyer, R. Roberts and D. Roy, RHEL preprint RPP/T/35.
25. G. C. Fox, ref. 23, p.
26. W. D. Walker, (private communication), Aspen Summer Study-1968, A.3-68-100.

PARTIAL WAVE ANALYSIS IN 2 → 3 BODY REACTIONS AND COMMENTS ON INELASTIC FINAL STATES IN $\pi\pi$ AND $K\pi$ SCATTERING*

R. J. Cashmore
Stanford Linear Accelerator Center
Stanford University, Stanford, California 94305

ABSTRACT

Multiparticle states derived from $\pi\pi$ and $K\pi$ scattering are discussed and the techniques for analysis of 3 body final states are described. The role that inelastic reactions play in determination of resonance parameters is investigated and the conclusion reached that they only provide information on the specific inelastic channel couplings. Thus the identification of resonances coupling to $\pi\pi$ or $K\pi$ systems is most easily made by observation of the elastic and charge exchange reactions.

INTRODUCTION

In the future analysis of $\pi\pi$ and $K\pi$ scattering will be performed at higher energies. This has two immediate consequences in that we expect (1) higher J^P states and (2) increased inelasticity. Thus we might expect that the inelastic channels will become an increasingly important tool in identifying new resonance states.

In this talk I want to (1) demonstrate that viable techniques exist for the analysis of 2 → 3 body reactions; (2) consider the available data in $K\pi \to K\pi\pi$ scattering and the application of such techniques; (3) discuss the contribution such inelastic information might make to $\pi\pi$ and $K\pi$ scattering; and finally (4) come to some conclusion about the best way of proceeding in the study of $\pi\pi$ and $K\pi$ scattering.

To illustrate many of these points I will draw on experience one has gained from studies[1] of

$$\pi N \to \pi\pi N \qquad (1)$$

INELASTIC REACTIONS IN $\pi\pi$ AND $K\pi$

In Table I, I have made a list (not exhaustive) of the inelastic reactions one might expect in $\pi\pi$ and $K\pi$ scattering, taking into account the restrictions of G-parity in the former case. In the case of $K\pi$ scattering, $K\pi\pi$ could well be an important state at low c.m. energies. Thus I will use this as an indication of what might be gained by observation and analysis of an inelastic reaction. The analysis methods for stable 2 body inelastic

---

*Work supported by the U. S. Atomic Energy Commission.

reactions $K\bar{K}$, $K\eta$ (I also include $\pi\omega$, $K\omega$) are well known and hence I will pay little attention to these.

I might note at this point that the 0^+ wave in $K\pi$ scattering can only lead to inelastic final states with ≥ 4 particles, a clear difference from the same J^P state in $\pi\pi$ scattering.

Table I Inelastic final states in $\pi\pi$ and $K\pi$ scattering

	$\pi\pi$	$K\pi$
2	$\pi\pi$	$K\pi$
	$K\bar{K}$	$K\eta$
3	$K\bar{K}\pi$	$K\pi\pi$ ($K^*\pi$, $K\rho$ ...)
		$KK\bar{K}$ (M > 1500)
4	$\pi\pi\pi\pi(\pi\omega)$	$K\pi\pi\pi(K\omega)$
etc.		

THE ANALYSIS OF 3 BODY FINAL STATES

<u>Formalism and method.</u> The formalism for this has been developed[2,3] to deal with

$$\pi N \to \pi\pi N \quad (2)$$

and later applied to

$$A_1, A_2 \ldots \to 3\pi . \quad (3)$$

The main ingredients are

(1) There exists strong resonance production in the final state, e.g., K^*, ρ. This is then introduced as an integral part of the model.

(2) The amplitude for the observation of a state JPM is written as a coherent sum over all of the possible intermediate quasi-two body states

$$A(K\pi\pi) = \sum_{JPM} f^{JPM}(\omega, t) \, G^{JPM}(\omega_1, \omega_2, \alpha, \beta, \gamma) \quad (4)$$

where $f^{JPM}(\omega, t)$ = amplitude for formation of the state JPM with mass ω and 4 momentum transfer t; $G^{JPM}(\omega_1, \omega_2, \alpha, \beta, \gamma)$ = amplitude for the decay of this state into $K\pi\pi$ ($\omega_1, \omega_2, \alpha, \beta, \gamma$ - the 5 variables necessary to describe the state). The decay amplitude is then

$$G^{JPM} = \sum_{Lj} X^{JPM}_{Lj} B_j(\omega_j) \quad (5)$$

where X^{JPM}_{Lj} contains all the angular momentum decomposition and $B_j(\omega_j)$ contains the final state enhancement factor for the pair of particles labelled j, e.g., Watson final state factor[5] or Breit-Wigner.

This is represented in Fig. 1.

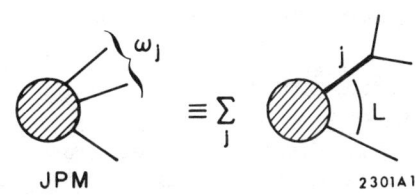

Fig. 1. Decay of state JPM to 3 bodies via intermediate isobar.

(3) We can then write the differential cross section as

$$d^5\sigma(\omega_1^2, \omega_2^2, \alpha, \beta, \gamma) \propto \sum_{\substack{JPM \\ J_1P_1M_1}} f^{JPM} G^{JPM} \begin{bmatrix} J_1P_1M_1 \\ f \end{bmatrix} G^{J_1P_1M_1} \end{bmatrix}^* \quad (6)$$

$$\propto \sum \rho_{MM_1}^{JP\,J_1P_1} G^{JPM} G^{J_1P_1M_1 *} \quad (7)$$

The object is then to determine

$$f^{JPM}(\omega, t) \quad \text{or} \quad \rho_{MM'}^{JP\,J'P'}(\omega, t) \quad (8)$$

and in order to make maximum use of the data this is usually done by the maximum likelihood method.

Results. The success of this method is best demonstrated in the results in $\pi\pi N$ partial wave analyses. There one finds that (1) Good fits to the data are obtained. However large numbers of events are necessary to obtain stable solutions, e.g., in $\pi\pi N$ we use ~5000-10,000 at each energy; and (2) The expected resonance structures are observed. This is seen in Fig. 2 where I show the F_{15} partial wave amplitudes — the F_{15} being clearly seen in the $\pi\Delta$, Nρ and Nσ channels.

I might also add that the analyses of $A \to 3\pi$ data have also clearly identified the resonant structure in the $J^PM = 2^+$, 1 waves (the A_2).

Conclusion. The conclusion is that one now has comparatively well understood methods of analyzing such reactions and one can consider applying them to the 3 body channels in $\pi\pi$ and $K\pi$ scattering.

THE DATA AND ITS UTILIZATION FROM $K\pi$ INELASTIC REACTIONS

The data. In Fig. 3, I show the data that exists on the reaction

$$K^+ p \to K^0 \pi^+ \pi^- \Delta^{++} \quad (9)$$

with incident 10 GeV/c K^+ mesons.[6] This is a large bubble chamber experiment and produces such meagre statistics.

The utilization of the data. In the spirit of $\pi\pi$ and $K\pi$ partial wave analyses one would optimistically like to follow one of two paths. (1) Amplitude extrapolation from the physical region to the pion pole — this means that one must determine the amplitudes at different t values. Now in inelastic final states one in general can have both natural and unnatural spin parity series (cf, $\pi\pi$ or $K\pi$ elastic scattering where one only has natural spin parity) so that the amplitude extraction is more difficult. From the $\pi\pi N$ experience this means $\gtrsim 1000$ events for each m, t bin. (2) Amplitudes from

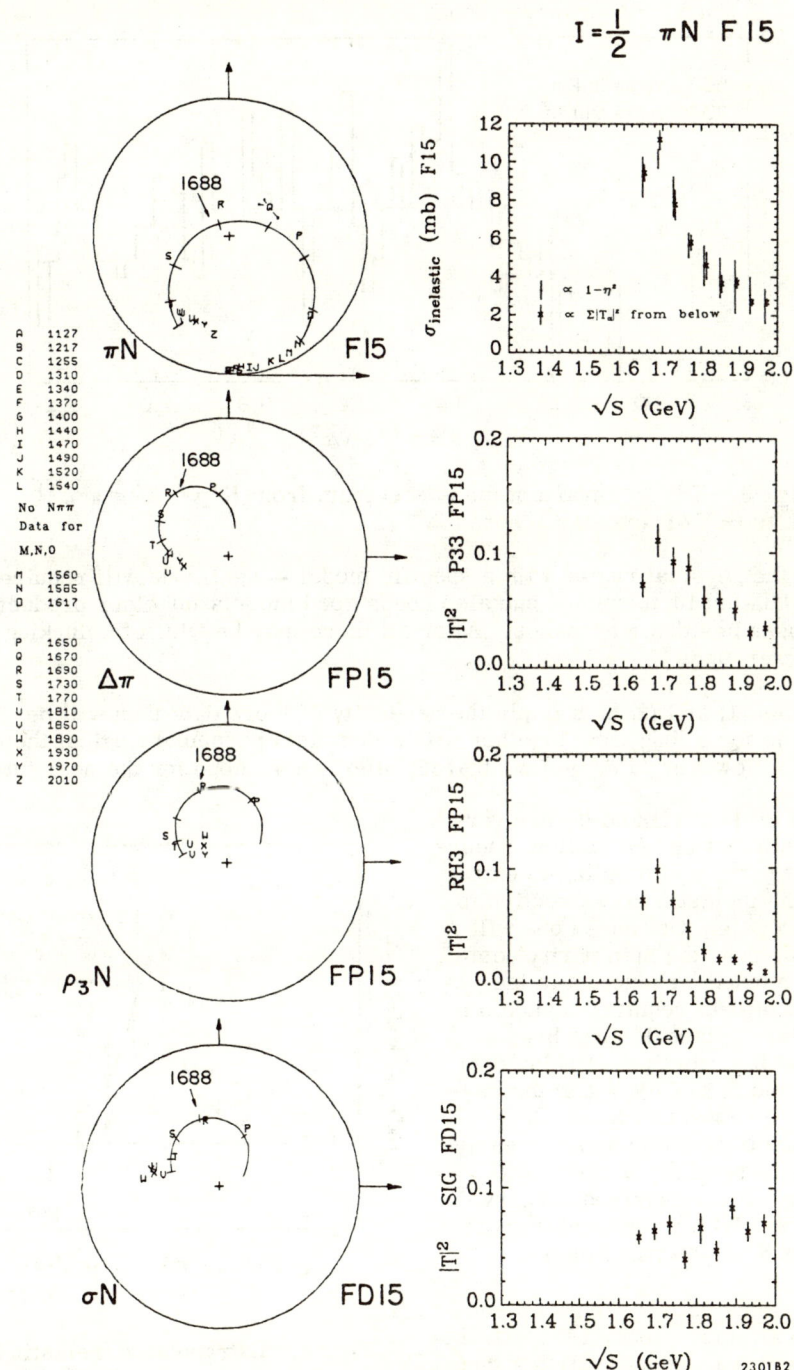

Fig. 2. The F_{15} partial wave.

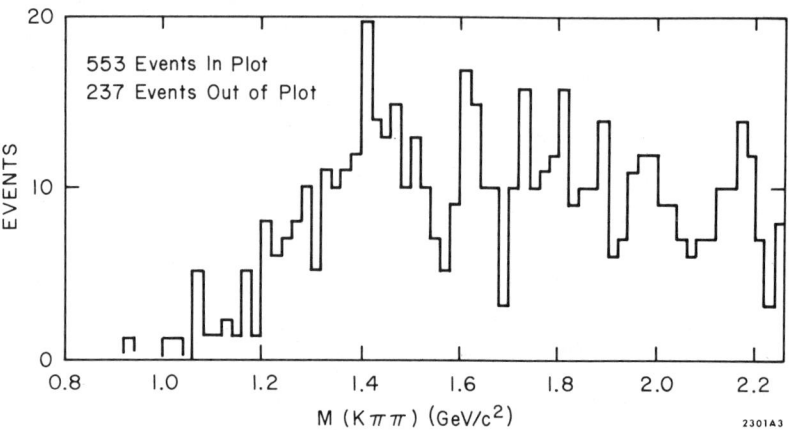

Fig. 3. $K^0\pi^+\pi^-$ invariant mass spectrum from $K^+p \to K^0\pi^+\pi^-\Delta^{++}$ at 10 GeV/c (other π^+p not in Δ^{++}).

fits in the physical region with a specific model — again one will require a lot of data but furthermore one also needs good models for other production processes besides π exchange. After all there may be Q's, etc. lurking in the background.

Thus (1) and (2) both imply the necessity of more data than will be available for a long time together with a complete phenomenological theory for (2). However, I do believe that (2) offers more hope for the near future.

At present (1) and (2) are very optimistic but one can follow a more modest path by attempting to extrapolate the inelastic cross section to the pion pole. Of course one will not then know the spin parity composition but the reduction in the amount of data required makes this feasible. In fact this has been attempted for reaction (9) using data at 5.0 and 8.25 GeV/c and the results are shown in Fig. 4. The presence of the $K^*(1400)$ is clearly seen on essentially a zero background. The superimposed curve is a Breit-Wigner derived from the current $K^*(1400)$ parameters.

Conclusion. It seems reasonable to attempt to obtain the extrapolated inelastic cross section although anything more ambitious will require at least an order of magnitude increase in the quantity of data.

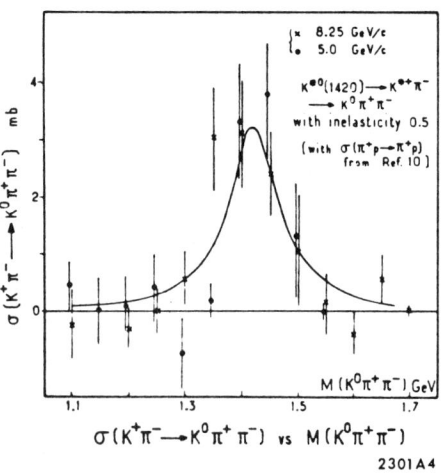

Fig. 4. Extrapolated inelastic $K\pi$ cross section versus $M(K^0\pi^+\pi^-)$ at 5.0 and 8.25 GeV/c using elastic (π^+p) cross section from a π-nucleon phase-shift analysis.

THE INFORMATION CONTAINED IN INELASTIC
PARTIAL WAVE AMPLITUDES

It is now important to understand the impact on analyses of the absence of detailed information on the inelastic partial wave amplitudes. Again as a guide I would like to take some of our work on the πN system in discussing this question.

<u>Resonance observation.</u> It is first worth noting that all of the resonances listed in the PDG tables have been observed first in stable 2 body partial wave analyses, i.e., $\pi N \to \pi N$, $\bar{K}N \to \bar{K}N$, $\Lambda\pi$, $\Sigma\pi$. Indeed our analysis has essentially only added one new state, the $D_{13}(1700)$ (already hinted at by EPSA) and removed another, the $P_{33}(\sim 1700)$.

<u>Resonance parameters</u> (pole position in the T-matrix). One might expect that the quantitative definition would improve by adding the inelastic channel information. We have been making coupled channel K-matrix analyses of the different πN partial waves and so we are in a position to investigate this question by performing the analysis with or without the inelastic data. Of course if the partial wave is inelastic sensible fits can only be obtained if an inelastic channel is introduced to preserve unitarity. In Fig. 5

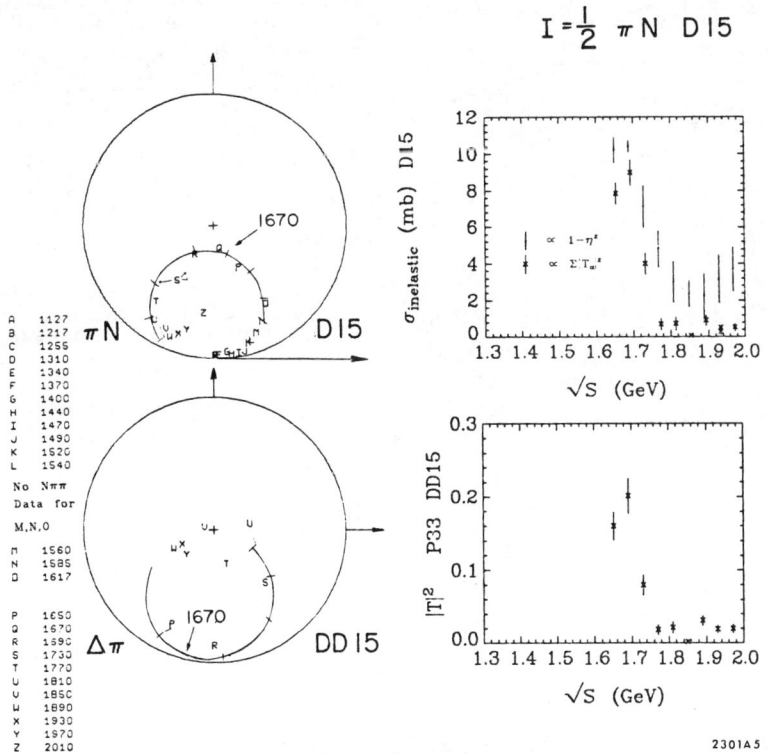

Fig. 5. D_{15} partial wave.

and Fig. 6 we see the Argand diagrams for the two partial waves D_{15} and F_{35}, both exhibiting large inelasticity but in two different channels, $\pi\Delta$ and $N\rho$ respectively. In analyzing just the elastic data we have considered the "junk" channel in both cases to be $\pi\Delta$, clearly a bad assumption for the F_{35}.

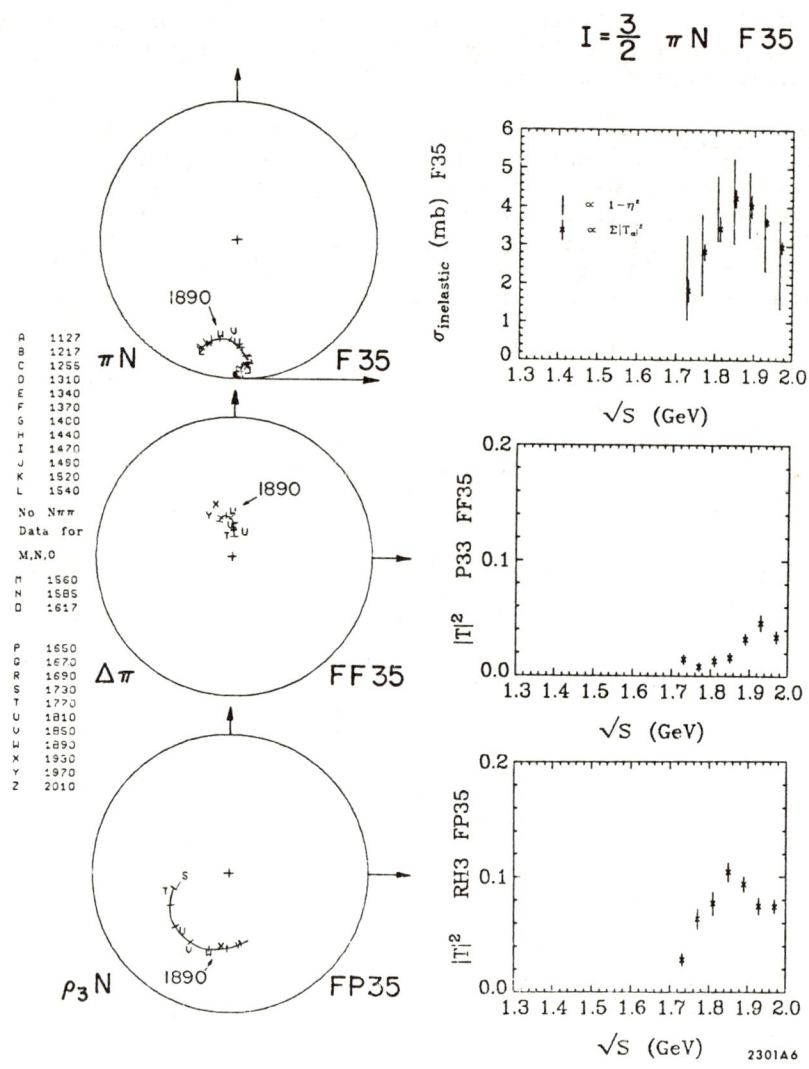

Fig. 6. F_{35} partial wave.

The results of these analyses is contained in Table II.

The conclusions from this work are, I think, obvious (1) Resonance pole positions are well determined by the elastic data alone; and (2) Little change is caused by the introduction of the inelastic data.

Table II Partial wave pole positions

Partial wave	Elastic data only	Elastic and inelastic
D_{15}	$1660 - i\,\frac{140}{2}$ (Junk = $\pi\Delta$)	$1666 - i\,\frac{159}{2}$ (Major channel = $\pi\Delta$)
F_{35}	$1810 - i\,\frac{275}{2}$ (Junk = $\pi\Delta$)	$1824 - i\,\frac{282}{2}$ (Major channel = $N\rho$)

<u>Resonance couplings</u>. Clearly one will never know the inelastic channel to which a resonance couples unless one measures inelastic channels. It is <u>very important</u> to know the couplings to specific decays for the evaluation of resonance classifications or symmetry schemes in general. However, one should realize that $\pi\pi N$ experience has taught us that this is really the only contribution.

Thus one is forced to conclude from these empirical observations that (1) The qualitative existence and quantitative evaluation of resonances which couple to stable 2 body channels are most easily determined by studying those channels; and (2) Other more complicated inelastic channels provide information only on the couplings to specific channels, although this in itself is very valuable data.

CONCLUSIONS

The following points are the conclusions we can draw: (1) The quantities of data required for partial wave analyses in multiparticle states derived from $\pi\pi$ and $K\pi$ scattering are large, making such projects unfeasible. We might be able to attempt such analyses in the future, first by assuming that certain resonances are present with a specific production mechanism and then secondly determining their coupling constants. (2) The extrapolation of the total inelastic cross section to the pion pole, which requires less data, may well provide a useful constraint in future partial wave analyses of the elastic scatterings. In this case it would be important to study

$$K^+ p \rightarrow (MM)\,\Delta^{++}$$

or

$$\pi^+ p \rightarrow (MM)\,\Delta^{++}$$

in order to obtain the total inelastic cross section. (3) Stable 2 body data and in particular the elastic channels — identify resonances and give good estimates of the pole parameters. This means that the value of the inelastic data lies in its ability to give the couplings to very specific final states.

Finally one should not infer that it is of little interest to study these higher multiplicity final states. Clearly one can never see $G = -1$, $S = 0$ boson resonances in 2π final states. However, if one wishes to identify resonances which do couple to the elastic channel in $\pi\pi$ or $K\pi$ scattering then studying those elastic channels will produce the most dramatic reward.

REFERENCES

1. D. J. Herndon et al., "A partial wave analysis of the reaction $\pi N \to \pi\pi N$ in the c.m. energy range 1300-2000 MeV," Report No. LBL-1065 and SLAC-PUB-1108 (1972); unpublished.
2. B. Deler and G. Valladas, Nuovo Cimento 45A, 559 (1966).
3. R. J. Cashmore, D. J. Herndon, and P. Soding, "The generalized isobar model," Report No. LBL-543.
4. D. V. Brockway, University of Illinois COO-1195-197 (unpublished).
5. K. M. Watson, Phys. Rev. 88, 1163 (1952).
6. K. W. Barnham et al., Nucl. Phys. B25, 49 (1970).
7. D. Mettel, Nucl. Phys. B42, 340 (1972).

Chapter 3. Production Amplitude Analysis and π-π Phase Shifts

AMPLITUDE ANALYSES OF HYPERCHARGE EXCHANGE REACTIONS[*,†]

R. D. Field
Brookhaven National Laboratory, Upton, New York 11973

ABSTRACT

The determination of the two-body production amplitudes for $Y=1$ exchange reactions is discussed. The pseudoscalar production line reversed reactions $K^-n \to \pi^-\Lambda$ and $\pi^-p \to K^0\Lambda$; the vector-meson production reactions $K^-p \to (\omega,\varphi)\Lambda$ and $\pi^-p \to K^{*0}\Lambda$; the reaction $K^-p \to \pi^-Y^{*+}(1385)$; and the photoproduction reaction $\gamma p \to K^+\Lambda$ are examined. Quark model, SU(3), and VDM comparisons are made and the EXD nature of the K^{**} and K^* Regge exchanges explored. In addition the behavior of the unnatural parity $Y=1$ exchanges (K, K_B, K_A) are investigated

I. INTRODUCTION

It has become increasingly evident throughout this conference that before substantial new progress can be made on the determination of π-π and K-π scattering one must have a better knowledge of the behavior of two-body production amplitudes. This is crucial in order to decide upon proper extrapolation procedures. Due to the self-analysing property of the Λ (and Σ) hyperon reactions with $Y=1$ exchange provide much information on the nature of production processes. Although the decay correlations from unpolarized initial particles are insufficient for a complete determination of the productions amplitudes, observing all the decay products affords one the opportunity of determining a nearly complete set.[1] By judiciously choosing the set of independent amplitudes (transversity type) the unknown quantities are one relative phase φ_R and one overall phase φ_0. In spite of the lack of knowledge of these two phases, one can learn a great deal about the behavior of these amplitudes.

In Sec. II the reactions $K^-n \to \pi^-\Lambda$ and $\pi^-p \to K^0\Lambda$ are discussed and a comparison of various models is presented. By observing the two-step decay of the $Y^*(1385)$ one can determine much about the amplitudes for the reaction $K^-p \to \pi^-Y^{*+}(1385)$. This is discussed and experimental results shown in Sec. III. In Sec. IV vector-meson production with $Y=1$ exchange is examined. New results for the amplitudes in $K^-p \to (\omega,\varphi)\Lambda$ are presented and compared with previous analyses for $\pi^-p \to K^{*0}\Lambda$. The similarities between the vector meson production reactions $K^-p \to (\omega,\varphi)\Lambda$ and the pseudoscalar production reactions $K^-n \to \pi^-\Lambda$ and $\pi^-p \to K^0\Lambda$ are discussed. In Sec. V SU(3)

---

*Invited talk presented at the International Conference on π-π Scattering and Associated Topics held at Florida State University, March 28-30, 1973.
†Work performed under the auspices of the U. S. Atomic Energy Commission.

and VDM are used to predict observables for the photoproduction reaction $\gamma p \to K^+\Lambda$ from the amplitudes for $K^- p \to (\omega,\varphi)\Lambda$. In Sec. VI we present a summary and discuss interesting future experimental observations. The definitions of transversity amplitudes and many useful relations including the connection between the transversity amplitudes and the s- and t-channel helicity amplitudes are presented in the Appendix.

II. PSEUDOSCALAR PRODUCTION WITH Y = 1 EXCHANGE
($K^-n \to \pi^-\Lambda$, $\pi^-p \to K^0\Lambda$)

The two body production process $PB \to P'B'$ can be described at each value of s and t by four complex transversity amplitudes $T_{\lambda'\lambda}(s,t)$, where $\lambda'(\lambda)$ corresponds to the component of spin of the outgoing baryon B' (target baryon B) along the transversity z-axis defined normal to the production plane (see Appendix). The two amplitudes T_{+-} and T_{-+} vanish by parity (A.3) leaving the two independent complex amplitudes $T_{++}(s,t)$, $T_{--}(s,t)$ (4 parameters at each s and t). These two transversity amplitudes are related to the helicity amplitudes as follows:

$$H_{++}(s,t) = T_{++}(s,t) + i\, T_{--}(s,t) \tag{2.1a}$$

$$H_{+-}(s,t) = T_{++}(s,t) - i\, T_{--}(s,t) \tag{2.1b}$$

and the helicity amplitudes are related to the observables $\frac{d\sigma}{dt}$ and P by

$$\frac{d\sigma}{dt} \propto |H_{++}|^2 + |H_{+-}|^2 \tag{2.2a}$$

$$P = \frac{2\,\mathrm{Im}\, H_{++} H_{+-}^*}{|H_{++}|^2 + |H_{+-}|^2} \quad . \tag{2.2b}$$

The observables are thus related to the transversity amplitudes by

$$\frac{d\sigma}{dt} \propto |T_{++}|^2 + |T_{--}|^2 \tag{2.3a}$$

$$P = \frac{|T_{++}|^2 - |T_{--}|^2}{|T_{++}|^2 + |T_{--}|^2} \quad . \tag{2.3b}$$

Thus by observing the Λ polarization (Λ decay) in addition to the differential cross section one can determine the magnitudes of the two transversity amplitudes for the reactions $K^-n \to \pi^-\Lambda$ and $\pi^-p \to K^0\Lambda$. One cannot determine the relative phase between the two amplitudes φ_R ($\varphi_R = \mathrm{Arg}(T_{++}) - \mathrm{Arg}(T_{--})$) nor can one determine the overall phase of the amplitudes φ_0. At each value of s and t one can determine two of the four parameters necessary to completely describe the reactions. The relative phase φ_R can only be determined by including polarized target information (A and R parameters).

From (2.1a,b) it is seen that without a knowledge of the relative phase φ_R the structures of the helicity amplitudes H_{++} and H_{+-} are unknown (both the magnitudes and phases of the helicity amplitudes depend on φ_R). Nevertheless the determination of $d\sigma/dt$ and P (i.e. $|T_{++}|^2$ and $|T_{--}|^2$) for the line reversed reactions $K^-n \to \pi^-\Lambda$ and $\pi^-p \to K^0\Lambda$ tells us much about the production process. (Strictly speaking $\pi^+n \to K^+\Lambda$ is the line reversed partner of the reaction $K^-n \to \pi^-\Lambda$ but $A(\pi^+n \to K^+\Lambda) = A(\pi^-p \to K^0\Lambda)$ by isospin.) The important features of the experimental data on $K^-n \to \pi^-\Lambda$ and $\pi^-p \to K^0\Lambda$ are:

(1) The differential cross sections for the two reactions are not equal. (Fig. 1) The differential cross section for $\pi^-p \to K^0\Lambda$ is steeper and smaller at large $|t|$. The two cross sections do appear to agree in the forward direction.

(2) The polarization for the two reactions are both non-zero and they have the same sign for $0 \lesssim |t| \lesssim 0.4$ (GeV/c)^2 (Fig. 2). Figure 2 also shows the magnitudes of the transversity amplitudes, where we have normalized so that $|T_{++}|^2 + |T_{--}|^2 = 1$.

The above experimental observations place great restrictions on high energy models attempting to describe reactions $K^-n \to \pi^-\Lambda$ and $\pi^-p \to K^0\Lambda$ which only have natural parity exchange (K^{**},K^*) in the t-channel. The high energy models for these reactions can be divided into three types. First, there are models that attempt to describe these reactions in terms of Regge poles (K^{**},K^*) only with no absorption:

Type 1a: Duality and EXD pole model - The duality diagram for $K^-n \to \pi^-\Lambda$ is non-planar (quark lines cross), whereas the diagram for $\pi^-p \to K^0\Lambda$ is planar (no crossed quark lines). Thus, the s-channel helicity amplitudes for $K^-n \to \pi^-\Lambda$ are predicted to be purely real and those for $\pi^-p \to K^0\Lambda$ are predicted to have a rotating phase $\exp(-i\pi\alpha(t))$.[2] These predictions are identical to the ones arrived at by assuming that the K^{**} and K^* Regge poles are strongly EXD. The model predicts

$$\frac{d\sigma}{dt}(K^-n \to \pi^-\Lambda) = \frac{d\sigma}{dt}(\pi^-p \to K^0\Lambda) \qquad (2.4a)$$

and

$$P(K^-n \to \pi^-\Lambda) = P(\pi^-p \to K^0\Lambda) = 0 \qquad (2.4b)$$

in obvious disagreement with the data.

Type 1b: Weakly EXD pole models - This model assumes only the trajectory functions of the K^{**} and K^* pole are equal and not the residues. This model predicts

$$\frac{d\sigma}{dt}(K^-n \to \pi^-\Lambda) = \frac{d\sigma}{dt}(\pi^-p \to K^0\Lambda) \qquad (2.5a)$$

and

$$P(K^-n \to \pi^-\Lambda) = -P(\pi^-p \to K^0\Lambda) \quad , \qquad (2.5b)$$

which disagrees with experiment.

Type 1c: Broken EXD pole model - In this model EXD is broken in both the residue and trajectory functions for the K^{**} and K^* pole. This model predicts

$$P \frac{d\sigma}{dt}(K^-n \to \pi^-\Lambda) = -P \frac{d\sigma}{dt}(\pi^-p \to K^0\Lambda) \quad . \tag{2.6}$$

This prediction is also in disagreement with the data since the polarization for $K^-n \to \pi^-\Lambda$ and $\pi^-p \to K^0\Lambda$ have the same sign for small $|t|$ (see Fig. 2). The K** and K* Regge poles are thus inadequate to describe these reactions no matter how much EXD is broken![3-4] One <u>must</u> introduce absorptive corrections and/or lower lying poles.

The second class of models are those that assume the bare K** and K* poles are EXD, and add absorptive corrections and/or lower lying poles in an attempt to explain the experimental data.

<u>Type 2a</u>: EXD poles plus "old" absorption[5]- This model assumes that the bare K** and K* Regge poles are EXD and that the EXD is broken by introducing absorptive corrections. The absorptive corrections (Regge cuts) are calculated using the "old" absorption model which calculates the partial wave projection of the cut H_{cut}^J using the Sopkovich prescription[6]

$$H_{cut}^J(s) = iH_{pole}^J(s) H_{elastic}^J(s) \tag{2.7}$$

where $H_{pole}^J(s)$ is the partial wave projection of the s-channel Regge pole helicity amplitude and $H_{elastic}^J(s)$ is the s-channel helicity partial wave projection of the elastic amplitude given by a Pomeron amplitude of the form

$$H_{Pomeron}(s,t) = is\sigma_{tot}(\frac{s}{s_0})^{\frac{-i\pi}{2}\alpha_P' t} e^{\frac{A}{2}t} \quad . \tag{2.8}$$

This type of model is able to explain the behavior of the polarizations in $K^-n \to \pi^-\Lambda$ and $\pi^-p \to K^0\Lambda$ but predicts

$$\frac{d\sigma}{dt}(\pi^-p \to K^0\Lambda) > \frac{d\sigma}{dt}(K^-n \to \pi^-\Lambda) \tag{2.9}$$

in disagreement with the data.

<u>Type 2b</u>: EXD poles plus "old" absorption plus daughters[7]- This model is identical to type 2a but with the addition of daughter trajectories. The daughter trajectories do not affect the model's ability to successfully fit the polarizations, but allows one to also fit the differential cross sections. Since the line reversal breaking of the differential cross sections is attributed to lower lying daughter trajectories, this model predicts

$$\frac{d\sigma}{dt}(K^-n \to \pi^-\Lambda) \to \frac{d\sigma}{dt}(\pi^-p \to K^0\Lambda)$$

as energy is increased.

Due to the inability of the "old" absorption model to successfully explain line reversal breaking, "new" absorption models were invented. The next class of models abandon the prescription of (2.7) and (2.8) in favor of a phenomenological prescription for calculating Regge cuts.

<u>Type 2c</u>: EXD poles plus phenomenological absorption[8-10] These models are constructed to absorb real and imaginary parts differently. One finds that to explain the data one must absorb the imaginary parts more than the real parts. Thus, for example, the real reaction ("1") $K^-n \to \pi^-\Lambda$ is absorbed less than the rotating reaction ($e^{-i\pi\alpha(t)}$)

$\pi^- p \to K^0 \Lambda$ providing

$$\frac{d\sigma}{dt}(K^- n \to \pi^- \Lambda) > \frac{d\sigma}{dt}(\pi^- p \to K^0 \Lambda)$$

in accordance with the data.[11,12]

<u>Type 2d</u>: T_{eff} absorption models - Strictly speaking this type should not be called a model. For this method one calculates the absorption according to

$$H^J_{cut}(s) = i \, H^J_{pole} \, T^J_{eff} \qquad (2.10)$$

and determines what kind of T^J_{eff} is necessary to fit the data. Different T^J_{eff}'s arise depending on the assumptions concerning the initial Regge poles. Barger and Martin[13] assume EXD K** and K* poles and no absorption in the s-channel flip amplitude. They then determine the structure of the non-flip amplitudes (pole + cut) necessary to fit the data. The results are interesting but depend, of course, on their initial assumptions.

The next type of model abandons the assumption of EXD bare Regge poles.

<u>Type 3a</u>: Non-EXD pole (no WSNZ) plus "old" absorption.[14] This is just the old SCRAM model. It assumes no wrong signature zero (WSNZ) of the Regge pole and produces dips by pole-cut interference. This model can fit the observables for $K^- n \to \pi^- \Lambda$ and $\pi^- p \to K^0 \Lambda$. The ability to fit the behavior of the differential cross sections is due to the relaxing of the EXD of the bare poles. There are, however, other reactions where this model fails badly (e.g., $\pi^- p \to \pi^0 n$).[15]

<u>Type 3b</u>: Non-EXD pole (no WSNZ) plus "fancy" Pomeron absorption - This model, recently proposed by G.L. Kane,[16] assumes broken EXD for the bare poles (no WSNZ). Absorption effects are calculated using (2.10) with $T_{eff} = P+D$, where P corresponds to the Pomeron parametrized by

$$P(s,t) = is \left[A e^{Bt} + A_0 e^{B_0 t} J_0(R_0 \sqrt{-t} \sqrt{\ell n s - i \frac{\pi}{2}}) \right].$$

The quantity D represents an attempt to include the contributions from the sum over intermediate states other than the elastic one, and is parameterized by

$$D_n(s,t) = is \, e^{Dt} J_n(R_0 \sqrt{-t} \sqrt{\ell n s - i \frac{\pi}{2}}),$$

where n is the net helicity flip. One feature of this T_{eff} is that it has a larger real part than previous models. Hartly and Kane[16] have successfully fit the systematics of many reactions using this model including the reactions $K^- n \to \pi^- \Lambda$ and $\pi^- p \to K^0 \Lambda$.[17]

In order to determine which of the above models is a correct description of two-body production, if any, or to decide which features of these models should be retained in future models, it will be necessary to know:

(1) The energy dependences of $d\sigma/dt$ and P for the line reversed reactions $K^- n \to \pi^- \Lambda$ and $\pi^- p \to K^0 \Lambda$.

(2) The relative phase φ_R which can be determined by polarized target experiments (with a component of polarization in the production plane).

Table I summarizes the various models applied to the reactions $K^- n \to \pi^- \Lambda$ and $\pi^- p \to K^0 \Lambda$.

III. THE REACTION $K^-p \to \pi^- Y^{*+}(1385)$

The reaction $K^-p \to \pi^- Y^{*+}(1385)$ is described by the transversity amplitudes $T_{2\lambda;2\lambda'}(s,t)$, where λ (λ') corresponds to the component of spin of the Y^* (target proton) along the transversity z-axis, which is normal to the production plane (see Appendix). Parity conservation (A.3) implies $T_{31} = T_{1\,-1} = T_{-1\,1} = T_{-3\,-1} = 0$. The remaining four complex amplitudes are related to the eight transversity density matrix elements as follows:

$$\rho_{33} = |T_{3\,-1}|^2 \tag{3.1a}$$

$$\rho_{11} = |T_{11}|^2 \tag{3.1b}$$

$$\rho_{-1\,-1} = |T_{-1\,-1}|^2 \tag{3.1c}$$

$$\rho_{-3\,-3} = |T_{-31}|^2 \tag{3.1d}$$

$$\mathrm{Re}\rho_{3\,-1} = \mathrm{Re}(T_{3\,-1} T^*_{-1\,-1}) = |T_{3\,-1}||T_{-1\,-1}|\cos\delta_1 \tag{3.1e}$$

$$\mathrm{Im}\rho_{3\,-1} = \mathrm{Im}(T_{3\,-1} T^*_{-1\,-1}) = |T_{3\,-1}||T_{-1\,-1}|\sin\delta_1 \tag{3.1f}$$

$$\mathrm{Re}\rho_{1\,-3} = \mathrm{Re}(T_{11} T^*_{-31}) = |T_{11}||T_{-31}|\cos\delta_2 \tag{3.1g}$$

$$\mathrm{Im}\rho_{1\,-3} = \mathrm{Im}(T_{11} T^*_{-31}) = |T_{11}||T_{-31}|\sin\delta_2 , \tag{3.1h}$$

where (A.5b) was used and where δ_1 [δ_2] is defined as the relative phase between the amplitudes $T_{3\,-1}$ and $T_{-1\,-1}$ [T_{11} and T_{-31}].

By observing the two step decay of the $Y^*(1385)$ (i.e., $Y^* \to \pi\Lambda$, $\Lambda \to \pi p$) in addition to the differential cross section for $K^-p \to \pi^- Y^{*+}$ one can determine all the transversity density matrix elements and thus the magnitudes of all four transversity amplitudes and the two relative phases δ_1 and δ_2. Without a polarized target one cannot determine the relative phase φ_R between the amplitudes with target proton transversity up ($\lambda' = 1/2$) and the amplitudes with target proton transversity down ($\lambda' = -1/2$). In addition one, of course, cannot determine the overall phase of the amplitudes φ_0. One can thus determine 6 out of the 8 parameters needed at each s and t value to completely specify the production process $K^-p \to \pi^- Y^{*+}$. The transversity density matrix elements and the phases δ_1 and δ_2 resulting from an analysis of $K^-p \to \pi^- Y^{*+}(1385)$[18] are displayed in Fig. 3, where the density matrix elements are normalized according to

$$\rho_{33} + \rho_{11} + \rho_{-1\,-1} + \rho_{-3\,-3} = 1 .$$

The simple non-relativistic additive quark model assumes that the transversity amplitudes for production process $K^-p \to \pi^- Y^*$ can be represented by a sum of quark-quark scattering amplitudes (Fig. 4a).[19] Transversity amplitudes that require a flip of more than one of the baryon quarks, such as $T_{3\,-1}$ (Fig. 4b), are not allowed and predicted to be zero. This model thus predicts

$$T_{3\,-1}(s,t) = T_{-31}(s,t) = 0 . \tag{3.2a}$$

It also predicts

$$T_{11}(s,t) = T_{-1\,-1}(s,t) \quad , \qquad (3.2b)$$

since these amplitudes are given by the same sum over quark-quark amplitudes. The model thus predicts that $\rho_{11} = \rho_{-1-1} = 1/2$ and that the remaining transversity density matrix elements vanish. As can be seen from the data in Fig. 3 these predictions do explain the gross features of the amplitudes, however, they are not satisfied exactly. The data clearly show non-zero values of ρ_{33}, $\text{Im}\rho_{3\,-1}$, and $\text{Re}\rho_{3\,-1}$ in the small momentum transfer region $|t_0-t| \lesssim 0.3$ $(\text{GeV}/c)^2$ indicating a non vanishing $T_{3\,-1}$ amplitude.

By the use of (A.4) the quark model predictions for the transversity amplitudes can be converted into a prediction about the s-channel helicity amplitudes. Namely,

$$H_{3/2;-1/2}(s,t) = H_{1/2;1/2}(s,t) = 0$$

and

$$H_{3/2;1/2}(s,t) = \sqrt{3}\; H_{1/2;-1/2}(s,t)$$

which implies (using (A.5a)) that $\rho_{33}^H = 3/8$, $\text{Re}\rho_{3\,-1}^H = \sqrt{3}/8$ and that the remaining s-channel density matrix elements vanish. These are the same as the Stodolsky-Sakurai predictions for ρ exchange in the reaction $\pi N \to \pi\Delta(1236)$. The s-channel density matrix elements for $K^-p \to \pi^-Y^{*+}$ are shown in Fig. 5. Again we see rough agreement with the quark model prediction.

IV. VECTOR MESON PRODUCTION WITH Y = 1 EXCHANGE

The twelve complex transversity amplitudes for the reaction $PB \to VB'$, where P and V are pseudoscalar and vector mesons, respectively, and B and B' are spin 1/2+ baryons are written as follows:

$$T_{\lambda'\lambda}^{\mu}(s,t) \quad ,$$

where λ, λ', and μ correspond to the components of spin of the target baryon B, baryon B', and vector meson V, respectively, along the transversity z-axis (see Appendix). Parity conservation implies $T_{\lambda'\lambda}^{\mu}(s,t)$ vanish when $(-1)^{\lambda'-\lambda+\mu}$ is odd. Thus six independent complex amplitudes are sufficient to describe the reaction $PB \to VB'$. These six transversity amplitudes are related to the helicity type amplitudes as follows:

$$T_{++}^{\;0} = -i\sqrt{2}\; A^+ \qquad T_{--}^{\;0} = -i\sqrt{2}\; A^-$$

$$T_{-+}^{\;1} = iB^+ + C^+ \qquad T_{+-}^{\;1} = iB^- + C^-$$

$$T_{-+}^{-1} = -iB^+ + C^+ \qquad T_{+-}^{-1} = -iB^- + C^- \quad , \qquad (4.1)$$

where

$$A^{\pm} = \frac{1}{2}\{H_{++}^{1} + H_{++}^{-1} \mp i(H_{-+}^{1} + H_{-+}^{-1})\}$$

$$B^{\pm} = \mp \frac{1}{2}\{H_{++}^{1} - H_{++}^{-1} \pm i(H_{-+}^{1} - H_{-+}^{-1})\}$$

$$C^{\pm} = \mp \frac{1}{\sqrt{2}}(H_{++}^{0} \pm iH_{-+}^{0}) \quad , \qquad (4.2)$$

and where the T's are helicity-transversity amplitudes (Jackson-transversity) when the corresponding H's are s-channel helicity (t-channel helicity) amplitudes. The $A^{\pm}$, $B^{\pm}$ and $C^{\pm}$ amplitudes are simply related to the amplitudes defined by Byers and Yang;[20] the superscripts $\pm$ refer to states with target baryon transversity $\lambda = \pm 1/2$.

Due to the self-analysing property of the Λ hyperon, the reactions $K^-p \to (\omega,\varphi)\Lambda$ and $\pi^-p \to K^{*0}\Lambda$ provide one the opportunity of learning much about the production amplitudes. Indeed observation of the joint density matrix elements between the vector meson V and Λ hyperon, in addition to the single density matrix elements of the vector meson, and the Λ polarization allows determination of the magnitudes of all six transversity amplitudes plus the determination of the relative phases between amplitudes with the same target baryon transversity (λ). The amplitudes can be divided into two sets of three amplitudes

$$\begin{pmatrix} T_{++}^{0} \\ T_{-+}^{1} \\ T_{-+}^{-1} \end{pmatrix} \qquad \begin{pmatrix} T_{--}^{0} \\ T_{+-}^{1} \\ T_{+-}^{-1} \end{pmatrix}$$

for the transversity amplitudes and

$$\begin{pmatrix} A^{+} \\ B^{+} \\ C^{+} \end{pmatrix}, \qquad \begin{pmatrix} A^{-} \\ B^{-} \\ C^{-} \end{pmatrix}$$

for the Byers and Yang type amplitudes. From the joint decay angular distributions of the vector meson V and Λ hyperon together with the differential cross section the amplitudes can be determined up to a relative phase φ_R between the target baryon transversity up ($\lambda = 1/2$) and down ($\lambda = -1/2$) groups and an overall phase φ_0. Thus out of the 12 parameters at each s and t necessary to completly specify the amplitudes, 10 can be determined.

The magnitudes of the transversity amplitudes for $K^-p \to (\omega,\varphi)\Lambda$ are shown in Fig. 6 as a function of momentum transfer, and where we have normalized

$$\sum_{\mu,\lambda',\lambda} |T_{\lambda'\lambda}^{\mu}|^2 = 1 \quad .$$

Figure 7 shows the relative phases

$$\theta_{\pm}^{+1} = \text{Arg}(T_{\mp\pm}^{+1}) - \text{Arg}(T_{\pm\pm}^{0})$$

for $K^-p \to (\omega,\varphi)\Lambda$ as a function of momentum transfer. As can be seen from Figs. 6 and 7 there are striking differences in the amplitude structure for $K^-p \to (\omega,\varphi)\Lambda$. In particular $K^-p \to \varphi\Lambda$ has large values of $|T_{--}^0|^2$ and small values of $|T_{\mp+}^0|^2$ while $K^-p \to \omega\Lambda$ has large values of the latter and small values of the former.

Arguments similar to those in Sec. III using the simple quark model lead to the prediction that[19]

$$T_{\lambda'\lambda}^\mu(K^-p \to \varphi\Lambda) = T_{\lambda'\lambda}^\mu(\pi^-p \to K^{*0}\Lambda) \quad ,$$

which also implies via (4.1) that the $A^{\pm}$, $B^{\pm}$, $C^{\pm}$ amplitudes are identical. Figure 8 shows a comparison of the magnitudes of the amplitudes determined from the following analyses:
(1) The reaction $K^-p \to \varphi\Lambda$ at 4.2 GeV/c[21] (Ref. 22)
(2) The reaction $K^-p \to \varphi\Lambda$ using the combined BNL and EP data (Ref. 23)
(3) The reaction $\pi^-p \to K^{*0}\Lambda$ at 4.5 GeV/c (Ref. 24)
(4) The reaction $\pi^-p \to K^{*0}\Lambda$ at 3.9 GeV/c (Ref. 25)

All these analyses are seen to give remarkably similar results in excellent agreement with the quark model prediction.[26]

The prediction of equality of the amplitudes for $K^-p \to \varphi\Lambda$ and $\pi^-p \to K^{*0}\Lambda$ is also arrived at using SU(3) and factorization on the various t-channel exchange amplitudes (see Table II). Thus it is not clear whether the experimental equality is a success for the quark model or for SU(3) plus factorization. The quark model, however, does not relate $K^-p \to \varphi\Lambda$ to $K^-p \to \omega\Lambda$, whereas SU(3) plus factorization does. From Table II it can be seen that if the K^{**} and K^* exchange amplitudes exhibit strong EXD then

$$\rho_n \frac{d\sigma}{dt}(K^-p \to \varphi\Lambda) = 2\rho_n \frac{d\sigma}{dt}(K^-p \to \omega\Lambda) \quad , \qquad (4.3)$$

where the differential cross sections are multiplied by $\rho_n = \rho_{11} + \rho_{1-1}$ to project out the natural parity (K^{**},K^*) part. In addition the duality diagram for $K^-p \to \varphi\Lambda$ is planar, whereas $K^-p \to \omega\Lambda$ has a nonplanar diagram (assuming ideal mixing for ω and φ). Thus duality diagram arguments predict the (K^{**},K^*) exchange to be purely real for $K^-p \to \omega\Lambda$ and have a rotating phase ($e^{-i\pi\alpha(t)}$) for $K^-p \to \varphi\Lambda$. Equation (4.3) is thus analogous (assuming SU(3)) to the line reversal relation for $K^-n \to \pi^-\Lambda$ and $\pi^-p \to K^0\Lambda$ (2.4a) except here one has the same incident particle (K^-) and hence (4.3) can be tested in the same experiment (this reduces normalization problems). The results for $K^-p \to (\omega,\varphi)\Lambda$ are shown in Fig. 9. The relation (4.3) is seen <u>not</u> to hold and just as occurred in pseudoscalar production (Fig. 1) the "rotating" reaction lies below the "real" reaction. Assuming SU(3) is good, one concludes the EXD of the K^{**} and K^* amplitudes is broken in a manner similar to that found in pseudoscalar production. It could be argued for $K^-n \to \pi^-\Lambda$ and $\pi^-p \to K^0\Lambda$ that the distention between "real" and "rotating" is meaningless and that the pion induced reaction is just smaller than the kion induced one. Here, however, both reactions are kion induced and the systematics are the same!

The reactions $K^-n \to \pi^-\Lambda$ and $\pi^-p \to K^0\Lambda$ have large non-zero polarizations produced by the K** and K* exchange amplitudes.[27] If systematics are indeed similar one expects large polarization effects in $K^-p \to (\omega,\varphi)\Lambda$. The Λ polarization in $K^-p \to (\omega,\varphi)\Lambda$ is, however, made up of the two pieces ($P = P_n + P_u$); one arising from the interference between natural parity t-channel amplitudes P_n and one arising from the interference between unnatural parity t-channel amplitudes P_u. By observing the joint correlations between the vector meson and the Λ decay in addition to single density matrix elements and Λ polarization one can determine these two pieces separately.[28] Figure 10 shows the results for $K^-p \to (\omega,\varphi)\Lambda$. It can be seen that the natural parity K** and K* amplitudes do indeed produce a large polarization P_n and that this polarization changes sign in going from ω to φ production. This sign change is predicted by the SU(3) coefficients in Table II. (The K* changes sign under $\omega \to \varphi$ whereas K** does not so $P_n \propto \mathrm{Im}[(K^{**})(K^*)^*]$ does change sign.)

The reactions $K^-p \, (\omega,\varphi)\Lambda$ and $\pi^-p \to K^{*0}\Lambda$ also provide a first look into the nature of unnatural parity Y=1 exchange (i.e., K, K_B, K_A exchange[29]). One striking fact about Fig. 10 is that P_u is not zero. The total polarization P receives a contribution from both natural parity exchange and a contribution from unnatural parity exchange. In particular for $K^-p \to \varphi\Lambda$ P_n and P_u have opposite signs for $|t| \lesssim 0.4$ GeV/c producing a small total polarization, whereas for $K^-p \to \omega\Lambda$ P_n and P_u have the same sign producing very large total polarization. The polarization P_u is made up of two terms[27]

$$P_u \propto \mathrm{Im}[(K)(K_A)^*] + \mathrm{Im}[(K_B)(K_A)^*] . \qquad (4.4)$$

The second term changes sign under $\varphi \to \omega$ production, whereas the first term does not (see Table II). In the absence of an amplitude with the quantum numbers of the K_A the polarization P_u would vanish. The experimental data for $K^-p \to (\omega,\varphi)\Lambda$ and also for $\pi^-p \to K^{*0}\Lambda$ seem to indicate the presence of K-K_A interference.[30] Certainly this was not expected.

To this point specific models for the behavior of the amplitudes have not been used. It is necessary, however, if one wants to compare $K^-p \to (\omega,\varphi)\Lambda$ using SU(3) to use a model for the behavior of the K**, K*, K, K_B and K_A exchanges listed in Table II. Figures 11, 12, and 13 show the results of a simple Regge pole fit to the observables for $K^-p \to (\omega,\varphi)\Lambda$ where SU(3) was used to relate the Regge exchanges in $K^-p \to \omega\Lambda$ to those in $K^-p \to \varphi\Lambda$ (Table II). Solution 1 (dashed curves) contains a K**, K*, K, and K_B pole whereas solution 2 (solid curve) also contains a K_A pole.[27] Figures 14 and 15 show the resulting fits to the transversity amplitudes for $K^-p \to (\omega,\varphi)\Lambda$. These effective pole fits illustrate the following points:

(1) Whereas the quark model has no prediction for the relationship between $K^-p \to (\omega,\varphi)\Lambda$, SU(3) successfully explains the relationships between these reactions. It explains the systematics of the relationships between the transversity amplitudes for $K^-p \to (\omega,\varphi)\Lambda$ (Figs. 15 and 16).[31]

(2) Large EXD breaking is required between the K** and K* effective poles. The fits require a K* without wrong signature nonsense α-factors (WSNZ).

(3) One gets a much better fit when a K_A pole is included and P_u is explained primarily by K-K_A interference (Fig. 13) since this quantity does not change sign under $\omega \to \varphi$. Whether one is really seeing an actual K_A pole or effects of Regge-Regge cuts and/or lower lying 1/s effects remains to be seen (studies of the energy dependences of the transversity amplitudes will help settle this point).

The question arises as to why effective K^{**} and K^* Regge poles are able to describe the observed natural parity behavior for $K^-p \to (\omega,\varphi)\Lambda$ and $\pi^-p \to K^{*0}\Lambda$ but are inadequate to explain the observables for $K^-n \to \pi^-\Lambda$ and $\pi^-p \to K^0\Lambda$. A possible explanation lies in the nature of absorptive corrections. These corrections tend to be smaller for "flip" type amplitudes (amplitudes vanishing in the forward directions) than for "non-flip" amplitudes (amplitudes peaking in the forward direction). Because the K^{**} and K^* couple only to vector meson equal ± 1 states and because of the evasive nature of these poles, all the natural parity (K^{**},K^*) amplitudes behave like "flip" amplitudes in $K^-p \to (\omega,\varphi)\Lambda$ and $\pi^-p \to K^{*0}\Lambda$. For the reactions $K^-n \to \pi^-\Lambda$ and $\pi^-p \to K^0\Lambda$, on the other hand, the K^{**} and K^* couple to both a "non-flip" and a "flip" amplitude. This is illustrated by Fig. 1 and Fig. 9 where it can be seen that $d\sigma/dt$ for $K^-n \to \pi^-\Lambda$ and $\pi^-p \to K^0\Lambda$ does not dip in the forward direction, whereas the natural parity contribution to $d\sigma/dt$ for $K^-p \to (\omega,\varphi)\Lambda$ dips in the forward direction. One can understand effective K^{**} and K^* poles working for $K^-p \to (\omega,\varphi)\Lambda$ and $\pi^-p \to K^{*0}\Lambda$ because absorptive correction are small for "flip" type amplitudes and not working for $K^-n \to \pi^-\Lambda$ and $\pi^-p \to K^0\Lambda$ because the large non-flip absorptive correction must be included.

V. PHOTOPRODUCTION WITH Y = 1 EXCHANGE

The vector meson dominance model (VDM) relates the amplitudes for the vector meson production reactions $K^-p \to (\rho,\omega,\varphi)\Lambda$ with those of the photoproduction reaction $\gamma p \to K^+\Lambda$ as follows:

$$A(\gamma p \to K^+\Lambda) \propto 3A(\rho p \to K^+\Lambda) + A(\omega p \to K^+\Lambda) - \sqrt{2}\, A(\varphi p \to K^+\Lambda), \tag{5.1}$$

where the ρ, ω, and φ have only ± 1 helicity. The amplitudes $A(\rho p \to K^+\Lambda)$, $A(\omega p \to K^+\Lambda)$, and $A(\varphi p \to K^+\Lambda)$ are related by line reversal to $A(K^-p \to \rho\Lambda)$, $A(K^-p \to \omega\Lambda)$, and $A(K^-p \to \varphi\Lambda)$, respectively. (Note the amplitudes $A(K^-p \to \rho\Lambda)$ and $A(K^-p \to \omega\Lambda)$ are predicted by the quark model or Table II to be equal.) Thus

$$A(\gamma p \to K^+\Lambda) \propto 4[A(K^-p \to \omega\Lambda)]_{l.r.} - \sqrt{2}\,[A(K^-p \to \varphi\Lambda)]_{l.r.}, \tag{5.2}$$

where l.r. means line reversed. Given a model that explains $K^-p \to (\omega,\varphi)\Lambda$ one can use (5.2) to predict the observables for the reaction $\gamma p \to K^+\Lambda$.

Figure 16 shows a comparison of the predicted Λ polarization for $\gamma p \to K^+\Lambda$ from the two Regge pole models of Sec. IV with the experimental values at 5.0 GeV/c.[32] It can be seen that the VDM predictions have the right sign although the magnitude is not as large as observed experimentally. A recent SLAC experiment has measured the polarized photon asymmetry Σ for $\gamma p \to K^+\Lambda$ and found it to be very large and approximately one for p_{lab} = 16.0 GeV/c.[33] The models of Sec. IV also predict this as is also shown in Fig. 16.

VI. SUMMARY AND FUTURE PROJECTS

In this section we present a summary and list interesting future experimental observations.

$K^-n \to \pi^-\Lambda$, $\pi^-p \to K^0\Lambda$ (2 amplitudes)

Using present data on $d\sigma/dt$ and P one can determine two out of the four parameters necessary to completely specify the amplitudes at each s and t. The data indicate that

(1) The K** and K* pole alone are inadequate to describe these reactions. One must include Regge cuts (absorption) corrections and/or lower lying exchanges.

(2) The amplitudes with quantum numbers of the K** and K* do not exhibit EXD. EXD is badly broken.

(3) The line reversal predictions from EXD poles (or duality diagram arguemtns) are badly broken at intermediate energies (3-8 GeV/c).

Interesting future projects include

(1) The determination of line reversal behavior at high energy (> 10 GeV/c). Do the line reversal predictions work better at very high energies?

(2) The determination of the relative phase φ_R. This will allow a complete determination of the production amplitudes (expect overall phase φ_0), but requires a polarized target experiment with a component of the polarization in the production plane (A and R parameters).

$K^-p \to \pi^-Y^{*+}(1385)$ (4 amplitudes)

By observing the two-step decay of the Y*(1385) together with $d\sigma/dt$ one can determine 6 out of the 8 parameters necessary to completely specify the amplitudes at each s and t. The data indicate that

(1) The naive quark model gives a good description of the transversity amplitudes, although the predictions do not hold exactly.

Interesting future projects include

(1) Better statistics so that $Im\rho_{ij}^H$ can be determined better and tests of duality predictions can be made.

(2) Perform amplitude analysis on $\pi^+p \to K^+Y^{*+}(1385)$[34] or $\pi^-p \to K^0Y^{*+}(1385)$. These reactions are related to $K^-p \to \pi^-Y^{*+}(1385)$ by line reversal and it would be interesting to make a comparison of the amplitudes.

$K^-p \to (\omega,\varphi)\Lambda$, $\pi^-p \to K^{*0}\Lambda$ (6 amplitudes)

By observing the joint correlations between the vector-meson and the Λ-hyperon in addition to the single density matrix elements, Λ polarization, and $d\sigma/dt$, one can determine 10 out of the 12 parameters necessary to completely specify the amplitudes at each s and t. The data indicate that

(1) The quark model (or SU(3)) prediction of equality of the amplitudes for $K^-p \to \varphi\Lambda$ and $\pi^-p \to K^{*0}\Lambda$ works remarkably well.

(2) Using SU(3) one can successfully relate the amplitudes for $K^-p \to \varphi\Lambda$ and $K^-p \to \omega\Lambda$.

(3) The amplitudes with quantum numbers of the K^{**} and K^* exchanges do not exhibit EXD and the systematics of this EXD breaking is similar to that found in $K^-n \to \pi^-\Lambda$ and $\pi^-p \to K^0\Lambda$.
(4) The amplitudes with the quantum numbers of the K_A trajectory are not zero at these energies. Whether this is due to an actual K_A pole or Regge-Regge cuts or $1/s$ effects remains to be seen. In any case, these reactions provide an excellent place to study the unnatural parity exchanges (K,K_A,K_B).

Interesting future projects include
(1) Determination of the energy dependences of the transversity amplitudes. If the unnatural parity exchanges (K,K_B,K_A) lie lower than the natural parity exchanges (K^{**},K^*) we should see the unnatural parity amplitudes die out faster than the natural parity ones.
(2) The determination of the transversity amplitudes for the corresponding Σ production reactions $(K^-p \to (\rho,\omega,\varphi)\Sigma, \pi^-p \to K^{*0}\Sigma)$. Unnatural parity exchanges should be less important here than in the corresponding Λ reactions.[35] There are several interesting predictions relating the Σ and Λ reactions.[31] For example, is it true that $P_\Sigma = -(P_n)_\Lambda$?
(3) The determination of the relative phase φ_R, which would allow the complete determination of the production amplitudes (except for an overall phase φ_0). This requires a polarized target experiment with a component of polarization in the production plane.

$\gamma p \to K^+\Lambda$ (4 amplitudes)

We find that VDM does not work very well in relating $K^-p \to (\omega,\varphi)\Lambda$ to $\gamma p \to K^+\Lambda$, although the correct sign of P_Λ is predicted and the polarized photon asymmetry (Σ) is predicted to be large and positive as is the case experimentally at 16 GeV/c.[33] Interesting future observations include the determination of the energy dependence of the polarized photon asymmetry. Experimental results indicate that the reaction proceeds via an almost pure natural parity (K^{**},K^*) exchange at 16 GeV/c. It would be interesting to determine the energy dependence of the unnatural parity contributions by measuring Σ at lower energies.

$K^-p \to (\rho,\omega,\varphi)Y^*(1385), \pi^-p \to K^{*0}Y^*(1385)$ (12 amplitudes)

By observing the two-step decay of the $Y^*(1385)$ in addition to the vector-meson decay (triple correlations!) and $d\sigma/dt$ one can determine 22 out of the 24 parameters necessary to completely specify the amplitudes at each s and t. There has been no such analysis to date but results should be forthcoming.[36] There are a number of quark model predictions to be tested also SU(3) relationships between
$K^-p \to \rho Y^*(1385)$ and $K^-p \to \varphi Y^*(1385)$ and between $K^-p \to \varphi Y^*(1385)$ and $\pi^-p \to K^{*0}Y^*(1385)$.

ACKNOWLEDGMENTS

I would like to thank all my collaborators: M. Aguilar-Benitez, F. Barreiro, A. Rouge, H. Videau, S.U. Chung, and especially Dr. R.L. Eisner. I thank H.A. Gordon, K.-W. Lai, and J.M. Scarr for interesting discussions concerning their results. I also thank Dr. N.P. Samios for his interest and support.

APPENDIX
DEFINITION AND USEFUL RELATIONS FOR TRANSVERSITY AMPLITUDES

The transversity amplitudes for the reaction ab → cd are written as follows:

$$T_{\lambda_c \lambda_d; \lambda_a \lambda_b}(s,t) \quad , \tag{A.1}$$

where $\lambda_a, \lambda_b, \lambda_c$ and λ_d correspond to the components of spin of the particles along the transversity z-axis. The transversity frame is defined with z-axis normal to the production plane (the normal is defined by $\hat{n} = \vec{p}_i \times \vec{p}_f / |\vec{p}_i \times \vec{p}_f|$, where $\vec{p}_i$ and $\vec{p}_f$ are the three momenta of the initial and final meson, respectively). The transversity y-axis is chosen either in the direction of the outgoing particle c (d) as seen in the c.m. frame (helicity-transversity frame) or in the direction of the incoming meson a (b) in the rest frame of the outgoing meson c (d) (Jackson-transversity frame); the x-axis is chosen so as to produce a right-handed coordinate system. Thus the s-channel helicity frame (H) and the helicity-transversity frame (HT) are related by

$$(-x_{HT}, y_{HT}, z_{HT}) = (x_H, z_H, y_H) \quad . \tag{A.2a}$$

Similarly the Jackson frame (J) (t-channel frame) and the Jackson-transversity frame (JT) are related by

$$(-x_{JT}, y_{JT}, z_{JT}) = (x_J, z_J, y_J) \quad , \tag{A.2b}$$

Parity conservation in the production process implies

$$T_{\lambda_c \lambda_d; \lambda_a \lambda_b}(s,t) = \eta_1 \eta_2 \eta_3^* \eta_4^* (-1)^{\lambda-\mu} T_{\lambda_c \lambda_d; \lambda_a \lambda_b}(s,t) \tag{A.3}$$

where η_i and s_i is the parity and spin of particle i and $\lambda = \lambda_1 + \lambda_2$, $\mu = \lambda_3 + \lambda_4$. In a given reaction half the transversity amplitudes vanish by parity.

From (A.2) it can be seen that the transversity type frames are related to the helicity type frames by a rotation $R = (\pi/2, \pi/2, \pi/2)$. Thus

$$T_{\lambda_3 \lambda_4; \lambda_1 \lambda_2}(s,t) = \sum_{\mu_1 \mu_2 \mu_3 \mu_4} H_{\mu_3 \mu_4; \mu_1 \mu_2}(s,t) D^{s_1}_{\mu_1 \lambda_1}(R) D^{s_2}_{\mu_2 \lambda_2}(R)$$

$$D^{s_3 *}_{\mu_3 \lambda_3}(R) D^{s_4 *}_{\mu_4 \lambda_4}(R) \quad , \tag{A.4}$$

where the amplitudes $T_{\lambda_3 \lambda_4; \lambda_1 \lambda_2}(s,t)$ correspond to helicity-transversity amplitudes (Jackson-transversity) when $H_{\mu_3 \mu_4; \mu_1 \mu_2}(s,t)$ are s-(t-) channel helicity amplitudes.

The joint density matrix elements are expressed in terms of the s-(t-) channel helicity amplitudes by the usual formula:

$$\rho_{nn'}^{mm'} = \sum_{\lambda_1 \lambda_2} H_{mn;\lambda_1\lambda_2} H^*_{m'n';\lambda_1\lambda_2} \bigg/ \Sigma, \quad (A.5a)$$

where $\Sigma = \sum_{\lambda_1\lambda_2\lambda_3\lambda_4} |H_{\lambda_3\lambda_4;\lambda_1\lambda_2}|^2$, and where $\rho_{nn'}^{mm'}$ are measured in the s-channel helicity (Jackson) frame. Similarly the joint density matrix elements measured in the transversity frame are related to the transversity amplitudes by

$$^T\rho_{nn'}^{mm'} = \sum_{\lambda_1\lambda_2} T_{mm';\lambda_1\lambda_2} T^*_{m'n';\lambda_1\lambda_2} \bigg/ \Sigma, \quad (A.5b)$$

where $\Sigma = \sum_{\lambda_1\lambda_2\lambda_3\lambda_4} |T_{\lambda_3\lambda_4;\lambda_1\lambda_2}|^2.$

Single density matrix elements are related to the corresponding joint density matrix elements by

$$\rho_{mm'} = \sum_\lambda \rho_{\lambda\lambda}^{mm'}. \quad (A.6)$$

From (A.5a), (A.5b), and (A.4) it is easy to derive the relation between $\rho_{nn'}^{mm'}$ and $^T\rho_{nn'}^{mm'}$.

Since rotation from the helicity-transversity frame (HT) to the Jackson-transversity frame (JT) corresponds to a rotation about the z-transversity axis, the transversity amplitudes have very simple properties under crossing. Namely,

$$^{HT}T_{\lambda_3\lambda_4;\lambda_1\lambda_2} = \exp[i\lambda_3 X_3(s,t) + i\lambda_4 X_4(s,t) - i\lambda_1 X_1(s,t) - i\lambda_2 X_2(s,t)]\ ^{JT}T_{\lambda_3\lambda_4;\lambda_1\lambda_2}, \quad (A.7)$$

where $X_i(s,t)$ is the s-t crossing angle of particle i.[37] As can be seen, there is only a phase change under crossing. The magnitudes of the transversity amplitudes are invariant under rotation from the helicity-transversity to the Jackson-tranversity frame.

Table I. Various Models for $K^-n \to \pi^-\Lambda$ $(d\sigma/dt, P)$ and $\pi^-p \to K^0\Lambda(d\bar{\sigma}/dt, \bar{P})$

Type of Model	Successful	Comment
(1a) Duality, EXD(K^{**}, K^*) poles	No	Predicts $d\sigma/dt = d\bar{\sigma}/dt$, $P = \bar{P} = 0$
(1b) Weak EXD (K^{**}, K^*) poles	No	Predicts $d\sigma/dt = d\bar{\sigma}/dt$, $P = -\bar{P}$
(1c) Broken EXD (K^{**}, K^*) poles	No	Predicts $P\, d\sigma/dt = -\bar{P}\, d\bar{\sigma}/dt$
(2a) EXD (K^{**}, K^*) poles plus "old" absorption	No	Predicts $d\bar{\sigma}/dt > d\sigma/dt$
(2b) EXD (K^{**}, K^*) poles plus "old" absorption plus daughters	Yes	Predicts $d\sigma/dt \to d\bar{\sigma}/dt$ as $s \to \infty$
(2c) EXD (K^{**}, K^*) poles plus phenomenological absorption	Yes	somewhat ad hoc
(2d) EXD (K^{**}, K^*) poles plus T_{eff} absorption	Yes	little predictive power
(3a) Non-EXD (K^{**}, K^*) poles (no WSNZ) plus strong absorption (SCRAM)	Yes	fails elsewhere (i.e., $\pi^-p \to \pi^0 n$)
(3b) Non-EXD (K^{**}, K^*) poles (no WSNZ) plus "fancy Pomeron" absorption	Yes	more parameters since EXD is abandoned.

Table II. SU(3) Clebsch-Gordon Coefficients for the Reactions
$K^-p \to (\omega,\varphi)\Lambda$ and $\pi^-p \to K^{*0}\Lambda$

Exchange	Type Coupling	$K^-p \to \omega\Lambda$	$K^-p \to \varphi\Lambda$	$\pi^-p \to K^{*0}\Lambda$
K^{**}	F-type	1	$\sqrt{2}$	$-\sqrt{2}$
K^*	D-type	-1	$\sqrt{2}$	$-\sqrt{2}$
K	F-type	1	$\sqrt{2}$	$-\sqrt{2}$
K_B	D-type	-1	$\sqrt{2}$	$-\sqrt{2}$
K_A	F-type	1	$\sqrt{2}$	$-\sqrt{2}$

REFERENCES

1. A.D. Martin, Proceedings of the Seventh <u>Rencontre de Moriond</u> (1972) ed. J. Tran Thanh Van.
2. H. Harari, Phys. Rev. Letters <u>22</u>, 562 (1969); J.L. Rosner, Phys. Rev. Letters <u>22</u>, 689 (1969).
3. R.D. Field and J.D. Jackson, Phys. Rev. <u>D4</u>, 693 (1971).
4. A. Krzywicki and J. Tran Thanh Van, Phys. Letters <u>30B</u>, 185 (1969).
5. A.C. Irving, A.D. Martin and C. Michael, Nucl. Phys. <u>B32</u>, 1 (1971).
6. N.J. Sopkovich, Nuovo Cimento <u>26</u>, 186 (1962); K. Gottfried and J.D. Jackson, Nuovo Cimento <u>34</u>, 735 (1964).
7. R.D. Field, Phys. Rev. <u>D5</u>, 86 (1972).
8. G.A. Ringland, R.G. Roberts, D.P. Roy and J. Tran Thanh Van, Nucl. Phys. <u>B44</u>, 395 (1972).
9. R.L. Thews, G.R. Goldstein and J.F. Owens, "A New Regge Absorption Model," University of Arizona preprint (1972).
10. D. Barkai and K.J.M. Moriarty, Nucl. Phys. <u>B50</u>, 354 (1972).
11. A. Martin and P.R. Stevens, Phys. Rev. <u>D5</u>, 147 (1972).
12. B. Sadoulet, Nucl. Phys. <u>B</u> (to be published).
13. V. Barger and A.D. Martin, Phys. Letters <u>39B</u>, 379 (1972).
14. F. Henyey, G.L. Kane, J. Pumplin and M.H. Ross, Phys. Rev. <u>182</u>, 1579 (1969); M. Ross, F.S. Henyey and G.L. Kane, Nucl. Phys. <u>B23</u>, 269 (1970).
15. R.D. Field, Lawrence Berkeley Laboratory, Report No. LBL-33 (1971) (unpublished).
16. B.J. Hartley and G.L. Kane (preprint).
17. J.S. Loos and J.A.J. Matthews, Phys. Rev. <u>D6</u>, 2467 (1972) have successfuly fit the hypercharge exchange reactions $K^-n \to \pi^-\Lambda$ and $\pi^-p \to K^0\Lambda$ using a dual absorptive type model (DAM). They must, however, make assumptions concerning the behavior of the real parts of the helicity amplitudes since DAM predicts the behavior of the imaginary parts only for K^{**} and K^* exchange.
18. M. Aguilar-Benitez, S.U. Chung, R.L. Eisner and R.D. Field, Phys. Rev. Letters <u>29</u>, 749 (1972).
19. A. Białas and K. Zalewski, Nucl. Phys. <u>B6</u>, 449 (1968); ibid <u>B6</u>, 465 (1968); ibid <u>B6</u>, 478 (1968).
20. N. Byers and C.N. Yang, Phys. Rev. <u>135</u>, B796 (1964).
21. We have combined data at 3.9 and 4.6 GeV/c and labelled it 4.2 GeV/c.
22. R.D. Field, R.L. Eisner, S.U. Chung and M. Aguilar-Benitez, Phys. Rev. <u>D</u> (to be published).
23. R.D. Field, R.L. Eisner, A. Rouge, H. Videau, M. Aguilar-Benitez, and F. Barreiro (to be published). We have combined the BNL data of Ref. 22 with Ecole Polytechnique data.
24. D. Crennell, H. Gordon, K.-W. Lai and J.M. Scarr, Phys. Rev. <u>D6</u>, 1220 (1972).
25. M. Abramovich, A.C. Irving, A.D. Martin and C. Michael, Phys. Letters <u>39B</u>, 353 (1972).
26. The $A^{\pm}$, $B^{\pm}$, and $C^{\pm}$ amplitudes defined here are related to the amplitudes of Ref. 25 by $A^{\pm} = -c_{\pm}$, $B^{\pm} = \mp ia_{\pm}$, $C^{\pm} = \mp b_{\pm}$; and to the amplitudes of Ref. 24 by $A^{\pm} = \underline{S}_{\pm}$, $B^{\pm} = \mp \underline{D}_{\mp}$, and $C^{\pm} = \mp \underline{A}_{\mp}$.

27. These poles are to be understood as "effective" poles into which the effects of Regge cuts and/or lower lying poles have been absorbed.
28. Having determined the amplitudes $A^{\pm}$, $B^{\pm}$, and $C^{\pm}$ it is an easy matter to calculate P_n and P_u. Namely, $P_n = (|A^+|^2 - |A^-|^2)/\Sigma$, $P_u = (|B^-|^2 - |B^+|^2 + |C^-|^2 - |C^+|^2)/\Sigma$ where $\Sigma = |A^+|^2 + |A^-|^2 + |B^+|^2 + |B^-|^2 + |C^+|^2 + |C^-|^2$.
29. The K_B is the $Y = \pm 1$ member of the octet to which the B meson belongs and the K_A is the $Y = \pm 1$ member of the octet to which the A_1 meson belongs.
30. To date there has been no evidence for the necessity of A_1 exchange in $\pi^- p \to \rho^0 n$. However, the question of the importance of A_1 exchange will not be answered until polarized proton target information is used to detect possible π-A_1 interference in $\pi^- p \to \rho^0 n$. Even if future experimental information substantiates the necessity of K_A exchange in $K^- p \to (\omega, \varphi) \Lambda$ and $\pi^- p \to K^{*0} \Lambda$ this would not necessarily imply that the A_1 is important in $\pi^- p \to \rho^0 n$. This would depend on the F/D ratio of the A_1-K_A octet.
31. One can successfully fit the observables for the vector meson production with $Y = 1$ reactions $K^- p \to (\rho, \omega, \varphi)(\Lambda, \Sigma)$ and $\pi^- p \to K^{*0}(\Lambda, \Sigma)$. See, R.D. Field, R.L. Eisner, and M. Aguilar-Benitez, Phys. Rev. D6, 1863 (1972).
32. G. Vogel, H. Burfeindt, G. Buschhorn, P. Heide, U. Kötz, K.-H. Mess, P. Schmuser, B. Sonne and B.H. Wilk, Phys. Letters 40B, 513 (1972).
33. R.H. Siemann (private communication).
34. Quark model predictions appear to be reasonably well satisfied for $\pi^+ p \to K^+ Y^{*+}(1385)$ at 8 GeV/c, although statistics are limited. J.V. Beaupre, H. Grässler, R. Speth, K.-F. Albrecht, M. Walter, G.T. Jones, V. Karimaki, W. Kottel, D. Sotiriou, R. Stroynowski, D. Kisielewska, P. Malecki, J. Zaorska, W. Zielinski, P.J. Dornan, K. Kumas, G. Otter, P. Schmid, H. Piotrowska, and R. Sosnowski, Nucl. Phys. B49, 405 (1972).
35. This is primarily due to the fact that $g^2_{NK\Sigma} \ll g^2_{NK\Lambda}$.
36. R.L. Eisner and M. Aguilar-Benitez (private communication).
37. See Ref. 31 for definitions of the crossing angles.
38. D.J. Crennell, Uri Karshon, Kwan Wu Lai, J.S. O'Neall, J.M. Scarr, R.M. Lea, T.G. Schumann and E.M. Urvator, Phys. Rev. Letters 23, 1347 (1969).
39. M. Abramovitch, H. Blumenfeld, V. Chaloupka, S.U. Chung, J. Diaz, L. Montanet, J. Pernego, S. Reucroft, J. Rubio and B. Sadoulet, Nucl. Phys. B27, 477 (1971).
40. R. Barloutaud, Duong Nhu Hoa, J. Griselin, D.W. Merrill, J.C. Scheuer, W. Hoogland, J.C. Kluyver, A. Minguzzi-Ranzi, A.M. Rossi, B. Haber, E. Hirsch, J. Goldberg and M. Laloum, Nucl. Phys. B9, 493 (1969).

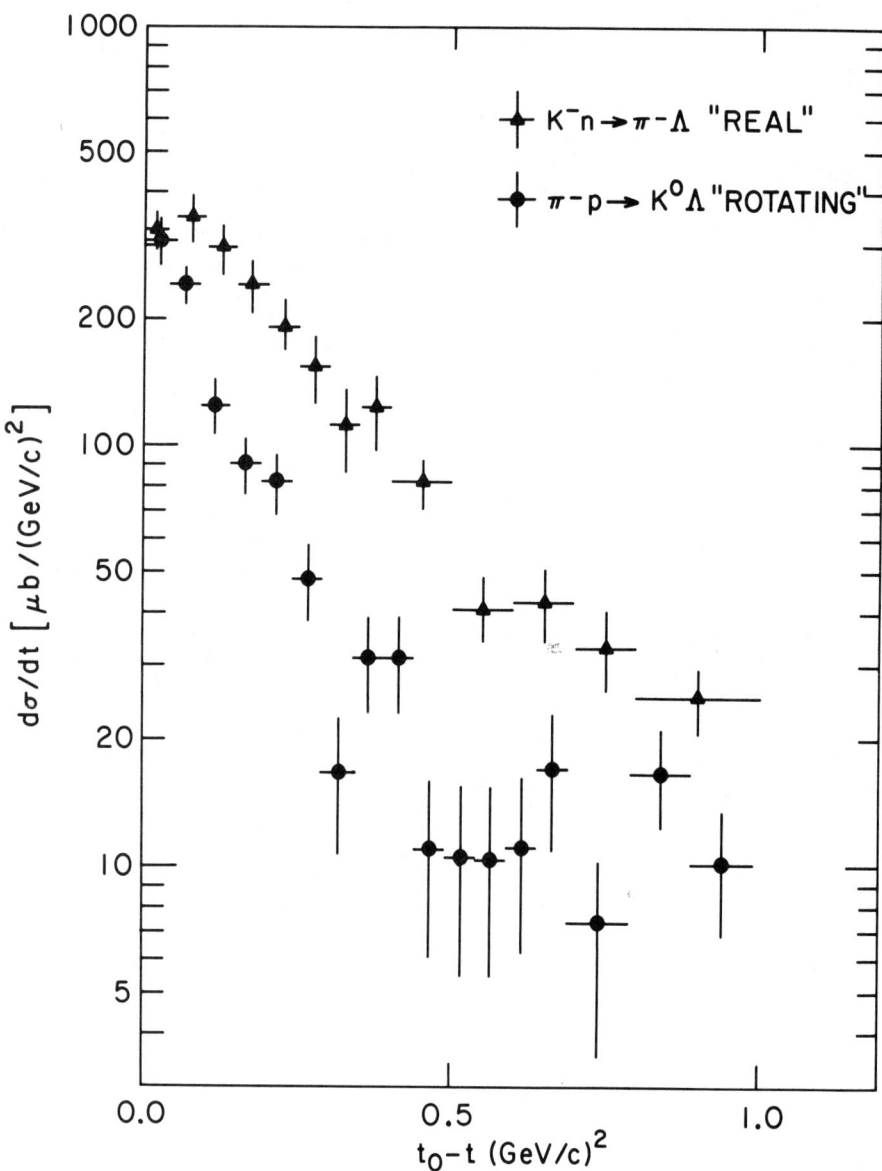

1. Experimental differential cross sections for the reaction $K^-n \to \pi^-\Lambda$ at 3.9 GeV/c (Ref. 38) and the line reversed reaction $\pi^-p \to K^0\Lambda$ at 3.9 GeV/c (Ref. 39). Duality diagrams arguments (or EXD) predict the amplitudes for $K^-n \to \pi^-\Lambda$ be purely real ("real") and those for $\pi^-p \to K^0\Lambda$ to have a rotating ("rotating") phase ($e^{-i\pi\alpha(t)}$).

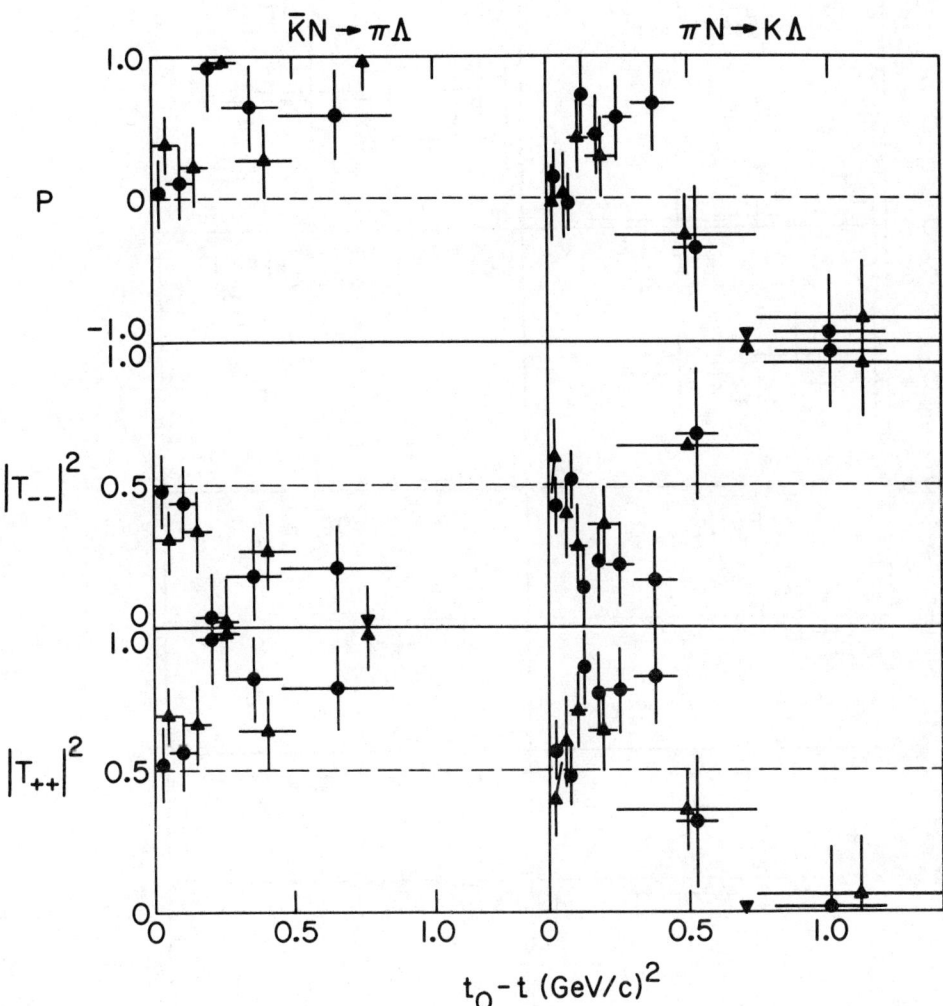

2. Experimental values of the Λ polarization P and magnitudes of the transversity amplitudes (normalized $|T_{++}|^2 + |T_{--}|^2 = 1$) for the reaction $K^-n \to \pi^-\Lambda$ at 3.0 GeV/c (Ref. 40, circles) and 3.9 GeV/c (Ref. 38, triangles) and for the line reversed reaction $\pi^-p \to K^0\Lambda$ at 3.9 GeV/c (Ref. 39, triangles) and 4.5 GeV/c (Ref. 24, circles).

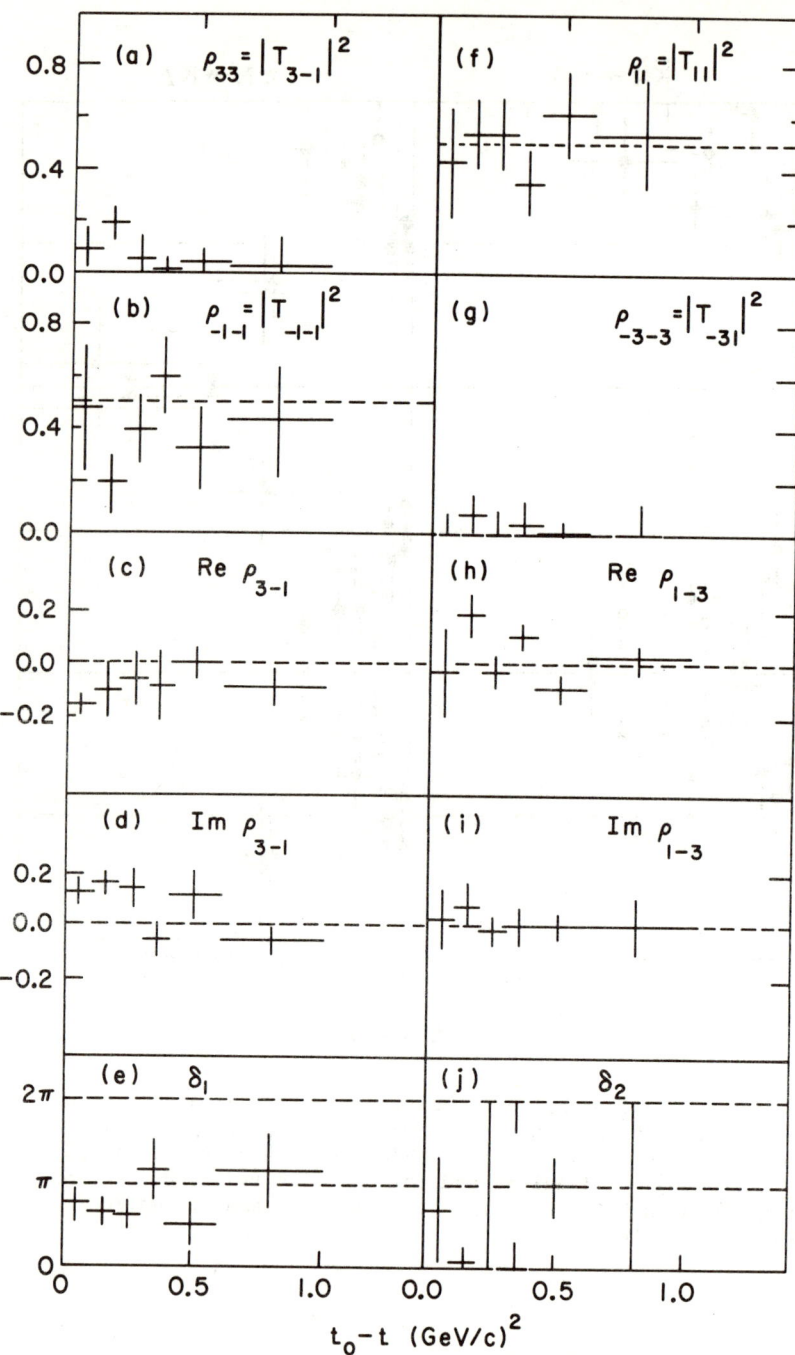

3. Transversity amplitudes and transversity density matrix elements for the reaction $K^-p \to \pi^-Y^{*+}(1385)$,[18] where δ_1 (δ_2) is the relative phase between T_{3-1} and T_{-1-1} (T_{11} and T_{-31}). The quark model predicts $T_{3-1} = T_{-31} = 0$ and $T_{11} = T_{-1-1} = 1/2$.

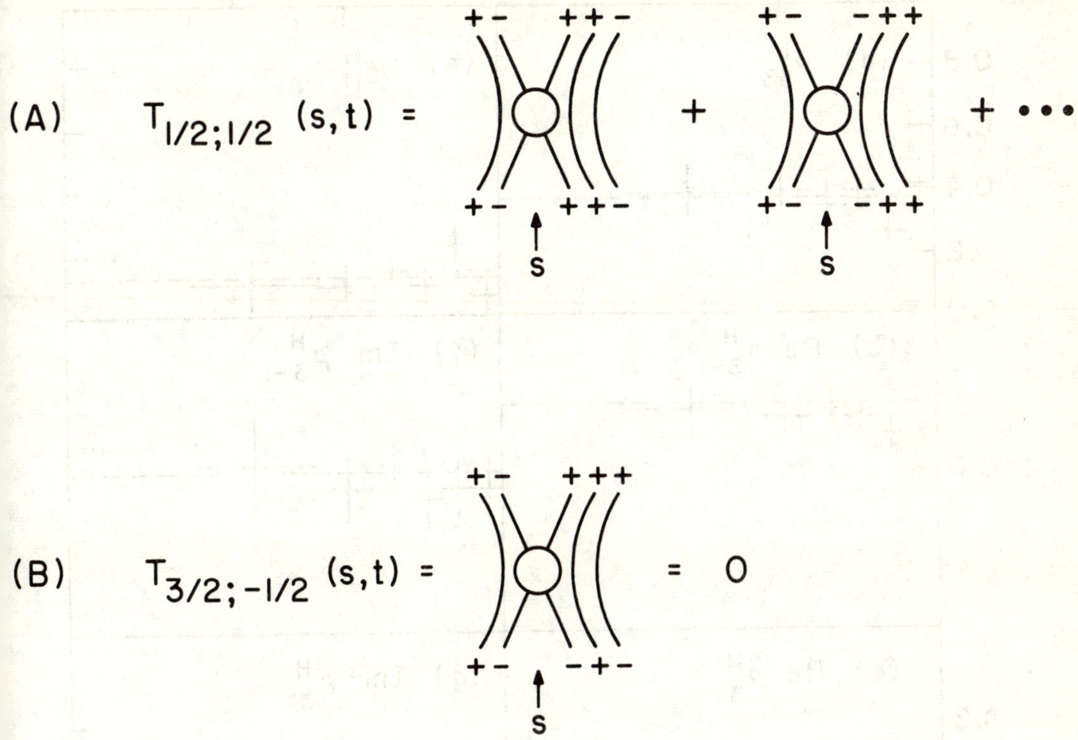

(A)
T1/2;1/2 (s,t) =
+ -
+ + -
+ -
- + +
+ •••
s
s
(B)
T3/2;-1/2 (s,t) =
+ -
+ + +
+ -
- + -
= 0
s

4. (A) Illustrates how T_{11} can be represented as a sum of quark-quark scattering amplitudes. The $\pm$ signs represent quark spin projections on the transversity z-axis (normal to production plane).
(B) Illustrates how $T_{3\,-1}$ requires that more than one of the baryon quarks flip sign. The quark model thus predicts that this amplitude vanish.

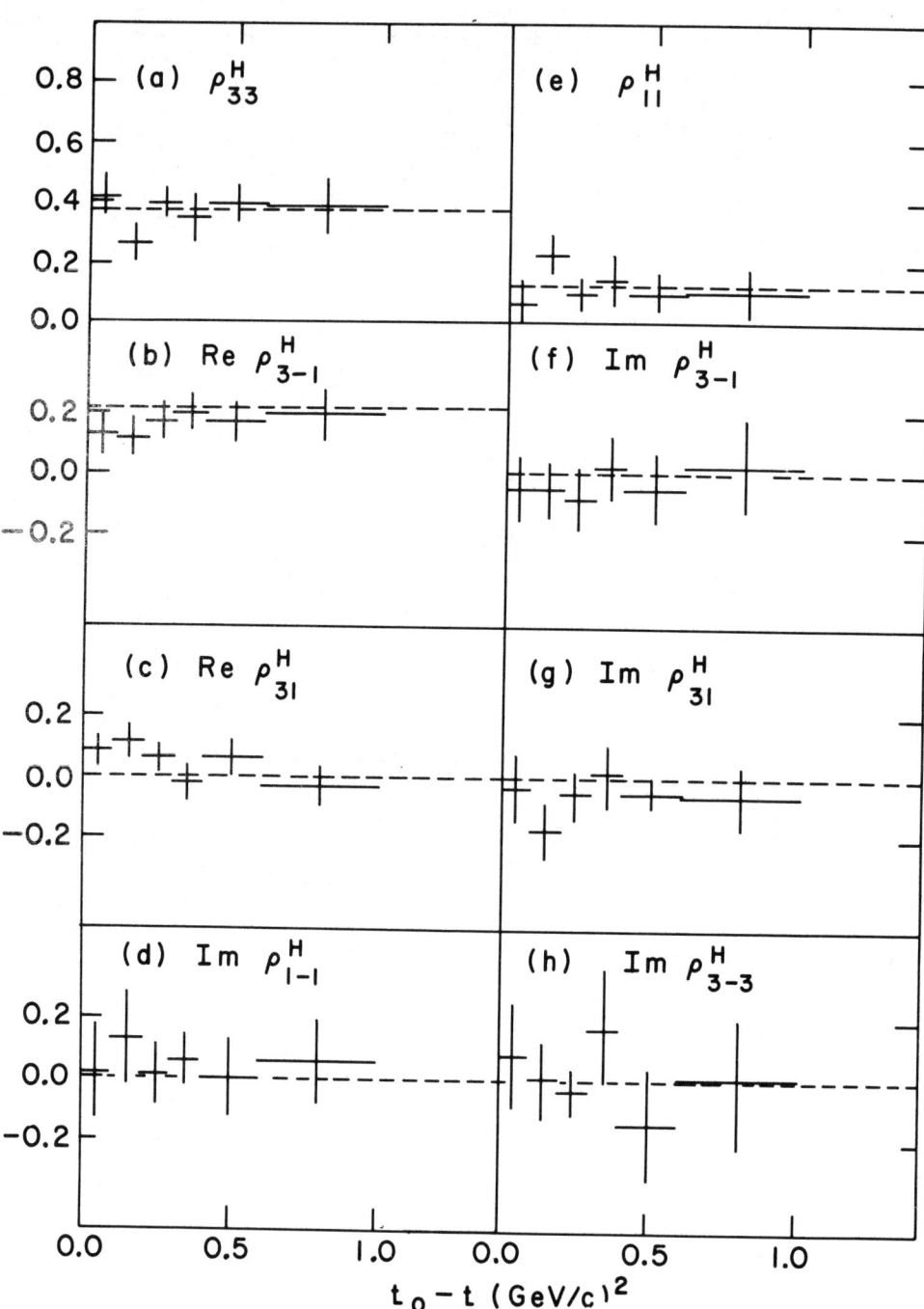

5. The s-channel helicity density matrix elements for the reaction $K^-p \to \pi^- Y^{*+}(1385)$.[18] The quark model predicts that $\rho^H_{33} = 3/8$, $\text{Re}\rho^H_{3-1} = \sqrt{3}/8$, and that the remaining elements vanish.

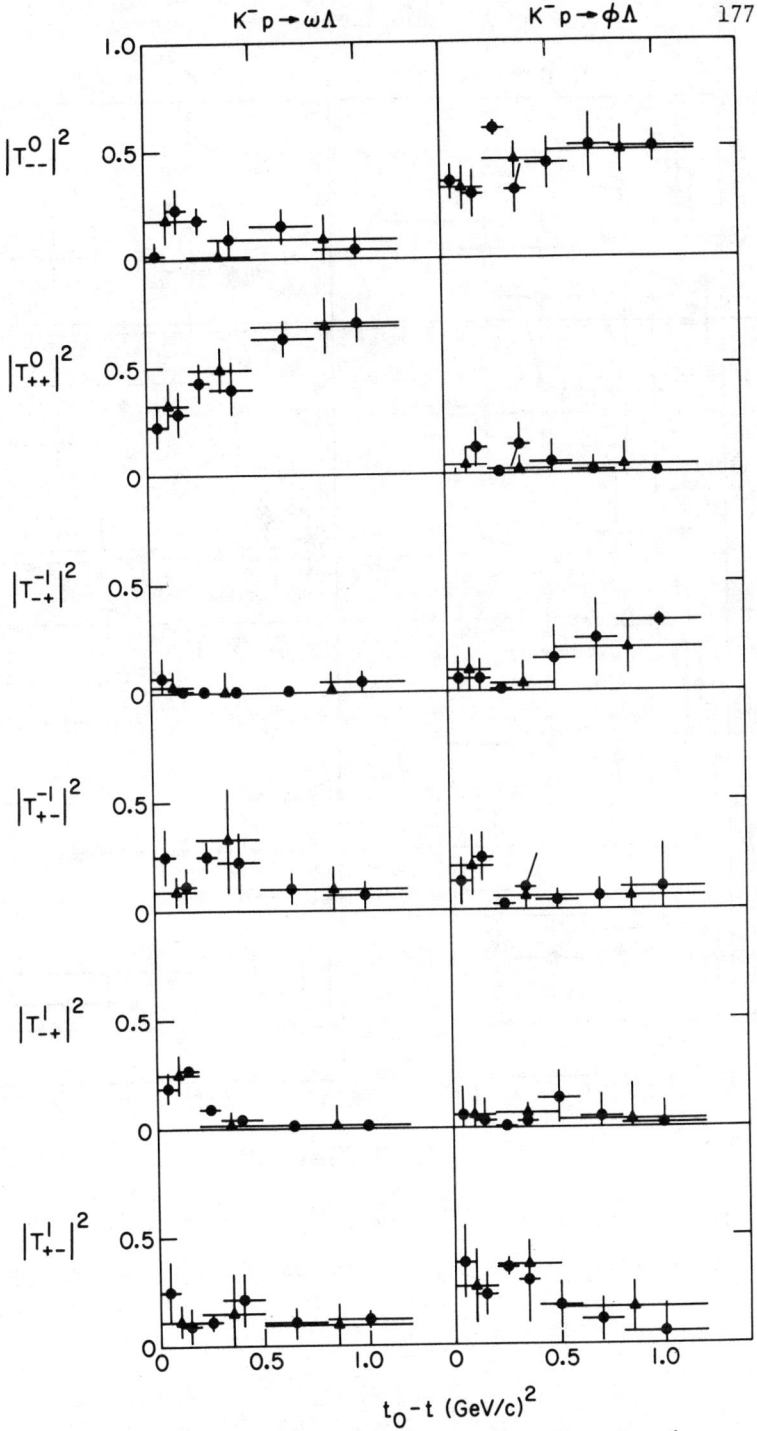

6. Magnitudes of the transversity amplitudes for the reactions $K^-p \to (\omega,\varphi)\Lambda$. The triangles are the combined 3.9 and 4.6 GeV/c BNL data (Ref. 22) and the circles are combined BNL and Ecole Polytechnique data (Ref. 23).

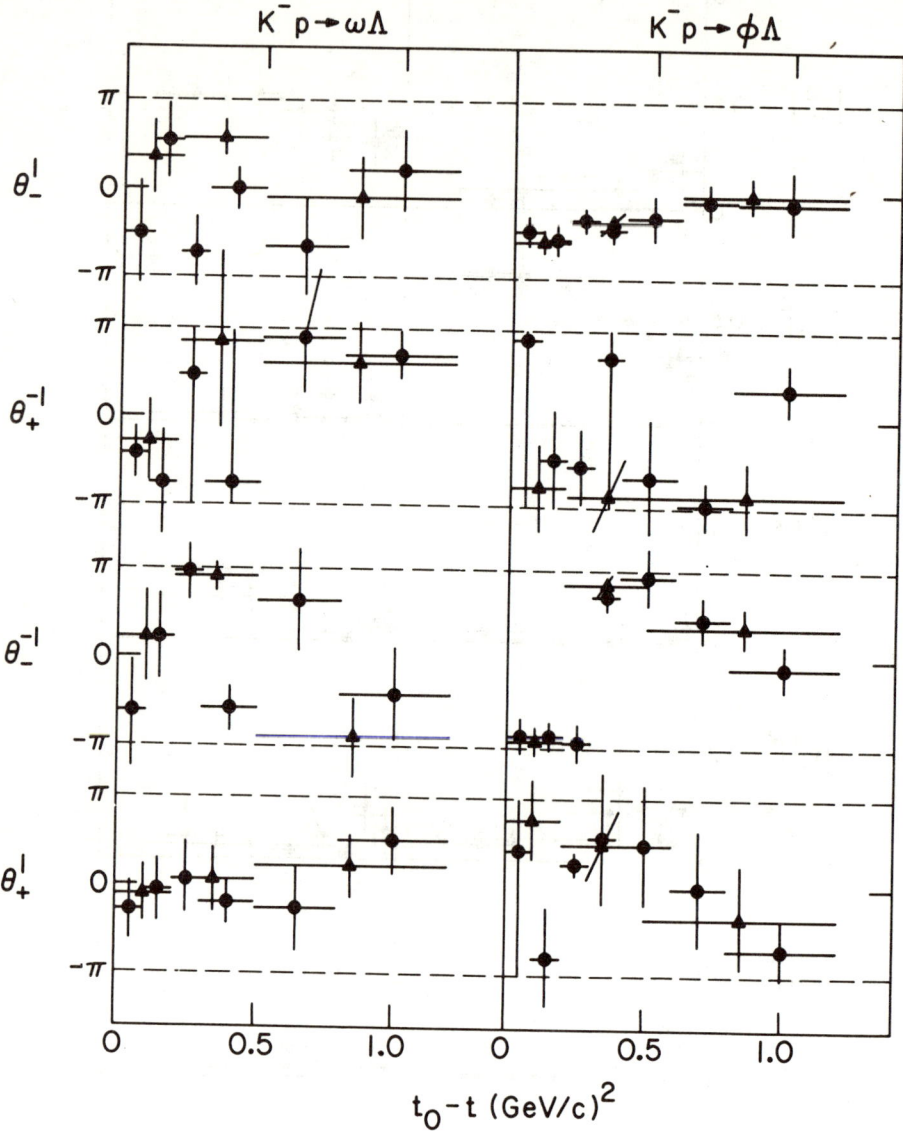

7. Relative phases of the transversity amplitudes, where $\theta^{\pm 1}_{\pm} = \text{Arg}(T^{\pm 1}_{\pm\pm}) - \text{Arg}(T^{0}_{\pm\pm})$. The data are the same as Fig. 6.

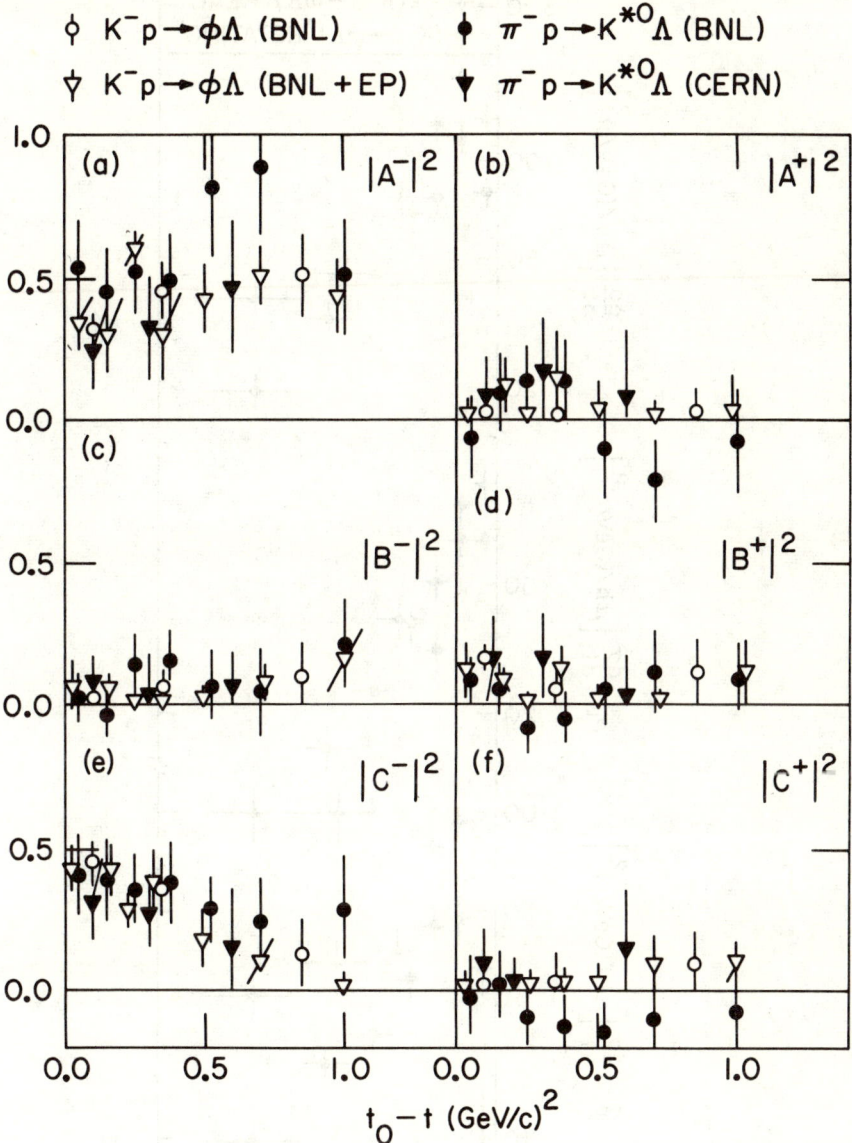

8. Comparison of the amplitudes for $K^-p \to \varphi\Lambda$ using the BNL data (Ref. 22, open circles) and the BNL + EP data (Ref. 23, open triangles) with the amplitudes for $\pi^-p \to K^{*0}\Lambda$ at 3.9 GeV/c Ref. 25, solid triangles) and at 4.5 GeV/c (Ref. 24, solid circles). The quark model predicts equality of the amplitudes for these two reactions.

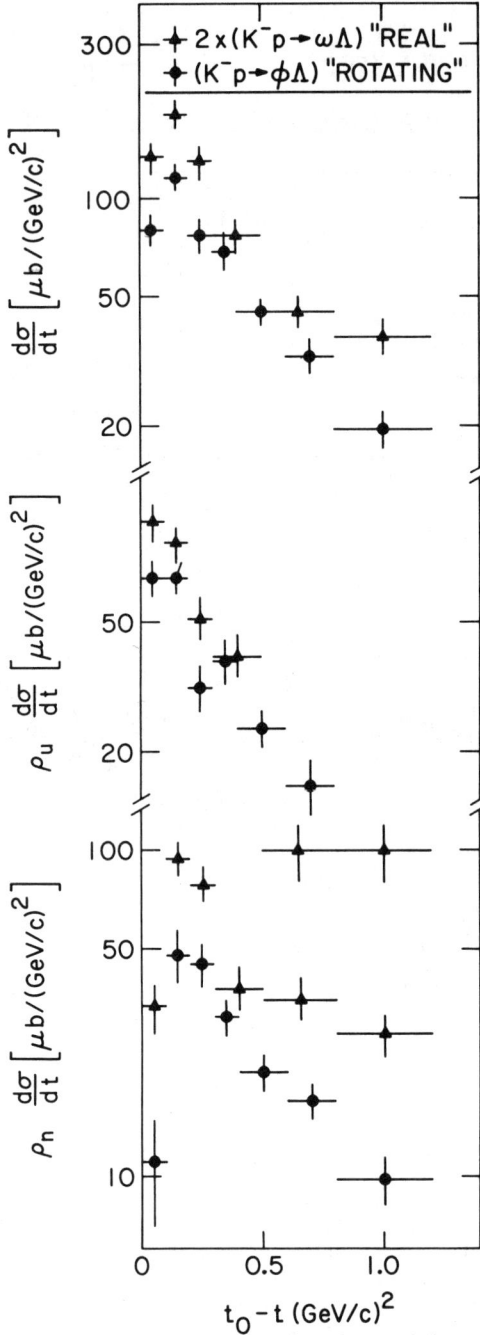

9. The natural parity ($\rho_n = \rho_{11}+\rho_{1\,-1}$), unnatural parity $\rho_u = \rho_{11}-\rho_{-1\,-1}$); and the total differential cross section for reaction $K^-p \to \varphi\Lambda$ and twice the reaction $K^-p \to \omega\Lambda$. The data are from Ref. 23. EXD plus SU(3) predicts the amplitudes for the natural parity part of the reaction $K^-p \to \omega\Lambda$ be purely real and those for $K^-p \to \varphi\Lambda$ have a rotating phase ($e^{-i\pi\alpha(t)}$).

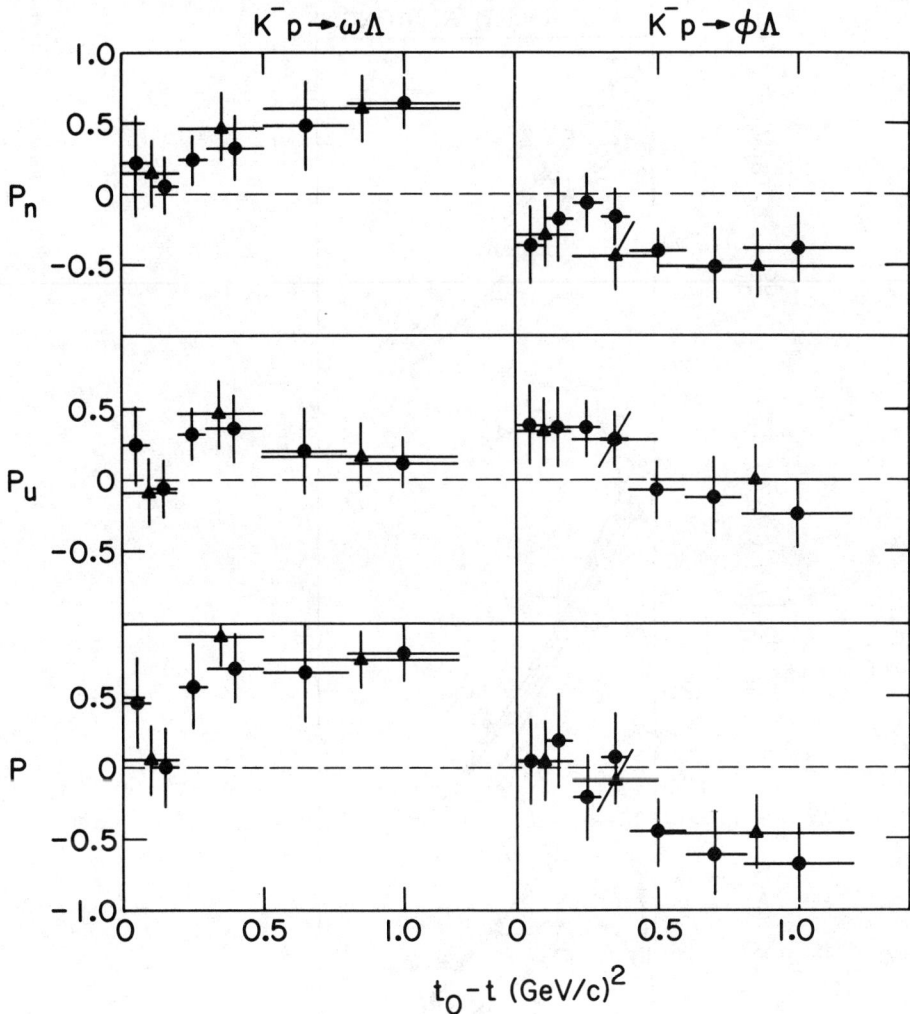

10. The natural parity polarization (P_n), unnatural parity polarization (P_u), and total polarization ($P = P_n + P_u$) for the reactions $K^-p \to (\omega,\varphi)\Lambda$ from Ref. 23.

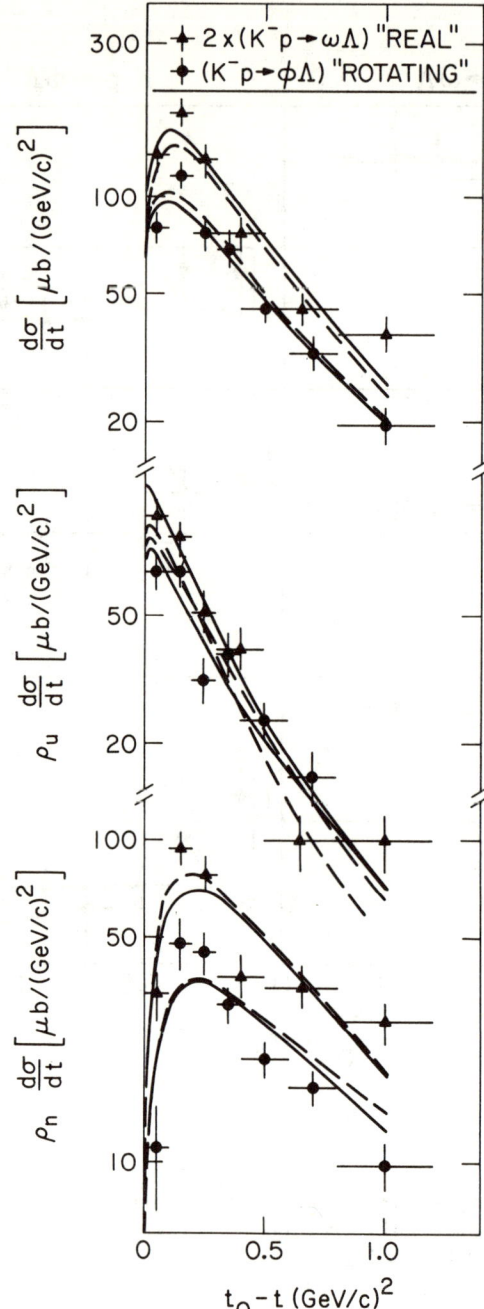

11. Shows an effective Regge pole plus SU(3) fit to the natural parity (ρ_n), unnatural parity (ρ_u) and total differential cross sections for $K^-p \to \phi\Lambda$ and twice $K^-p \to \omega\Lambda$. The dashed curves are from a model that includes a K^{**}, K^*, K, and K_B effective pole, whereas the solid curves are from a model that also includes a K_A effective pole. SU(3) is used to relate the two reactions. The data are from Ref. 23.

183

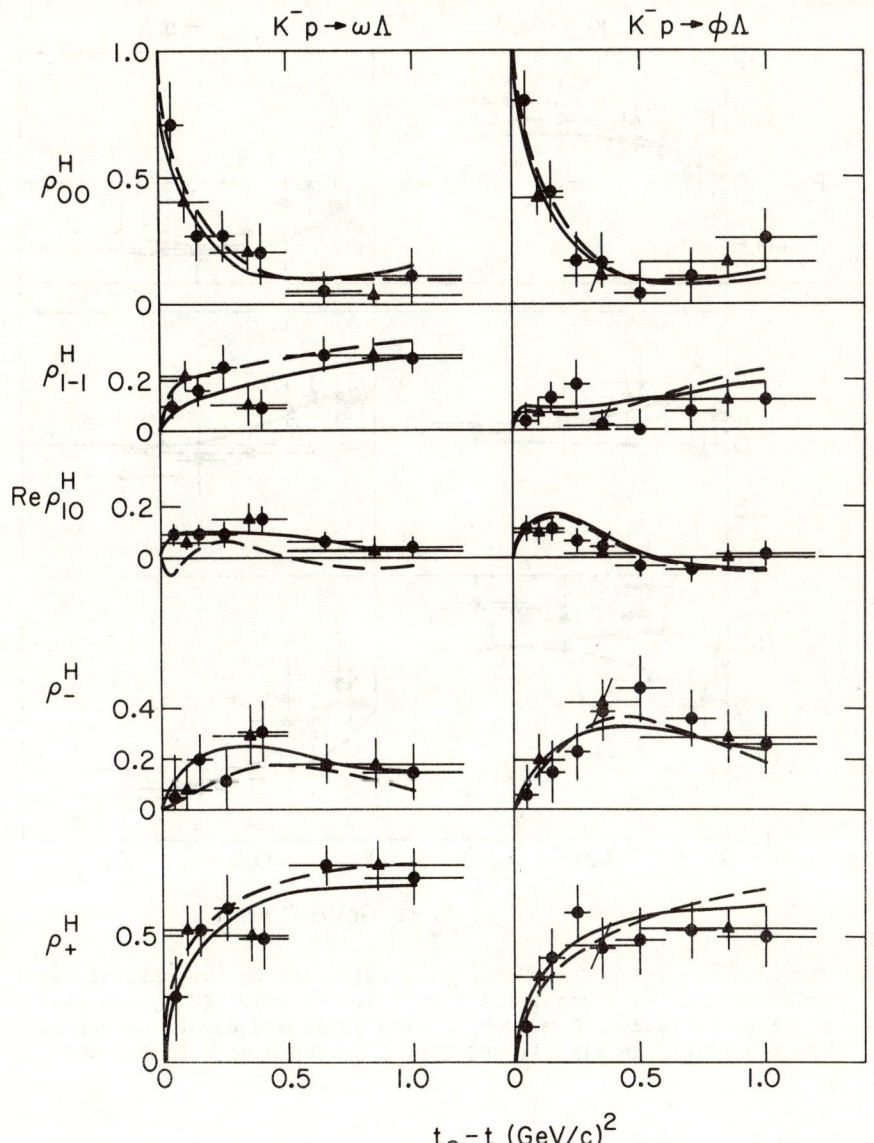

12. Shows an effective Regge pole plus SU(3) fit to the single density matrix elements in the helicity frame for $K^-p \to (\omega,\varphi)\Lambda$, where $\rho_\pm^H = \rho_{11}^H \pm \rho_{1-1}^H$. The dashed curves are from a model that includes a K^{**}, K^*, K, and K_B effective pole, whereas the solid curves are from a model that also includes a K_A effective pole. SU(3) is used to relate the two reactions. The data are from Ref. 23.

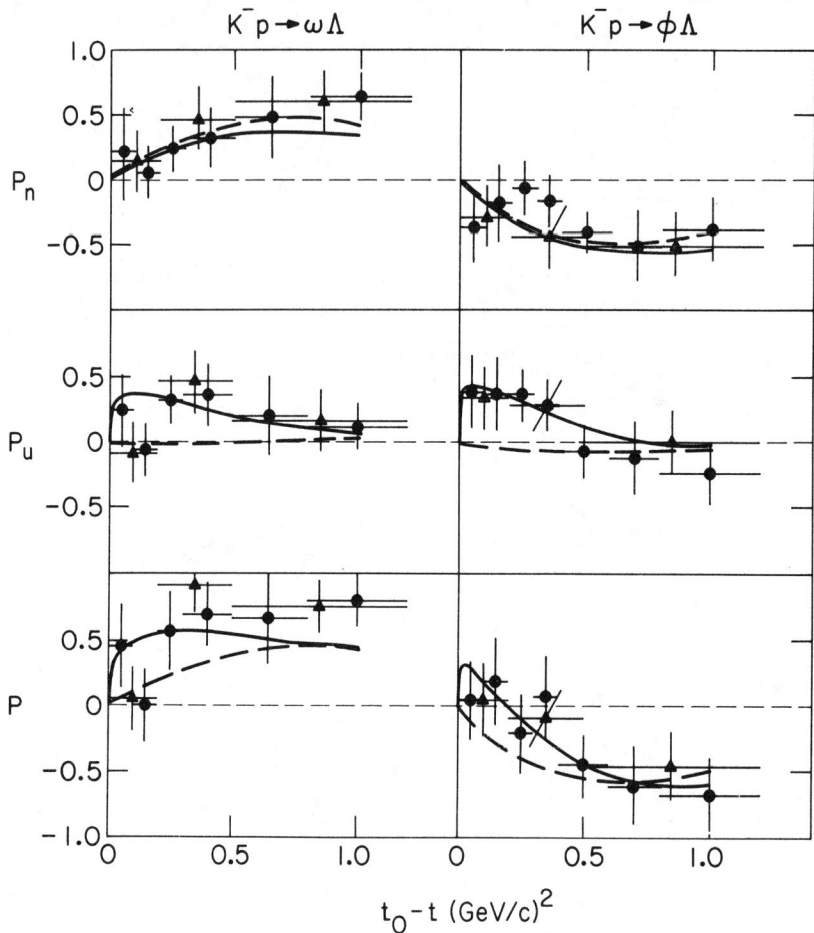

13. Shows an effective Regge pole plus SU(3) fit to the natural parity polarization (P_n), unnatural parity polarization (P_u), and the total polarization ($P = P_n + P_u$). The solid and dashed curves are the two models in Fig. 11 and Fig. 12. The data are from Ref. 23.

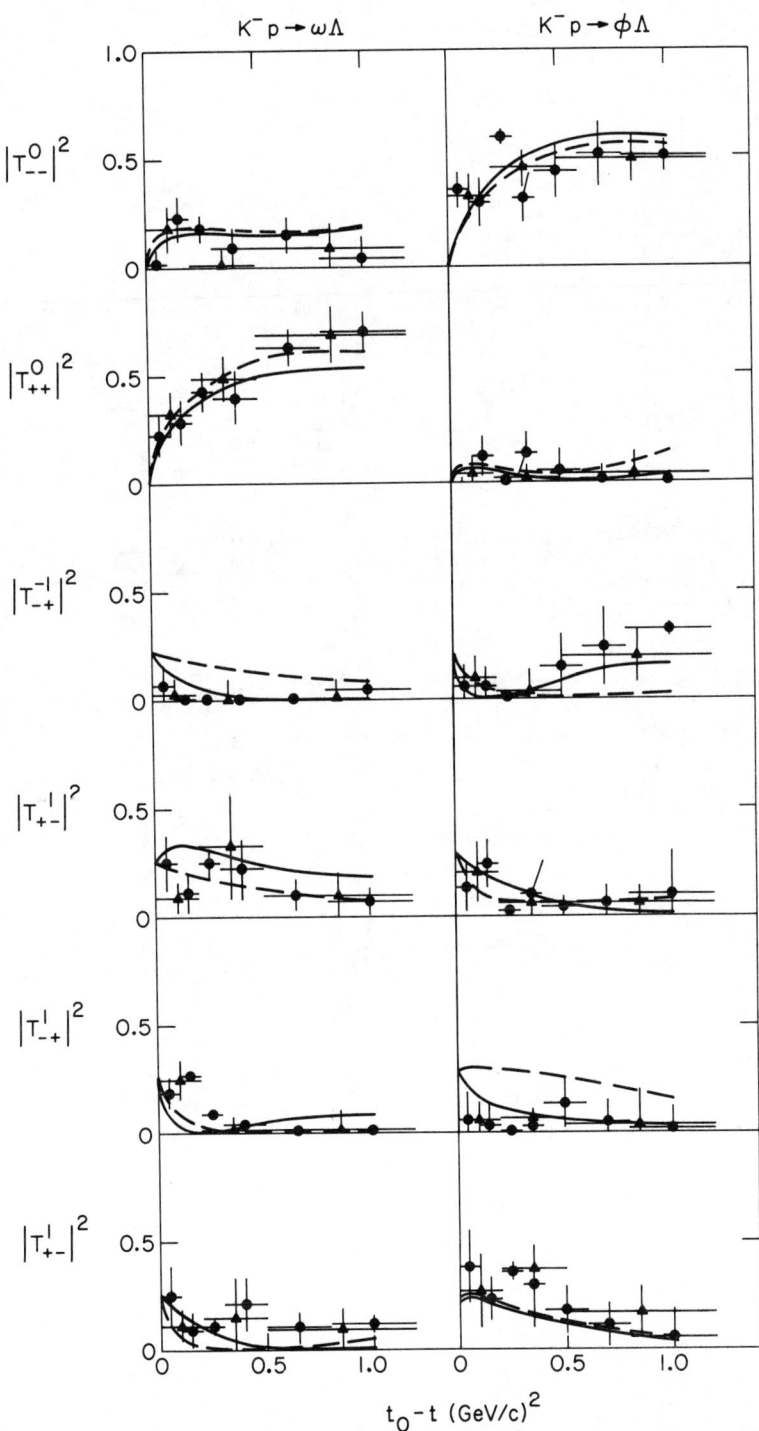

14. Comparison of the predictions of the two effective pole plus SU(3) models shown in Figs. 11, 12, and 13 with the experimentally determined transversity amplitudes of Fig. 6.

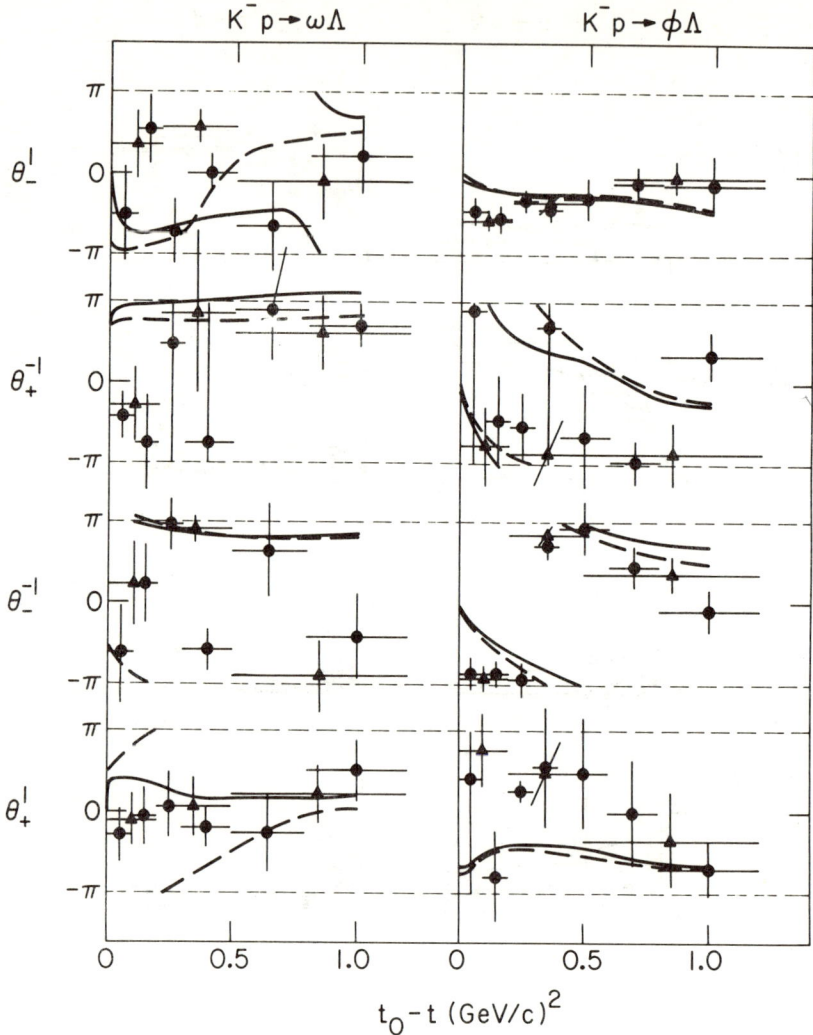

15. Comparison of the predictions of the two effective pole plus SU(3) models shown in Figs. 11, 12, and 13 with the relative phases of Fig. 7.

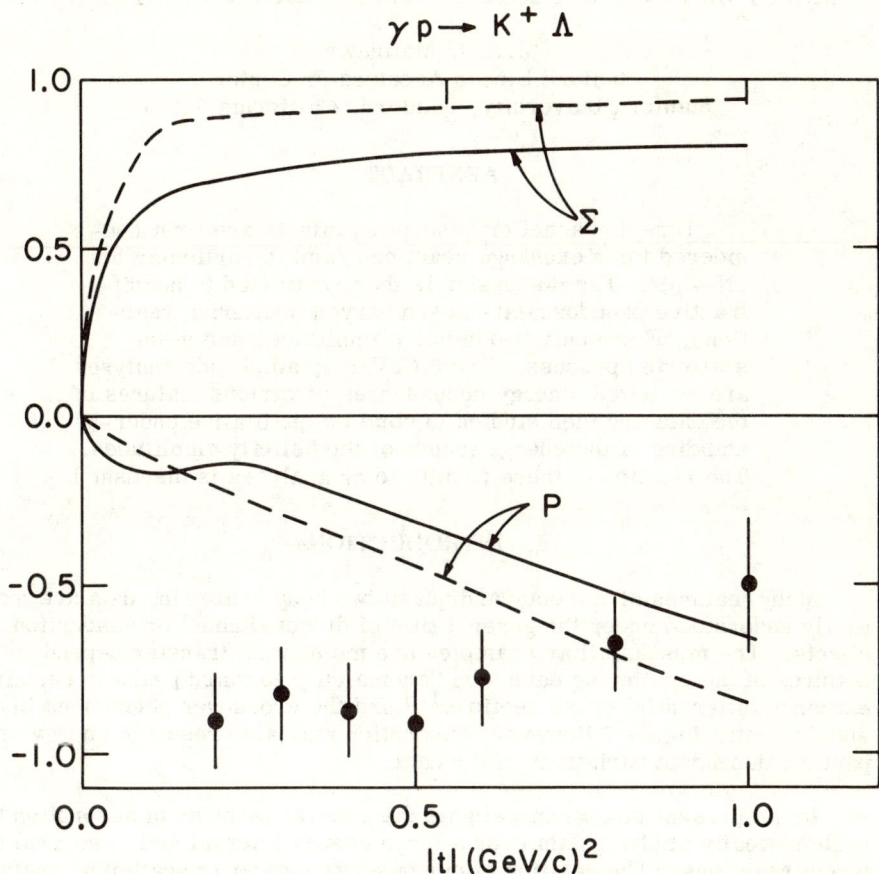

16. Shows the predictions of the two effective pole plus SU(3) models shown in Figs. 11, 12, and 13 for the Λ polarization (P) and polarized photon asymmetry (Σ) for the reaction $\gamma p \to K^+\Lambda$, where VDM is used to relate the photon to the vector mesons ρ, ω, and φ. The data are Λ polarization at 5.0 GeV/c from Ref. 32.

DIRECT CHANNEL EFFECTS IN NONDIFFRACTIVE SCATTERING*

J.A.J. Matthews
Stanford Linear Accelerator Center
Stanford University, Stanford, California 94305

ABSTRACT

Direct channel or absorption effects are first considered for π exchange reactions, and in particular for $\pi N \to \rho N$. The discussion is then restricted to nondiffractive pseudoscalar-meson baryon scattering reactions, where only two helicity amplitudes define the scattering process. The 6 GeV/c πp amplitude analyses are reviewed; energy dependences of various features of the data are then studied to obtain a qualitative understanding of the energy trends of the helicity amplitudes. The relation of these results to $\pi\pi$ analyses is discussed.

I. INTRODUCTION

Many features of two body and quasi-two body scattering data are most easily understood under the general title of direct channel or absorption effects. The most familiar examples are momentum transfer dependent features of the scattering data: the "anomalous" forward peaks in certain π exchange differential cross sections[1-5] and the crossover phenomena in elastic scattering.[6,7] However, absorption may also result in energy or particle dependent variations in the data.

In the present talk we investigate the general features of absorption by systematically studying data over a large energy interval and in several different reactions. The relation of our observations to $\pi\pi$ scattering analyses is emphasized.

The qualitative ideas associated with absorption are introduced in Sec. II using the familiar reaction $\pi N \to \rho N$. The study is then restricted in Sec. III to the simpler pseudoscalar-meson baryon scattering reactions, described by two helicity amplitudes, and having only natural parity exchanges in the t channel. Only nondiffractive reactions are considered.

The πp amplitude analyses at 6 GeV/c are discussed first, providing an introduction to the t dependence of the scattering amplitudes. To extend the results of the amplitude analyses, data in several reactions are studied to determine qualitative trends in the energy dependences of the scattering amplitudes. In particular data in πp and Kp reactions are contrasted to reveal possible direct channel effects.

The observations are summarized in Sec. IV.

---

*Work supported by the U. S. Atomic Energy Commission.

II. INTRODUCTION TO ABSORPTION

For many years absorption has been known to provide a simple explanation of the $\pi^- p \to \rho^0 n$ data at small values of momentum transfer. Historically, simple one pion exchange (OPE) failed in ρ^0 production, predicting $\rho_{00}^{GH} = 1$. In contrast the Gottfried-Jackson OPE plus absorption model[8] successfully reproduced the ρ^0 differential cross section and density matrix elements.

More recently it has been emphasized that in π exchange reactions an unambiguous signature of absorption occurs for $-t < m_\pi^2$. As discussed by Kane,[9] ρ_{00}^{GJ} and $d\sigma/dt$ for ρ^0 production should turn over in the forward direction, but ρ_{11}^H $d\sigma/dt$ should have a sharp forward peak. This has since been observed at 15 GeV/c,[4,10] see Fig. 1, and more recently at 17 GeV/c.[11] Analogous features are seen in the forward cross sections for the reactions $\gamma p \to \pi^+ n$,[1] $p\bar{p} \to n\bar{n}$,[3] and np charge exchange.[2]

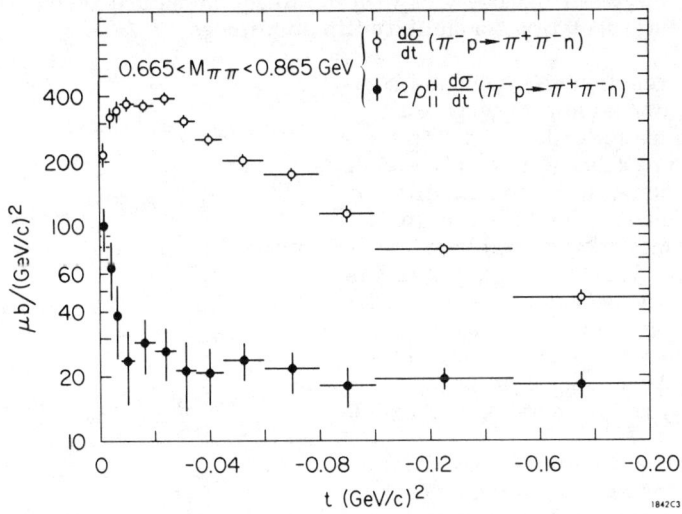

Fig. 1. Differential cross sections from the reaction $\pi^- p \to \rho^0 n$ at 15 GeV/c.

In the absorption model the explanation for this behavior is straightforward. Absorption "smooths" the amplitudes, thus in the forward direction helicity amplitudes possess only the minimal t dependence, $(t_{min}-t)^{\Delta\lambda}$, consistent with conservation of angular momentum. The net helicity flip, $\Delta\lambda$, is defined in the s channel or helicity frame. In the limit of large energies, the dominant π exchange amplitudes flip the nucleon s channel helicity, resulting in π exchange contributions to $\pi N \to \rho N$ of the form:

$$H_{+-}^{\lambda\rho} = f_{\Delta\lambda}(t) \sim \frac{(t_{min}-t)^{\Delta\lambda}}{(t-m_\pi^2)} \tag{1}$$

where λ_ρ is the ρ helicity, and $\Delta\lambda = \lambda_\rho - 1$. For scattering near $t \sim t_{min}$, the $\lambda_\rho = 1$ amplitude dominates yielding $\rho_{11}^H \sim 0.5$ as observed in Fig. 1. A popular model that embodies the features of absorption is the Williams OPE-δ [12] or poor man's absorption model;[13] this has been successfully compared to the high energy ρ^0 production data.[10]

The unique features of the data discussed above are a special argument for absorption however, relying as much on the proximity of the π pole to the physical region as on absorption itself. More generally absorption or geometrical models suggest that s channel helicity amplitudes have the approximate form[4]:

$$f_{\Delta\lambda}(t) \sim J_{\Delta\lambda}(r\sqrt{-t}) \qquad (2)$$

where r, the radius in impact parameter space where the amplitude is maximum, is approximately $r \sim 1\,\text{fm} \sim 5\,\text{GeV}^{-1}$. Thus $\Delta\lambda = 0, 1$ amplitudes are predicted to have minima (or zeros) at $-t \sim 0.2, 0.6\,\text{GeV}^2$, respectively. The former zero is responsible for the crossover effect in elastic scattering reactions,[7] the latter usually vies with the Regge signature zero as the more basic interpretation for helicity flip amplitudes.

For the reaction $\pi N \to \rho N$, the combination $\rho_{00}^H \, d\sigma/dt$ isolates unnatural parity exchange to leading order in the energy. If A_1 exchange is small[15] then $\rho_{00}^H \, d\sigma/dt$ isolates the π contribution to the single s channel helicity amplitude $f_{\Delta\lambda=1}$. Recent experimental results (Refs. 16-18) suggest that $\rho_{00}^H \, d\sigma/dt$ has a change in t dependence near $-t \sim 0.6\,\text{GeV}^2$, perhaps even possessing a dip in this momentum transfer region. The most optimistic evidence of this type (Refs. 17, 18) is shown in Figs. 2 and 3.

One explanation for the $\rho_{00}^H \, d\sigma/dt$ data associates only the forward peak with π exchange; this decreases rapidly becoming less than an approximately t independent background near $-t \sim 0.6\,\text{GeV}^2$.[19] Alternatively the data may suggest that the π amplitude has a minimum near $-t \sim 0.6\,\text{GeV}^2$, similar to the ρ amplitude in $\pi^- p \to \pi^0 n$. For the ρ amplitude the dip in $\pi^- p \to \pi^0 n$ can be interpreted as either an absorption effect (Eq. (2)) or a manifestation of the Regge signature factor:

$$f_{\Delta\lambda}(t) \sim \tilde{f}(t)\left(1 - e^{-i\pi\alpha_\rho(t)}\right) \qquad (3)$$

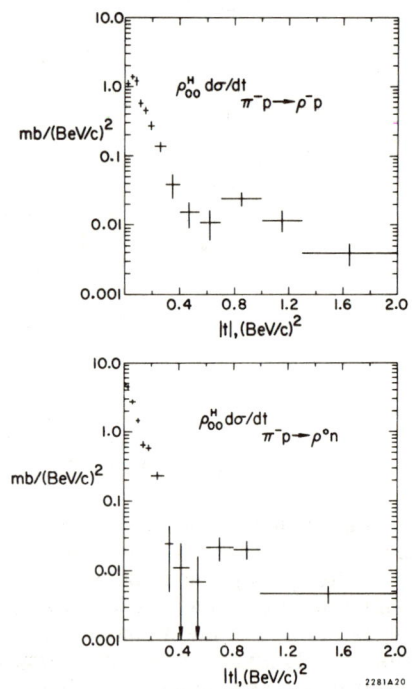

Fig. 2. Differential cross sections $\rho_{00}^H \, d\sigma/dt$ at 4.42 GeV/c from Ref. 17.

since $\alpha_\rho(t) = 0$ at $-t \sim 0.6$ GeV.^{2} However, for π exchange the signature zero should occur at $-t \sim 1.0$ GeV^2, in disagreement with the data. The ρ_{00}^H dσ/dt data may therefore provide the first realistic comparison of absorption and Regge model explanations for the structure of helicity flip amplitudes.

III. PSEUDOSCALAR-MESON BARYON SCATTERING

To continue the study of absorption, we now restrict the scope of this talk to nondiffractive pseudoscalar-meson baryon scattering reactions — the simplest class of reactions for which a substantial library of data presently exists. Features of the data will be related to the s channel helicity amplitudes:

NONFLIP $\equiv f_{\Delta\lambda=0} \equiv f_{++}$

and (4)

FLIP $\equiv f_{\Delta\lambda=1} \equiv f_{+-}$

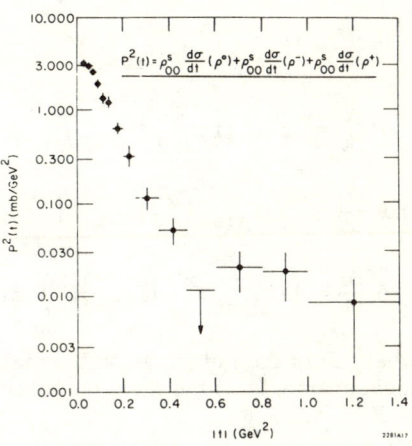

Fig. 3. Combined differential cross sections ρ_{00}^H dσ/dt from the three channels $\pi^-p \to \rho^-p$, $\pi^-p \to \rho^0n$ and $\pi^+p \to \rho^+p$ at 6 GeV/c, from Ref. 18.

The amplitudes are briefly reviewed in Table I.

Table I s channel helicity amplitudes

Amplitude	$|f_{\Delta\lambda}|^2$	Typical Reactions	Dominant t channel quantum numbers
$f_{\Delta\lambda=0}$	dσ/dt vs t (sharp forward peak)	$K_L^0 p \to K_S^0 p$ $\pi p \to K(\Lambda,\Sigma)$ $\bar{K}p \to \pi(\Lambda,\Sigma)$	ω^0 K^*, K^{**} $\bigg\} I_t = 0, \tfrac{1}{2}$
$f_{\Delta\lambda=1}$	dσ/dt vs t (turnover near t=0)	$\pi^-p \to \pi^0 n$ $\pi^-p \to \eta^0 n$	ρ A_2 $\bigg\} I_t = 1$

Measurable quantities then have the following forms:

$$\left. \begin{array}{l} \dfrac{d\sigma}{dt} = |f_{\Delta\lambda=0}|^2 + |f_{\Delta\lambda=1}|^2 \\[6pt] P\dfrac{d\sigma}{dt} = -2\,\text{Im}\left(f_{\Delta\lambda=0}\,f^*_{\Delta\lambda=1}\right) \\[6pt] R\dfrac{d\sigma}{dt} = -\left(|f_{\Delta\lambda=0}|^2 - |f_{\Delta\lambda=1}|^2\right)\cos\theta_{\text{lab}} - 2\,\text{Re}\left(f_{\Delta\lambda=0}\,f^*_{\Delta\lambda=1}\right)\sin\theta_{\text{lab}} \\[6pt] A\dfrac{d\sigma}{dt} = \left(|f_{\Delta\lambda=0}|^2 - |f_{\Delta\lambda=1}|^2\right)\sin\theta_{\text{lab}} - 2\,\text{Re}\left(f_{\Delta\lambda=0}\,f^*_{\Delta\lambda=1}\right)\cos\theta_{\text{lab}} \end{array} \right\} \quad (5)$$

where P is the polarization normal to the scattering plane, and R, A are measurements of the nucleon polarization in the scattering plane, as shown in Fig. 4.

Having measured a sufficient number of quantities in Eq. (5), the detailed structure of the scattering amplitudes can be determined in a model independent manner. In $\pi N \to \pi N$ for example, there are four complex amplitudes: t channel isospin $I_t=0, 1$ and helicities $\Delta\lambda=0, 1$; for comparison in $\pi N \to \rho N$ there are twelve complex amplitudes: $I_t=0, 1$ and six helicity amplitudes. Such an amplitude analysis for $\pi N \to \pi N$ has recently been possible at 6 GeV/c; the results of the Argonne analysis[20] are shown in Fig. 5. Qualitatively we

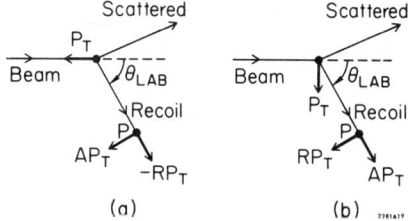

Fig. 4. Scattering geometries for the measurement of the R and A polarization parameters. Initial target polarization, P_T, is in the scattering plane.

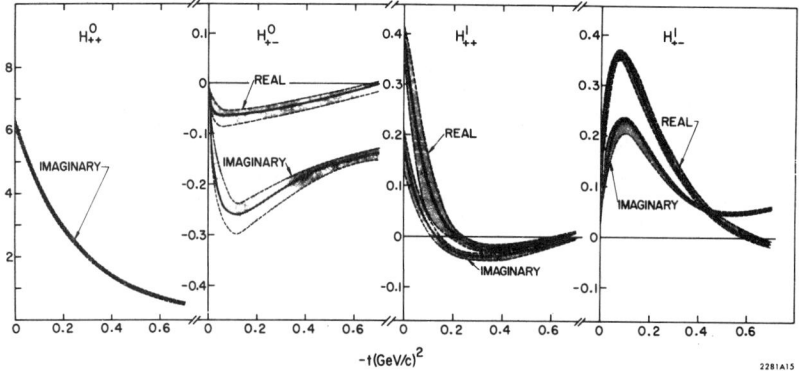

Fig. 5. πN s channel helicity amplitudes, $H^{I_t}_{\Delta\lambda}$, determined at 6 GeV/c from Ref. 20.

observe:

(a) $I_t=0$ amplitudes, since the Pomeron dominates the helicity nonflip (and possibly also the helicity flip) amplitude, no model independent information is obtained for the f^0 exchange amplitudes; and

(b) $I_t=1$ amplitudes, the ρ exchange amplitude is consistent with absorption model predictions for the helicity nonflip amplitude (cf. Eq. (2)), and with absorption or simple Regge model predictions for the helicity flip amplitude (cf. Eqs. (2) and (3)).

To test the sensitivity of these results to possible systematic effects in the data or to different analysis techniques, the Argonne[20] and Saclay[21] solutions are compared in Fig. 6. The agreement is good, except possibly for $-t \gtrsim 0.4$ GeV^2 where some deviations are observed in the $I_t=1$ amplitudes. These amplitudes will be taken therefore as a guide to our further study of meson baryon scattering reactions.

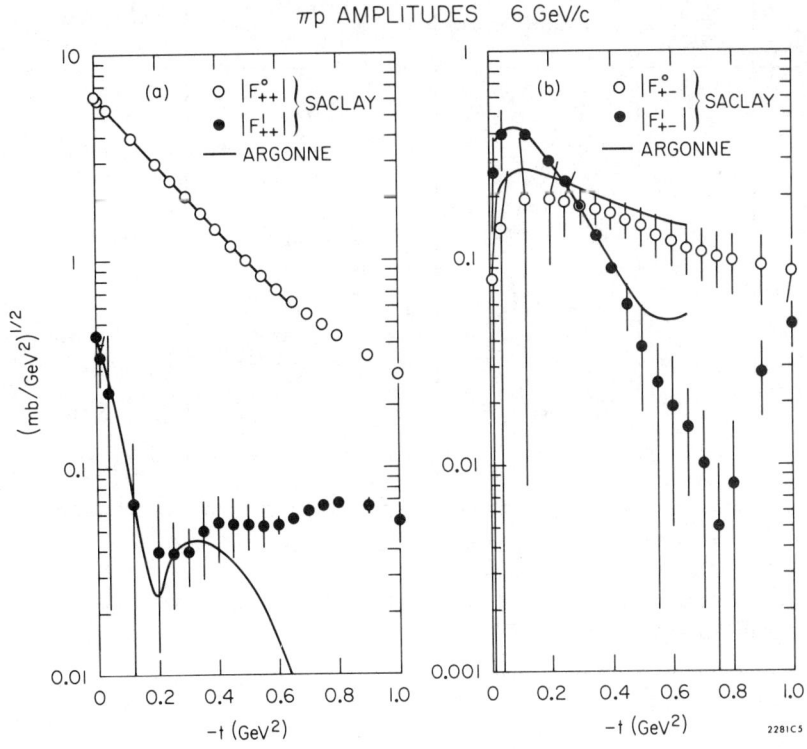

Fig. 6. Moduli of the 6 GeV/c πN helicity amplitudes from the Argonne analysis, Ref. 20, and the Saclay analysis, Ref. 21.

To extend the results of the 6 GeV/c πN amplitude analyses to different energies and reactions, we now consider six signposts which may lead to qualitative, if not quantitative, extrapolations.

Special Channels

If high energy scattering amplitudes can be described by the t channel exchanges involved, then reactions having only one known t channel exchange should provide the best means to systematize our study of the data. This logic has motivated the many analyses of $\pi^- p \to \pi^0 n$ and $\pi^- p \to \eta^0 n$ reactions; recently data has also begun to accumulate in several new channels:

(a) $K_L^0 p \to K_S^0 p$ — ω^0 exchange dominates the forward cross section;

(b) $K^- p \to \eta^0 \Lambda$ — K^* exchange dominates; and

(c) $K^- p \to \eta' \Lambda$ — K^{**} exchange dominates.

These reactions are listed with the "old faithfuls" in Table II.

Table II Special channels of interest

Old faithfuls	New allies	Reaction	SU(3) amplitudes (a)	t channel quantum number exchange
X		$\pi^- p \to \pi^0 n$	$-\sqrt{2}\, V$	ρ
X		$\pi^- p \to \eta_8 n$	$\sqrt{2/3}\, T$	A_2
		$\pi^- p \to \eta_1 n$	$\sqrt{2/3}\, S_T T$	A_2
	X	$K_L^0 p \to K_S^0 p$	$\frac{1}{2}[(4F-1)\omega - \rho]^{(b)}$ $\equiv (2F-1)V$	ω, ρ
X		$K^- p \to \eta_8 \Lambda^0$	$-1/6(2F+1)(T+3V)$	$K^*(K^{**})$
X		$K^- p \to \eta_1 \Lambda^0$	$1/3(2F+1) S_T T$	K^{**}

(a) Note that F is defined such that F+D=1, and experimentally $F_{\Delta\lambda=0} \sim 1.25$, $F_{\Delta\lambda=1} \sim 0.25$.

(b) It is assumed that ϕ exchange is negligible, having zero coupling at the nucleon vertex.

The physical states η^0 and η' are dominantly SU(3) octet and singlet components:

$$\eta^0 = \eta_8 \cos\theta - \eta_1 \sin\theta$$

$$\eta' = \eta_8 \sin\theta + \eta_1 \cos\theta$$

where θ is typically in the range $-11 \gtrsim \theta \gtrsim -23°$, the limits of the quadratic and linear SU(3) mass formulae.[22] A compilation[23] of the $K^-p \to (\eta^0, \eta')\Lambda$ data is shown in Fig. 7. The data in the two channels are quite different in structure, and may suggest that absorption differs for vector and tensor exchange reactions.[24]

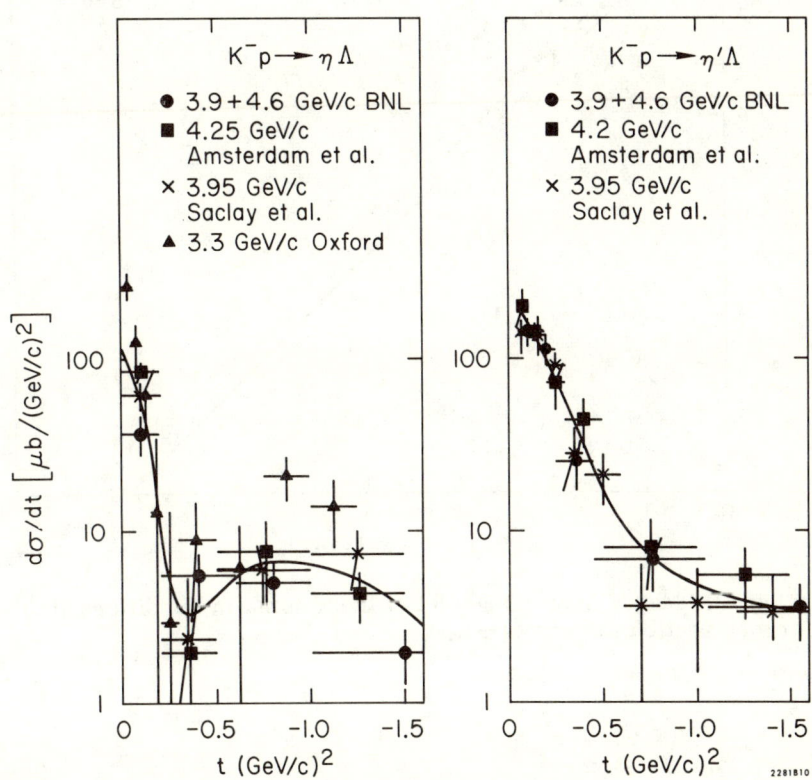

Fig. 7. Compilation of data in the reactions $K^-p \to \eta^0 \Lambda$ and $K^-p \to \eta'\Lambda$.

Amplitude zeros and minima in differential cross sections

An intriguing feature of many differential cross sections is the existence of minima at approximately fixed values of t (or u), independent of the reaction energy. This is shown in Figs. 8 and 9 where the locations of minima in the differential cross sections for $\pi^-p \to \pi^0 n$ and $\pi^-p \to \eta^0 n$ are recorded,[25] fixed t dips are observed at $-t \sim 0.55$ GeV^2 and $-t \sim 1.65$ GeV^2 respectively.

Similar fixed t dependences are found in the contributions of s channel resonances to the imaginary parts of the s channel helicity amplitudes. This is shown for πp scattering[26] in Fig. 10, where the locations of the first zero in the contribution of the dominant resonances to helicity flip and nonflip amplitudes are plotted.

196 J.A.J. Matthews

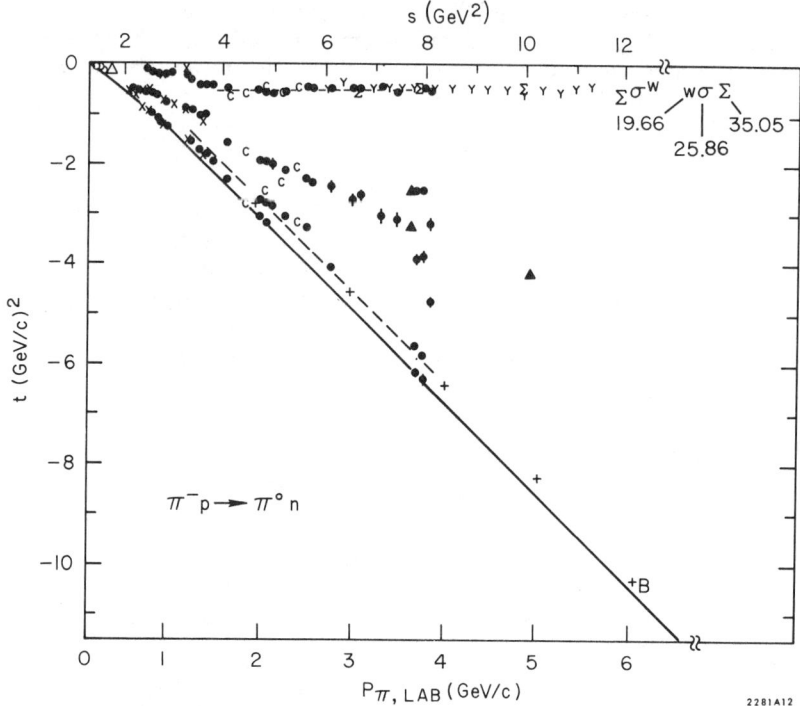

Fig. 8. Mandelstam plane plot of the minima in the differential cross section for $\pi^- p \to \pi^0 n$.

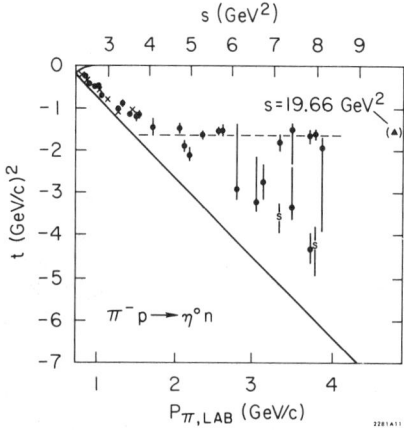

Fig. 9. Mandelstam plane plot of the minima in the differential cross section for $\pi^- p \to \eta^0 n$.

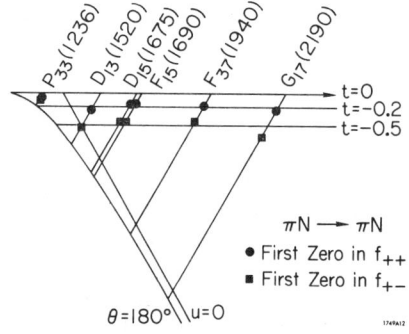

Fig. 10. Location of the first zero in the contribution of prominent resonances to the s-channel helicity amplitudes for πp scattering.

For K⁻p scattering a more complete separation[27] has been done yielding s channel helicity amplitudes with definite t channel isospin $I_t=0,1$. The results are shown in Fig. 11. For the $I_t=0$ helicity nonflip amplitude, and the $I_t=1$ helicity flip amplitude, the fixed t zero structures at $-t \sim 0.2$ GeV² and $-t \sim 0.5$ GeV² respectively, are again observed. The other two amplitudes are smaller in magnitude than the first two mentioned (cf. Fig. 6), thus the random structure of zeros in these results, Fig. 11b,c, may only reflect uncertainties in the analysis and in the resonance parameterizations used.

The data suggest therefore that: (a) zeros or pronounced minima occur at values of momenta transfer that change only slowly, if at all, with beam momentum; and (b) several of these features exist virtually from reaction threshold to the highest energies presently measured. We note that these results also carry over to $\pi\pi$ scattering where recent analyses[28] reveal similar fixed t zero structures in $\pi\pi$ amplitudes with well defined t channel isospin.

Polarization changes with energy
___

In channels with one t channel exchange, or with two exchanges thought to be exchange degenerate (EXD) polarization provides a possible means to observe different relative energy dependences of helicity flip and nonflip amplitudes (see Eq. (5)). For example, a large class of absorption models modify or "absorb" the nonflip amplitude to a much greater extent than the flip amplitude. If the absorption is then energy dependent, the nonflip amplitude will vary with energy more rapidly than the helicity flip amplitude, resulting in possible changes in the polarization.

To determine the energy trends in the data, polarization results for the reactions $\pi^- p \to \pi^0 n$,[29] $\bar{K}^0 p \to \pi^+ \Lambda^0$,[30] and $\pi^+ p \to K^+ \Sigma^+$,[31] are plotted in Fig. 12

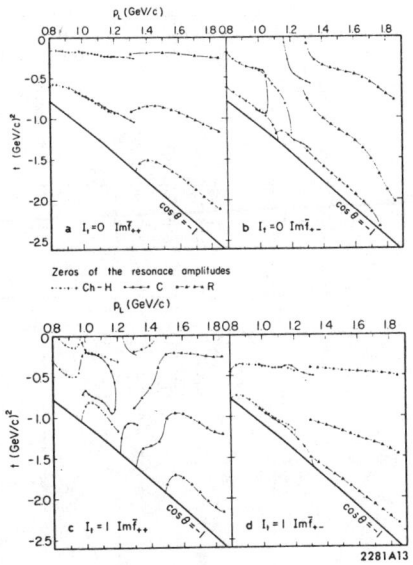

Fig. 11. Location of zeros in the imaginary parts of the s channel helicity amplitudes for K⁻p scattering.

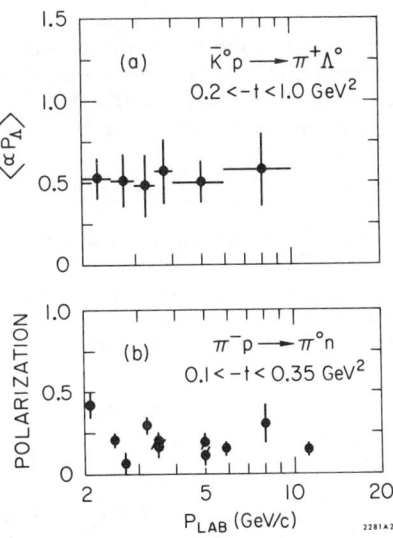

Fig. 12. Polarization data plotted as a function of beam momentum.

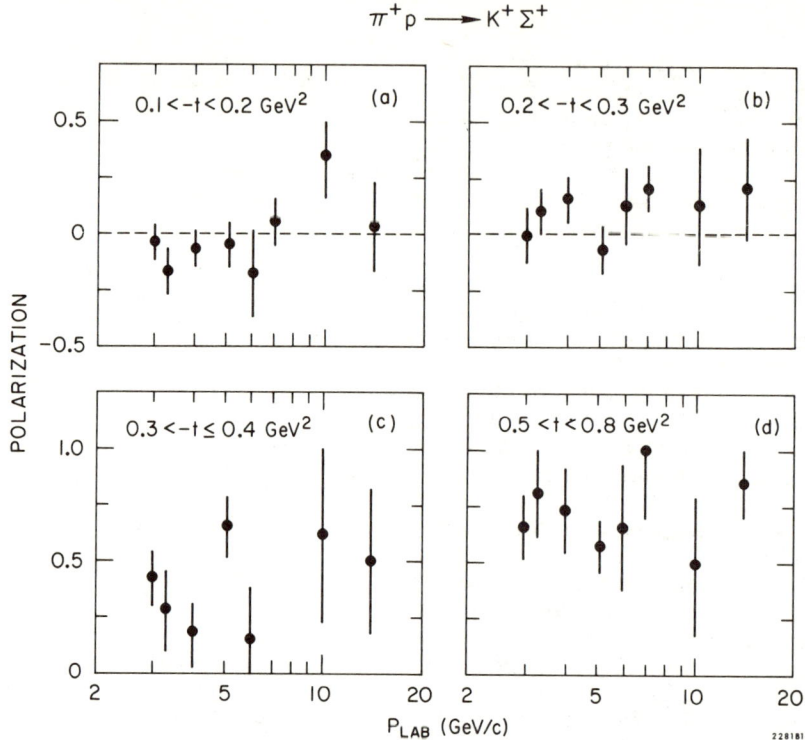

Fig. 13. Polarization data for $\pi^+ p \to K^+ \Sigma^+$ plotted as a function of momentum transfer and beam momentum.

and Fig. 13 for beam momenta $\gtrsim 2$ GeV/c. Momentum transfer intervals are chosen where the polarization is only slowly varying in t. The data are observed to be consistent with little or no energy dependence. This suggests therefore that helicity flip and nonflip amplitudes have similar energy dependences in the momentum interval ~ 2 GeV/c to ~ 14 GeV/c.

Phases of the amplitudes at t=0
─────────────────────────────

Although amplitude analyses typically require a prohibitive experimental effort, this is not the case at t=0 for many pseudoscalar-meson baryon scattering reactions. That is, the magnitude of the helicity nonflip amplitude is obtained directly from the forward differential cross section, and the imaginary part of the amplitude is provided by the optical theorem, for example:

$$\text{Im}\,(\pi^- p \to \pi^0 n) = -\frac{k}{4\sqrt{2}\pi}\left(\sigma^T_{\pi^- p} - \sigma^T_{\pi^+ p}\right)$$

and

$$\text{Im}\,(K^0_L p \to K^0_S p) = -\frac{k}{8\pi}\left(\sigma^T_{K^- n} - \sigma^T_{K^+ n}\right)$$

(6)

Recent results for the phase of the forward amplitude for $K_L^0 p \to K_S^0 p$,[32] are shown in Fig. 14. The data are consistent with having a constant phase, $\phi = -133.4 \pm 3.3°$, over the momentum interval 1.5 to 50 GeV/c, in remarkable agreement with the naive Regge model prediction, $\phi = -135°$ for $\alpha_\omega(0) = 0.5$. The curves on the figure result from using the optical theorem:

$$\phi = -\tan^{-1}\left\{\left(\frac{\left(\frac{d\sigma}{dt}\right)_0}{\left(\frac{d\sigma}{dt}\right)_{optical}}\right)^{-\frac{1}{2}} - 1\right\} \quad (7)$$

and parameterizing the data with the power law form:

$$\left(\frac{d\sigma}{dt}\right) = A\, p_{lab}^{-n}$$

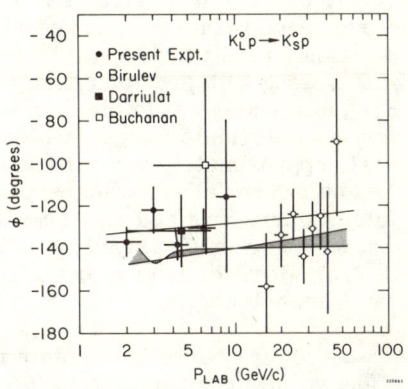

Fig. 14. Phase of the scattering amplitudes at t=0 for the process $K_L^0 p \to K_S^0 p$.

The solid and shaded curves compare the phases of the forward amplitudes for $K_L^0 p \to K_S^0 p$ and $\pi^- p \to \pi^0 n$,[33] respectively. The uncertainties in the curves are $\sim \pm 6°$ for $\pi^- p \to \pi^0 n$ and $\sim \pm 8°$ for $K_L^0 p \to K_S^0 p$.

If the phases of the forward amplitudes are in approximate agreement with the Regge phase, then equal forward cross sections should be observed for those processes related by s-u crossing and dominated by EXD t channel exchanges.[34] Near equality of the t=0 cross sections is in fact observed in the channels $\pi p \to K\Sigma$ and $\bar{K}p \to \pi\Sigma$,[35] see Fig. 15, as well as for Kp charge exchange[36] and $\pi p \to K\Lambda$, $\bar{K}p \to \pi\Lambda$ data.[35]

In summary it is observed that: (a) the forward phase for $K_L^0 p \to K_S^0 p$ is consistent with being energy independent; and (b) the phases of the amplitudes at t=0 are consistent with simple Regge predictions.

<u>Shrinkage</u>

Using the parameterizations

$$\frac{d\sigma}{dt} \propto e^{b(s)t} \quad (8)$$

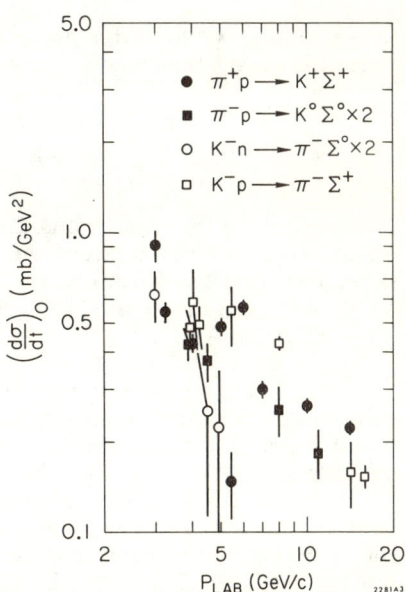

Fig. 15. Differential cross sections at t=0 for the reactions $\pi p \to K\Sigma$ and $\bar{K}p \to \pi\Sigma$.

or

$$\frac{d\sigma}{dt} \propto p_{lab}^{2\alpha(t)-2} \qquad (9)$$

the energy dependence of the scattering amplitudes can be studied as a function of momentum transfer. A recent Serpukhov result for $\pi^- p \to \pi^0 n$ is shown along with previous data[37] in Fig. 16. The curve in the figure represents the simple Regge prediction, $\alpha' = 1$, approximately normalized to the data below 20 GeV/c. The Serpukhov data is interesting and may suggest that shrinkage has stopped by ~ 20 GeV/c. However, the evidence is not yet overwhelming.

At lower energies we obtain the effective Regge trajectory, $\alpha_{eff}(t)$, shown in Fig. 17 for the SLAC $K_L^0 p \to K_S^0 p$ data,[32] and for the reaction $\pi^- p \to \pi^0 n$.[38] The solid curve in the figure gives the canonical ρ, ω trajectory $\alpha(t) = 0.5 + t$.

The $K_L^0 p \to K_S^0 p$ and $\pi^- p \to \pi^0 n$ data (Fig. 17) do show shrinkage, $\alpha = \alpha(t)$, however

$$\alpha_{K_L^0 p \to K_S^0 p} < \alpha_{\pi^- p \to \pi^0 n}$$

for $-t \lesssim 0.4$ GeV^2. Analogous differences are also found between the energy dependences of the reactions $\bar{K}p \to \pi\Lambda$ or $\pi p \to K\Lambda$ and the related Σ reactions, $\bar{K}p \to \pi\Sigma$ or $\pi p \to K\Sigma$.[24,30]

Simple Regge models would erroneously predict similar energy dependences for $K_L^0 p \to K_S^0 p$ and $\pi^- p \to \pi^0 n$ reactions, and for Λ and Σ reactions. By contrast, it was observed in the previous section that the phases of the forward amplitudes were in good agreement with simple Regge predictions. For example, the $K_L^0 p \to K_S^0 p$ data[32] yield $\alpha_{eff}(0) = 0.48 \pm 0.04$ as determined from the phase of the forward amplitude, but $\alpha_{eff}(0) = 0.30 \pm 0.03$

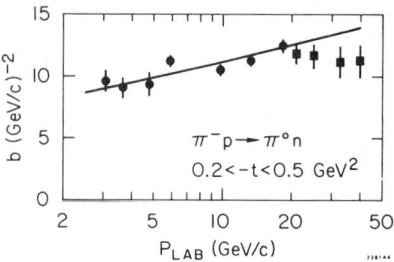

Fig. 16. Slope parameter for the reaction $\pi^- p \to \pi^0 n$.

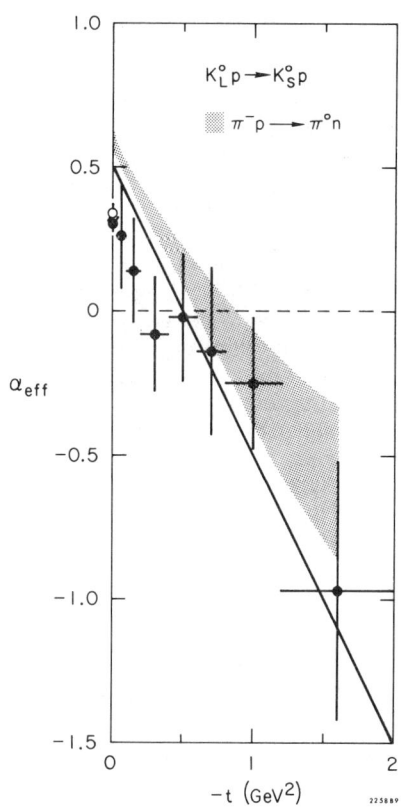

Fig. 17. Energy dependence of the process $K_L^0 p \to K_S^0 p$; $\pi^- p \to \pi^0 n$ data[38] is shown shaded in the figure.

from the energy dependence of the forward cross sections (see Fig. 17). The Regge model relationship between the phase and the energy dependence of the scattering amplitudes fails therefore at t=0; a similar conclusion for the helicity nonflip amplitude at momentum transfers -t > 0 can be inferred from the πp amplitude analyses at 6 GeV/c.[20,21]

Cross sections at t=0

In factorizable models (see for example Table II) the reactions $\pi^- p \to \pi^0 n$ and $K_L^0 p \to K_S^0 p$ are related by a single constant, assuming ρ and ω exchange amplitudes have similar energy dependences. This prediction disagrees with the data, as discussed in the last section; the magnitude of the discrepancy is observed by comparing the forward cross sections in Figs. 18 and 19.[32] An analogous comparison can be made using total cross sections differences;[39] where $\pi^- p \to \pi^0 n$ is replaced by

$$\Delta \sigma_{\pi^\pm p} = \sigma^T_{\pi^- p} - \sigma^T_{\pi^+ p}$$

and

$$K_L^0 p \to K_S^0 p \quad \text{by} \quad \Delta \sigma_{K^\pm p}$$

(see Eq. (6)). The results are shown in Fig. 20. Again the Kp data are observed to decrease with energy more rapidly than the πp data.

One conclusion is that ω and ρ exchanges, dominating $K_L^0 p \to K_S^0 p$ and $\pi^- p \to \pi^0 n$ respectively, are just intrinsically different. Alternatively, the comparison of forward $K_L^0 p \to K_S^0 p$ and Kp charge exchange cross sections[40] shown in Fig. 21 reveals that the magnitudes of the cross sections (pure coincidence?) as well as their momentum dependence are in excellent agreement. This result suggests that the t channel exchanges in these processes (ρ, ω, A_2) are consistent with EXD, and that direct channel effects cause the disagreement in the energy dependences

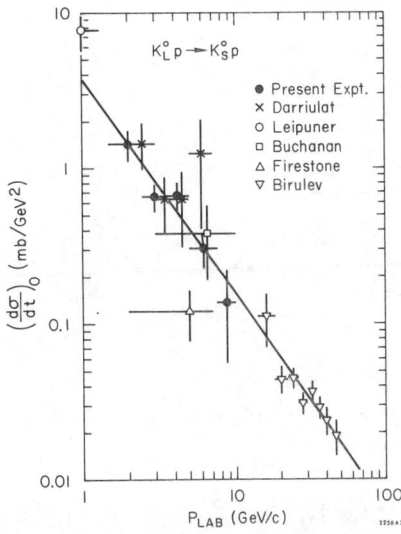

Fig. 18. Differential cross sections at t = 0 for the reaction $K_L^0 p \to K_S^0 p$.

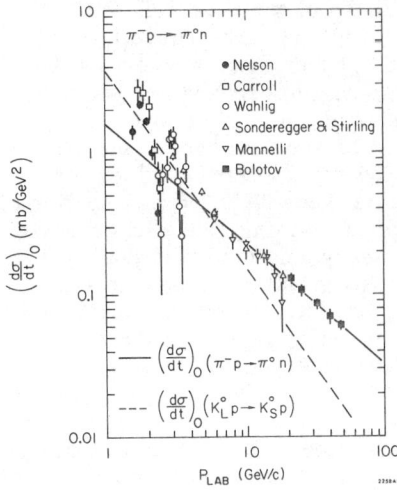

Fig. 19. Differential cross sections at t=0 for $\pi^- p \to \pi^0 n$; the solid curve is a power law fit to the data above 5 GeV/c, the dashed line represents the forward cross sections for $K_L^0 p \to K_S^0 p$.

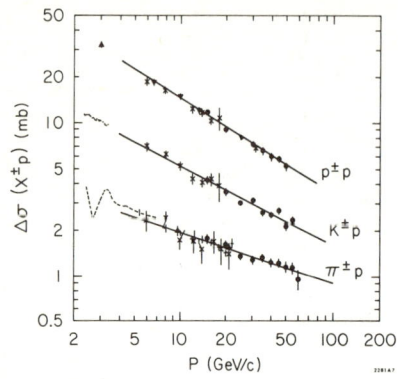

Fig. 20. Differences of total cross sections.

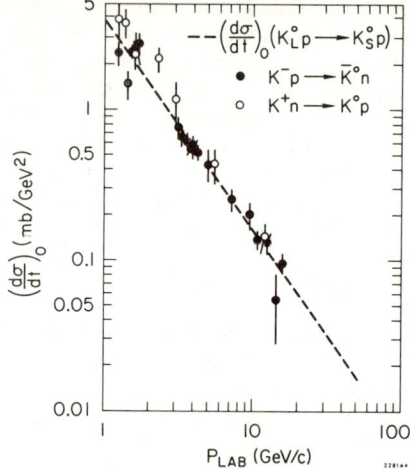

Fig. 21. Differential cross sections at t = 0 for the reactions $K^-p \to \bar{K}^0 n$ and $K^+ n \to K^0 p$; the dashed line represents the forward cross sections for $K_L^0 p \to K_S^0 p$.

of the $K_L^0 p \to K_S^0 p$ and $\pi^- p \to \pi^0 n$ forward cross sections.

To check this conjecture we note that the extrapolated $K^-p \to \bar{K}^0 n$ cross section, Fig. 21, provides an upper bound to the total cross section difference $K^-N = \sigma^T_{K^-p} - \sigma^T_{K^-n}$, shown as the solid curve on the "K^-N" data in Fig. 22. The curve falls below the Serpukhov data, suggesting that the Kp charge exchange cross section should infact lie above the $K_L^0 p \to K_S^0 p$ cross section at momenta $\gtrsim 20$ GeV/c. However, negative values for the Serpukhov "K^+N" cross section differences, Fig. 22, disagree with lower energy data (and with duality)[41] suggesting that small systematic effects in the total cross sections may be causing problems in these cross section differences.

Thus, if ρ and ω trajectories are assumed EXD, simple t channel factorization is in disagreement with the $K_L^0 p \to K_S^0 p$ and $\pi^- p \to \pi^0 n$ data. The comparison of $K_L^0 p \to K_S^0 p$ and Kp charge exchange data then suggest that factorization is infact broken by direct channel effects.[42]

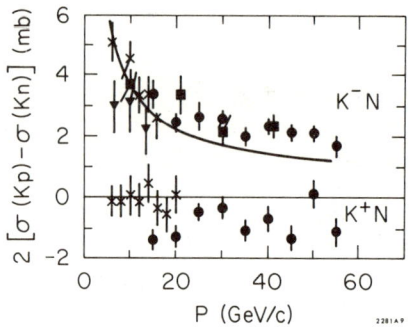

Fig. 22. Differences of total cross sections for the reactions $K^-p - K^-n$ and $K^+p - K^+n$.

IV. SUMMARY

Our approach in this talk was first to indicate that absorption or direct channel effects are important in scattering reactions, then to systematically investigate energy trends in the data. This latter study presented evidence for disagreements with t channel factorization, and we conjectured that this was additional evidence for direct channel effects.

Generally, the energy trends in the data are consistent with shrinkage of the forward differential cross sections, but indicate that many features are essentially energy independent:

(a) the positions of minima in the differential cross sections;

(b) the location of zeros in the imaginary parts of some s channel helicity amplitudes; and

(c) the polarization, and phase of the t=0 scattering amplitude for those reactions with a limited number of possible exchanges in the t channel.

These observations provide qualitative as well as quantitative constraints on the energy dependence of the actual scattering amplitudes.

The pseudoscalar-meson baryon scattering results suggest that similar features may also exist in the $\pi\pi(K\pi)$ scattering amplitudes. For example, although amplitude zeros have been used in selecting phase shift solutions, our present observations suggest that approximately fixed t zeros occur in those amplitudes isolating a known meson exchange in the t channel. Similarly the observed energy independence of the phase of the forward meson-baryon scattering amplitudes may provide a guide in selecting $\pi\pi(K\pi)$ phase shift solutions at relatively high mass. Finally, the different direct channel effects in πp and Kp data indicate that π extrapolations may have different characteristics in $\pi\pi$ and $K\pi$ analyses, and perhaps also in analyses at different $\pi\pi$ or $K\pi$ masses.

V. ACKNOWLEDGEMENTS

I would like to thank S. Barish, K.-W. Lai, W. Michael, R. K. Yamamoto and A. Yokosawa for assistance with their data, R. Diebold for a discussion of their $\pi\pi$ results, and G. Brandenburg, M. Davier, D. Leith, and J. Loos for helpful discussions. I also wish to thank D. Leith for his support and interest.

REFERENCES

1. A. M. Boyarski et al., Phys. Rev. Letters 20, 300 (1968).
2. M. B. Davis et al., Phys. Rev. Letters 29, 139 (1972).
3. J. G. Lee et al., Nucl. Phys. B52, 292 (1973).
4. F. Bulos et al., Phys. Rev. Letters 26, 1453 (1971).
5. F. Henyey, G. L. Kane, J. Pumplin and M. H. Ross, Phys. Rev. 182, 1579 (1969).
6. A. B. Wicklund et al., Phys. Rev. Letters 29, 1415 (1972).
7. M. Davier and H. Harari, Phys. Letters 35B, 239 (1971).
8. K. Gottfried and J. D. Jackson, Nuovo Cimento 34, 735 (1964).
9. G. L. Kane, Experimental Meson Spectroscopy, ed. C. Baltay and A. H. Rosenfeld (Columbia University Press, New York, 1970).
10. P. Baillon et al., Phys. Letters 35B, 453 (1971).

11. G. Gayer et al., "$\pi\pi$ phase shifts, amplitude analysis and vector dominance test in the ρ-region of $\pi^-p \to \pi^-\pi^+n$ at 17.2 GeV/c" CERN preprint (1972).
12. P. K. Williams, Phys. Rev. 181, 1963 (1969).
13. G. Fox, Argonne Workshop on Meson Spectroscopy (1971).
14. M. Ross, F. S. Henyey and G. L. Kane, Nucl. Phys. B23, 269 (1970); H. Harari, Ann. Phys. (N.Y.) 63, 432 (1971); J.A.J. Matthews, Proceedings of Canadian Institute of Particle Physics Summer School, ed. R. Henzi and B. Margolis (McGill University Press, Montreal, 1972).
15. P. Estabrooks and A. D. Martin, Phys. Letters 41B, 350 (1972).
16. J.A.J. Matthews et al., Nucl. Phys. B32, 366 (1971); Y. Williamson et al., Phys. Rev. Letters 29, 1353 (1972); W. Michael and G. Gidal, Phys. Rev. Letters 28, 1475 (1972); P. Estabrooks and A. D. Martin, Phys. Letters 42B, 229 (1972); R. Diebold, private communication.
17. S. Barish and W. Selove, private communication.
18. J. M. Scarr and K.-W. Lai, Phys. Rev. Letters 29, 310 (1972).
19. The presence of a confirmed dip in ρ_{00}^H $d\sigma/dt$ at several incident momenta would presumably rule out background as an explanation.
20. P. Johnson et al., Phys. Rev. Letters 30, 242 (1973).
21. G. Cozzika et al., Phys. Letters 40B, 281 (1972).
22. A. D. Martin and C. Michael, Phys. Letters 37B, 513 (1971); F. D. Gault, H. F. Jones and M. D. Scadron, Nucl. Phys. B51, 353 (1973); E. Fischback, M. M. Nieto, H. Primakoff and C. K. Scott, Phys. Rev. Letters 29, 1046 (1972).
23. C. Michael, 16th International Conference on High Energy Physics, CERN preprint REF. TH 1567-CERN (1972).
24. A. C. Irving, A. D. Martin and V. Barger, "Analysis of data for hypercharge exchange reactions," CERN preprint REF. TH 1585-CERN (1972).
25. R. K. Yamamoto et al., "Negative pion charge exchange and η^0 production from 1.3 to 3.9 GeV/c," MIT preprint (1972).
26. H. Harari, SLAC preprint SLAC-PUB-837 (1970).
27. M. Fukugita and T. Inami, Nucl. Phys. B44, 490 (1972).
28. M. Pennington, "$\pi\pi$ amplitudes, structure and zeros," International Conference on $\pi\pi$ Scattering and Associated Topics, Tallahassee (1973).
29. P. Bonamy et al., Nucl. Phys. B16, 335 (1970); D. D. Brobnis et al., Phys. Rev. Letters 20, 274 (1968); P. Bonamy et al., Amsterdam International Conference on Elementary Particles (1971); D. Hill et al., Phys. Rev. Letters 30, 239 (1973).
30. R. J. Yamartino, Ph.D. thesis, Stanford University (1973).
31. S. M. Pruss et al., Phys. Rev. Letters 23, 189 (1969); A. Bashian et al., Phys. Rev. D4, 2667 (1971).
32. G. W. Brandenburg et al., to be published.
33. G. Höhler and R. Strauss, "Tables of pion-nucleon forward amplitudes," Karlsruhe preprint (1971).
34. F. J. Gilman, Phys. Letters 29B, 673 (1969).
35. J. S. Loos and J.A.J. Matthews, Phys. Rev. D6, 2463 (1972).
36. D. Cline, J. Matos, and D. D. Reeder, Phys. Rev. Letters 23, 1318 (1969).
37. V. N. Bolotov et al., "A study of $\pi^-p \to \pi^0n$ charge-exchange in the momentum range 20 to 50 GeV/c," Serpukhov preprint (1972);

P. Sonderegger et al., Phys. Letters 20, 75 (1966);
A. V. Stirling et al., Phys. Rev. Letters 14, 763 (1965).

38. G. Höhler, J. Baacke, H. Schlaile and P. Sonderegger, Phys. Letters 20, 79 (1966).

39. S. P. Denison et al., "Differences of total cross sections for momenta up to 65 GeV/c," Serpukhov preprint (1972).

40. A. A. Hirata et al., Nucl. Phys. B30, 157 (1971);
M. Aguilar-Benitez, R. L. Eisner and J. B. Kinson, Phys. Rev. D4, 2583 (1971);
I. Butterworth et al., Phys. Rev. Letters 15, 734 (1965);
C. G. Wohl, Ph.D. thesis UCRL-16288 (1965);
Y. Goldschmidt-Clermont et al., Phys. Letters 27B, 602 (1968);
L. Moscoso et al., Phys. Letters 32B, 513 (1970);
P. Astbury et al., Phys. Letters 23, 396 (1966);
D. Cline, J. Penn and D. D. Reeder, Nucl. Phys. B22, 247 (1970);
A. Firestone et al., Phys. Rev. Letters 25, 958 (1970);
G. C. Mason and C. G. Wohl, "Reactions $K^-p \to \bar{K}^0 n$, $K^-p \to \Lambda\pi^0$ and $K^-p \to \Lambda\eta$ at 3.13, 3.30, and 3.59 GeV/c," Oxford University preprint (1973);
R. Blokzijl et al., Nucl. Phys. B51, 535 (1973);
R. J. Miller et al., "Two body final states in K^-p interactions at 14.3 GeV/c," Rutherford preprint RPP/H/102 (1972);
E. H. Willen et al., "High energy differential cross sections for K^0 production," BNL preprint BNL 16681 (1972).

41. H. Harari, Phys. Rev. Letters 20, 1395 (1968);
P.G.O. Freund, ibid 20, 235 (1968).

42. The explanation for the difference in the energy trends of the forward cross sections for $K^0_L p \to K^0_S p$ and $\pi^-p \to \pi^0 n$ is not clear, however. Typically absorption models correct simple Regge predictions in proportion to the total cross sections of the particles involved. However, $\sigma^T_{Kp} < \sigma^T_{\pi p} < \sigma^T_{pp}$ in contrast to the energy dependences of the data, see Fig. 20, where $\Delta\sigma_{pp}$ and $\Delta\sigma_{Kp}$ data are quite similar, and substantially different from the energy dependence of the $\Delta\sigma_{\pi p}$ data. Similar problems arise in discussing the difference in energy dependence of the channels $\bar{K}p \to \pi\Lambda$ and $\bar{K}p \to \pi\Sigma$.

ππ PHASE-SHIFT ANALYSIS FROM 600 to 1900 MeV

B. Hyams, C. Jones and P. Weilhammer

CERN, Geneva, Switzerland

W. Blum, H. Dietl, G. Grayer, W. Koch, E. Lorenz,
G. Lütjens, W. Männer, J. Meissburger, W. Ochs,
U. Stierlin and F. Wagner

Max-Planck-Institut für Physik und Astrophysik, München, Germany

ABSTRACT

We have performed an energy-dependent and an energy-independent analysis of elastic ππ scattering, using data for the reaction $\pi^-p \to \pi^-\pi^+n$ at 17.2 GeV/c. The ππ energy covers the range from 600 MeV to 1900 MeV. Apart from the well-known resonances ρ, f and g, we find a strong S-wave in the ρ and also in the f-meson region. A P-wave resonance occurs in both analyses at ~ 1600 MeV with a total width of 180 MeV and an elasticity of 0.25, which can be identified with the ρ' meson in its ππ decay mode. The zeros in the complex cos θ plane of the scattering amplitude are studied as a function of the energy. The position of the real parts of the zeros are in qualitative agreement with the prediction of the Veneziano model. Furthermore, we show that an analogous relation holds also for the imaginary parts of the zero trajectories.

1. INTRODUCTION

The main source of information about ππ elastic scattering comes from one pion exchange (OPE) dominated reactions, for example the reaction

$$\pi^-p \to \pi^-\pi^+n \qquad (1)$$

at small momentum transfer t of the proton to the neutron. The simple OPE model needs to be corrected in order to explain the data in detail. It has been shown recently with high statistics[1,2] that ρ production in reaction (1) follows closely an OPE model with absorptive corrections as proposed by P.K. Williams[3]. Therefore reaction (1) is a good candidate for accurate determination of $\pi\pi$ amplitudes.

So far such calculations[4] of $\pi\pi$ interactions have been carried out mainly in the energy region of the ρ meson and the $K\bar{K}$ threshold in order to establish a unique S-wave and to understand the cusp phenomenon related to the strong coupling of $K\bar{K}$ to $\pi\pi$. In this paper we present phase-shift results for a $\pi\pi$ energy range from 600-1900 MeV, which includes also the second (f) and third (g) resonance in $\pi\pi$. The data for reaction (1) come from our spark chamber experiment at 17.2 GeV/c. Some details about the data are given in Section 2.

Instead of performing an extrapolation in t to the π pole, we calculate the $\pi\pi$ amplitudes from the $\pi\pi$ density matrix elements integrated over a small t range near t = 0. Under certain assumptions these averaged density matrix elements can be obtained from the data. This procedure is described in Section 3.

In order to determine the $\pi\pi$ amplitudes as functions of the $\pi\pi$ energy in an energy-dependent analysis we use two-dimensional K matrices for the $\pi\pi$ partial-wave amplitudes. The details of this parametrization are given in Section 4.

To check the stability of our solution we determined also the $\pi\pi$ phases in each energy bin independently. The results of both analyses are discussed in Section 5.

Section 6 is devoted to a study of the zeros of the $\pi\pi$ scattering amplitudes.

2. THE DATA

Each event of reaction (1) is characterized by the $\pi\pi$ mass $m_{\pi\pi}$, the momentum transfer t, and two angles of the π^- direction in the $\pi\pi$ rest frame. After the necessary corrections for the limited acceptance of our apparatus, for each bin in $m_{\pi\pi}$ and t, there exist a certain number of spherical harmonic moments $\langle Y_L^M \rangle$ as discussed elsewhere[5]. Some plots of moments have already been presented[6].

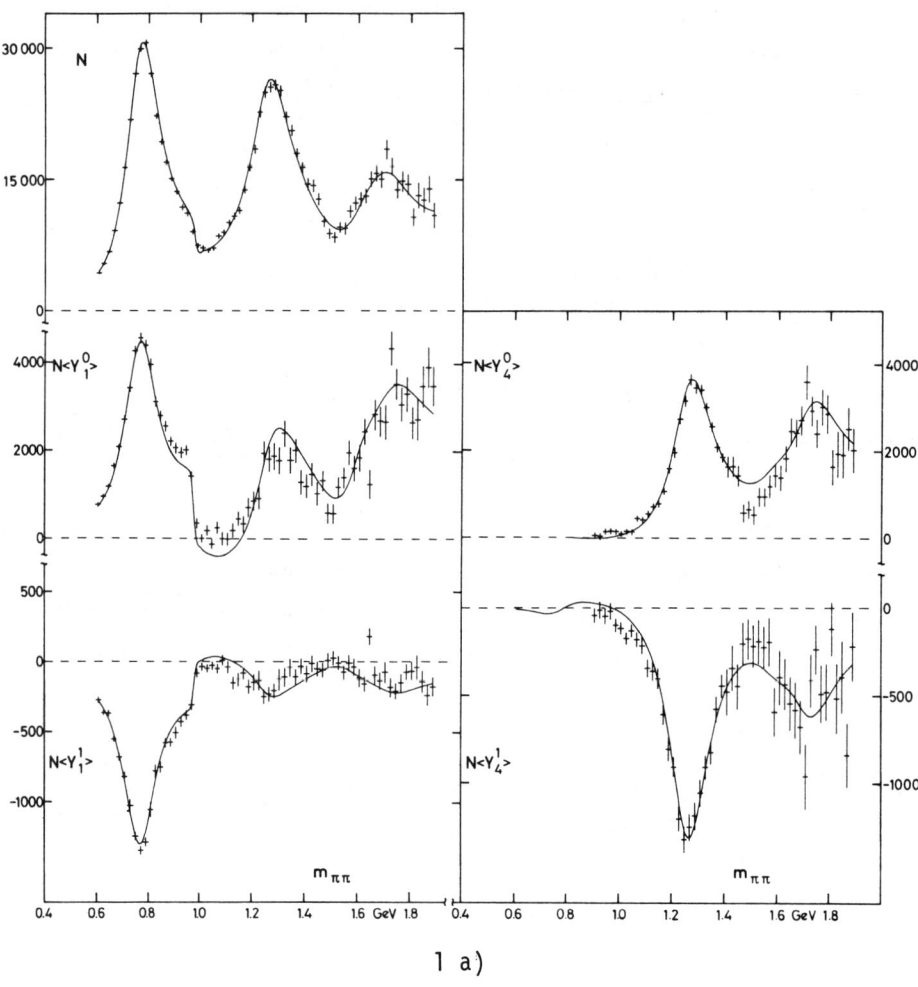

Fig. 1 The unnormalized moments of the $\pi\pi$ decay angular distributions as measured in the $n\pi^+\pi^-$ final state at 17.2 GeV/c (Gottfried-Jackson system; $0.01 \leq |t| \leq 0.15$ GeV^2). The curves represent the result of the energy-dependent fit over the mass region between 600 and 1900 MeV.

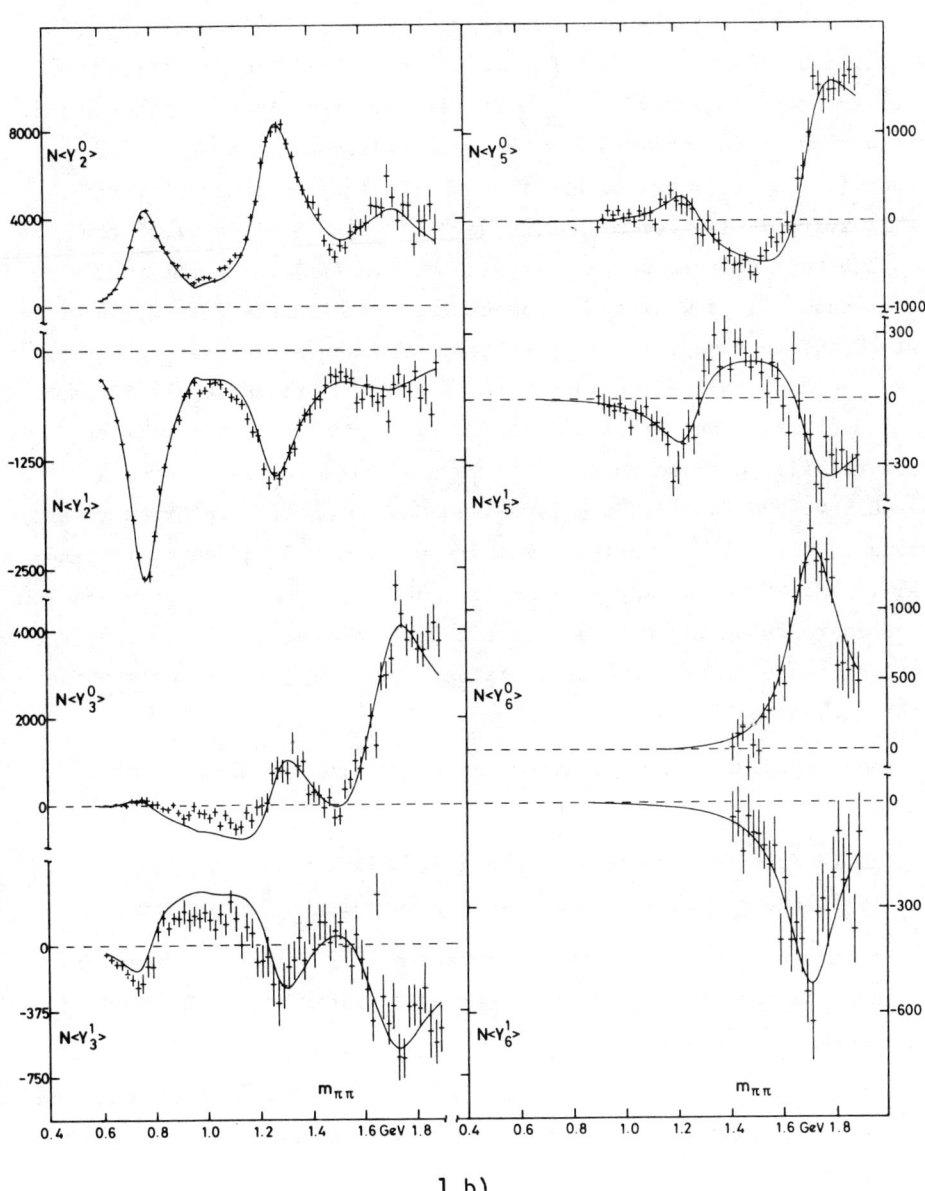

1 b)

For the present analysis we use moments integrated over the t interval $0.01 \leq |t| \leq 0.15$ (GeV/c)^2. The lower limit has been chosen in order to exclude t_{min} effects, the upper to minimize non-OPE contributions. The moments have been calculated in the Gottfried-Jackson (GJ) system, as one observes the following simple properties in this system: (i) For $m_{\pi\pi} \geq 600$ MeV and $|t| \leq 0.15$ (GeV/c)^2 the moments with $M \geq 2$ are compatible with zero. (ii) The normalized moments show only a weak t dependence (see for example Ref. 2) in contrast to the s-channel helicity system. For our phase-shift analysis we use the unnormalized moments $N\langle Y_L^0 \rangle$ and $N\langle Y_L^1 \rangle$ between $m_{\pi\pi} = 600$ and $m_{\pi\pi} = 1900$ MeV. They are shown in Fig. 1. Details of the fitting procedure will be given in Ref. 5. In the three energy intervals 600-900 MeV, 900-1400 MeV and 1400-1900 MeV (hereafter referred to as regions I, II, III) we included moments up to L = 3, 5 and 6, respectively. The higher moments have been found negligible. In the g-meson region there is an indication of small L = 7 moments. However, as the determination of these moments is rather poor, we did not include them in the present analysis.

Some immediate conclusions can be drawn from the moments of Fig. 1:

i) The leading resonances ρ, f and g are clearly visible in $d\sigma/dm_{\pi\pi}$ and in the L = 2, 4, 6 moments corresponding to their spin.

ii) The moments with L = 3 in the ρ region (L = 5 in the f region) show the presence of small D- (F-waves) interfering with the leading resonance.

iii) The M = 1 moments forbidden in OPE show the same structure as the M = 0 moments. Therefore it is desirable to include both in the analysis.

iv) Whereas the gross features of the mass dependence in the M = 0 moments account qualitatively for the interfering ππ amplitudes, a quantitative interpretation may lead to difficulties. For example, taking naïvely the M = 0 moments as ππ "on shell" moments one concludes from the L = 0 and L = 2 moments at 790 MeV (see Fig. 1) for the ratio R of the S-wave to P-wave cross-section:

R = 0.752. For a pure imaginary elastic P-wave and a negligible isospin 2, S-wave unitarity demands R ≤ 0.148. This "experimental" S-wave cross-section would exceed the unitarity limit by a factor of 5.

3. METHOD FOR THE PHASE-SHIFT ANALYSIS

In Sub-section 3.1 we will describe an analysis under assumptions proposed by one of us[7]. An important feature of this method is that the assumption allows certain experimental tests independent of the ππ phases. This will be discussed in Sub-section 3.2.

3.1 *Amplitude analysis and ππ on-shell amplitudes*

A complete determination of the ππ density matrix $\rho_{mm'}^{\ell\ell'}$, is impossible without information from polarized target experiments. (ℓ denotes the spin of the ππ system, m its z component in the GJ system). The basic idea of the method is to determine the density matrix elements $\rho_{00}^{\ell\ell'}$ under plausible assumptions and to calculate the ππ amplitudes from the $\rho_{00}^{\ell\ell'}$ elements.

In the GJ frame the OPE model contributes only to $\rho_{00}^{\ell\ell'}$. Additional mechanisms, such as absorption and/or A_2 exchange, generate $\rho_{mm'}^{\ell\ell'}$, with m,m' ≠ 0 and may modify the $\rho_{00}^{\ell\ell'}$ from OPE. A_2 decouples from $\rho_{00}^{\ell\ell'}$, and the absorptive corrections to $\rho_{00}^{\ell\ell'}$ as well as a possible A_1 exchange can be assumed to be negligible compared to the dominant OPE contribution.

The two assumptions to calculate $\rho_{00}^{\ell\ell'}$ from the data are the following:

(A) Helicity amplitudes with nucleon flip in the S-channel helicity system dominate at our high value of the beam momentum.

(B) Helicity amplitudes for production of a ππ system with spin ℓ are in phase for each natural and unnatural exchange amplitude separately. The relative phase between natural and unnatural exchange amplitudes is independent of ℓ.

These assumptions are motivated mainly by the success of the OPE model with absorptive corrections of Ref. 3 to explain ρ production at small t. This model would lead to complete phase coherence of the amplitudes with the same ℓ. The weaker assumption (B) allows also for an A_2 Regge-pole contribution with a production phase independent of ℓ. Assumption (B) is similar to the factorization assumption, which was proposed first by Schlein[8], to determine production amplitudes and ππ phases. His method, however, needs further assumption about the mass-dependence in order to get unique results. Assumptions (A) and (B) have been discussed also by Frogatt and Morgan[9] and applied to an amplitude analysis of ρ production. Recently Estabrooks and Martin[10] reported an amplitude analysis and presented a solution with an additional phase between the m = 0 and m = 1 amplitudes. This possibility will be discussed in Sub-section 3.2.

In order to get expressions for the density matrix elements in the GJ system in terms of amplitudes we use the fact that all moments $\langle Y_L^M \rangle$ with M ≥ 2 are zero for $|t| \leq 0.15$ $(GeV/c)^2$. Together with assumptions (A) and (B) a straightforward calculation shows that all amplitudes with helicity m ≥ 2 have to vanish and the moduli of the m = 1 amplitudes g_+^ℓ with natural and g_-^ℓ with unnatural exchange have to be equal (in the following we omit the nucleon indices)

$$|g_+^\ell| = |g_-^\ell|. \tag{2}$$

Since both exchanges add incoherently in the density matrix, we can eliminate g_+^ℓ with Eq. (2) and get the following non-zero density matrix elements in terms of the amplitudes:

$$\begin{aligned}
\rho_{00}^{\ell\ell'} &= g_0^\ell g_0^{\ell'*} \\
\rho_{01}^{\ell\ell'} &= g_0^\ell g_-^{\ell'*} \\
\rho_{11}^{\ell\ell'} &= 2 g_-^\ell g_-^{\ell'*}.
\end{aligned} \tag{3}$$

g_0^ℓ denotes the helicity m = 0 amplitude. The density matrix ρ of Eq. (3) is normalized to the cross-section, i.e.

$$\frac{d^2\sigma}{dt\,dm_{\pi\pi}} = \sum_{\ell} \left(\rho_{00}^{\ell\ell} + 2\rho_{11}^{\ell\ell} \right) . \tag{4}$$

From (B) it follows that r_ℓ in the relation

$$g_0^\ell(t,m_{\pi\pi}) = r_\ell(t,m_{\pi\pi}) \, g_-^\ell(t,m_{\pi\pi}) \tag{5}$$

has to be real. For $\ell \leq 3$ in each t and $m_{\pi\pi}$ bin there are 13 measured moments $N\langle Y_L^M \rangle$, from which we have to determine the seven independent real parameters in g_0^ℓ and the three real ratios r_ℓ. Therefore one has a three-constraint fit, which allows a test of our assumptions (A) and (B).

If one extrapolates g_0^ℓ to the π pole, the $\pi\pi$ on-shell-amplitudes can be obtained from the residue. If only OPE contributes to g_0^ℓ, we can determine this residue also from the average $\bar{g}_0^\ell$ in the t interval $0.1 \leq |t| \leq 0.15$ (GeV/c)^2. In this t range the GJ moments are approximately independent of t. Therefore, we use the parametrization (3) also for a calculation of the averaged density matrix elements in terms of average values $\bar{g}_{0,-}^\ell$ from the integrated moments. This procedure can be also justified in a Williams-type model as shown in the Appendix.

The $m_{\pi\pi}$ dependence of $\bar{g}_0^\ell$ is taken from the OPE model:

$$\bar{g}_0^\ell = c_N \, m_{\pi\pi} \sqrt{\frac{2\ell+1}{q}} \, T_\ell(m_{\pi\pi}) . \tag{6}$$

T_ℓ denotes the $\pi^+\pi^-$ partial wave amplitude (normalized to $|T_\ell - i/2| \leq 1/2$) and q the $\pi\pi$ c.m. momentum. Assumption (B) holds for the average values

$$\bar{g}_0^\ell = \bar{r}_\ell(m_{\pi\pi}) \, \bar{g}_-^\ell(m_{\pi\pi}) . \tag{7}$$

The normalization constant c_N, which is independent of $m_{\pi\pi}$ and ℓ, is adjusted such that $|T_1| = 1$ holds in the ρ meson peak. For our later energy-dependent fit we need $\bar{r}_\ell$ in Eq. (7) as a function of $m_{\pi\pi}$. The nearly mirror-symmetric pattern of $N\langle Y_L^0 \rangle$ and $N\langle Y_L^1 \rangle$ suggests a smooth

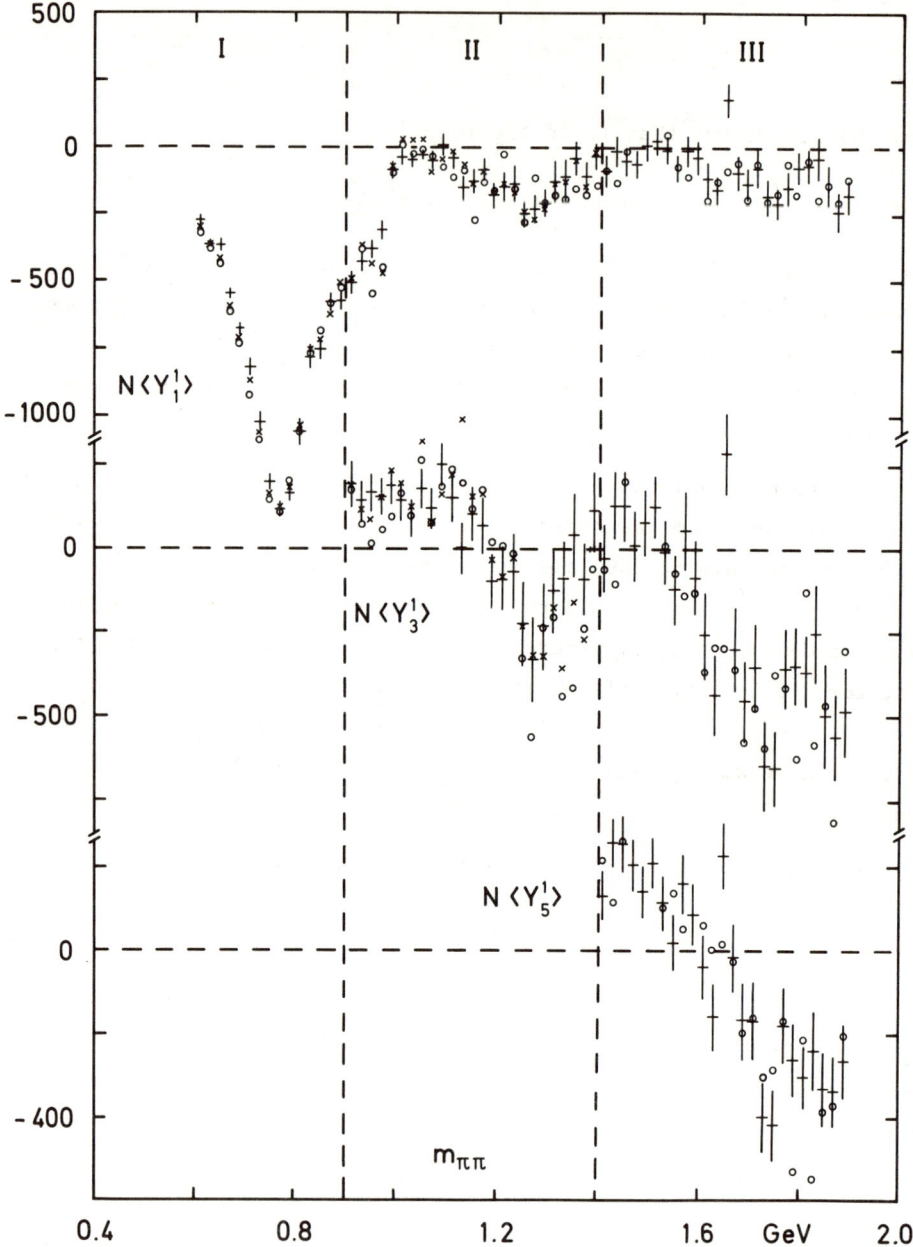

Fig. 2 Comparison between the moments $N\langle Y^1_{2L+1}\rangle$ as measured and as predicted from the assumptions of absent nucleon non-flip amplitudes and phase coherence of the amplitudes with the same ππ spin. Calculation with dominant waves only: O. Calculation including a correction from the next higher partial wave: ×.

215

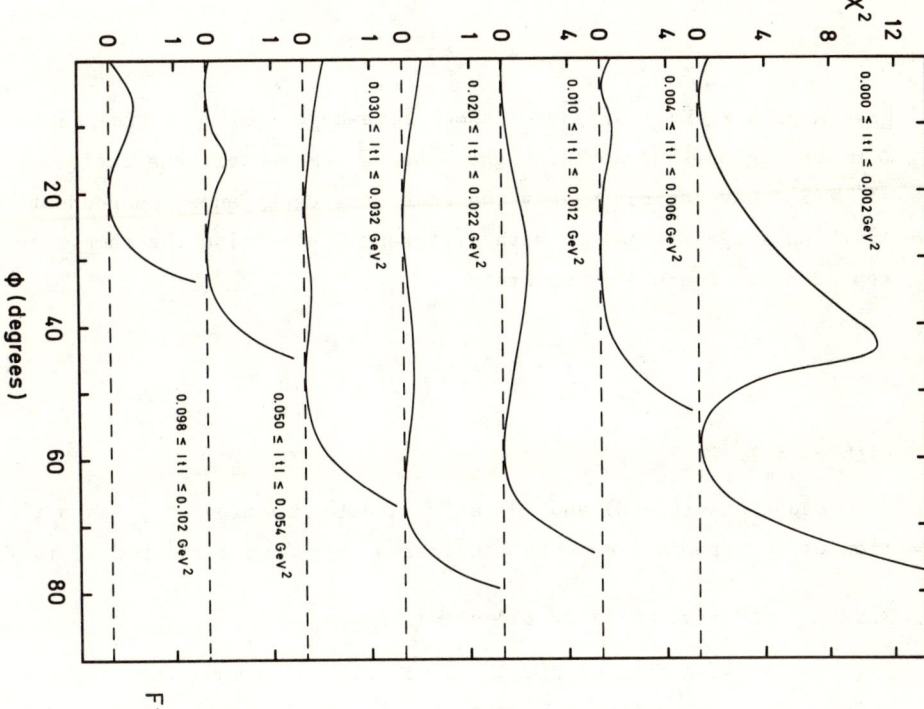

Fig. 3 χ^2 profiles for the phase ϕ between the S-channel ρ-production amplitudes (0.71 ≤ $m_{\pi\pi}$ ≤ 0.83) g_0^1 and g_-^1 for some t-intervals. These plots show that from the $\pi^+\pi^-$ data alone the value of ϕ cannot be determined within a rather wide range.

Fig. 4 χ^2 distributions for the energy-independent fits in the three mass regions. χ^2_{exp} measures the deviation of the fit from the experimental moments, χ^2_{ED} the deviation of the amplitudes from the energy-dependent fit.

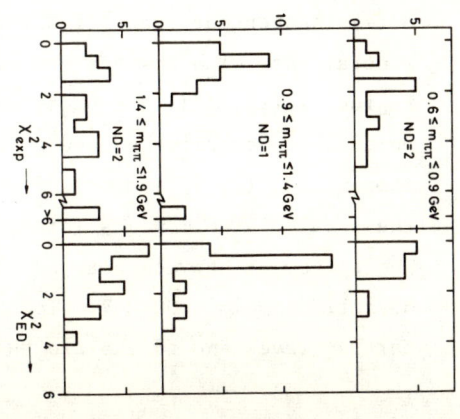

$m_{\pi\pi}$ dependence of $\bar{r}_\ell$. The Williams model would predict

$$\bar{r}_\ell \sim \frac{m_{\pi\pi}}{\sqrt{\ell(\ell+1)}}$$

[see Appendix, Eq. (A.7)]. As shown elsewhere[11] this is true up to the ρ meson region, but not at higher masses. Moreover, the ratios $\langle Y_L^1 \rangle / \langle Y_L^0 \rangle$ have approximately the same mass dependence independent of L, which suggests that $\bar{r}_\ell$ with different ℓ have also the same mass dependence. Therefore, we write

$$\bar{r}_\ell = c_\ell \sum_{\nu=0}^{3} a_\nu m_{\pi\pi}^\nu \qquad (8)$$

with $c_1 = 1$.

Equations (6)-(8) and (3) allow a determination of T_ℓ as a function of $m_{\pi\pi}$ from the moments integrated over $0.01 \leq |t| \leq 0.15$ (GeV/c)^2.

3.2 Experimental tests of the assumptions

In this reaction we discuss some tests of assumptions (A) and (B) which can be made without knowing the on-shell amplitudes T_ℓ. If assumption (A) holds, the rank of $\rho_{mm'}^{\ell\ell'}$ cannot exceed 2. This has been verified experimentally for the special case of ρ production[12]. The most important test is the fact that the assumptions (A) and (B) lead to an overconstrained system of equations for the unknown amplitudes $\bar{g}_0^\ell$ and $\bar{g}_-^\ell$. For $\ell_{max} = 3$ we have 13 equations for 10 unknowns in $\bar{g}_{0,-}^\ell$. For a test in region I we consider only S- and P-waves, in region II we include also D-waves. In region I (II) the same arguments lead to 3 (7) independent quantities for 4 (9) observables. In region III the general case with F-waves has to be considered. Therefore in the regions I (II, III), respectively, we can use 3 (7, 10) of the moments to determine $\bar{g}_0^\ell$ and $\bar{g}_-^\ell$ up to a common phase and predict the remaining 1 (2, 3) moments of the form $N\langle Y_{2L+1}^1 \rangle$. The comparison of these calculated moments (circles) with the data is shown in Fig. 2. The agreement between both is quite good, except for some systematic deviations below 800 MeV. To check stability against including high partial waves and to explain the non-zero L = 3 (5) moments in region

I (II) we added to the moments the terms linear in the D (F)-wave. These additional terms do not change the number of constraints. The moments predicted by this calculation are indicated by crosses in Fig. 2. The effect of these high partial waves is small, but they improve the agreement below the ρ meson. We conclude from Fig. 2 that the assumptions (A) and (B) are compatible with the data. In Ref. 10 an amplitude analysis for ρ production has been performed and an additional production phase between the amplitudes g_0^ℓ and g_-^ℓ in the s-channel helicity frame has been found. Figure 3 shows the χ^2 profile as a function of this phase ϕ for various t bins in the ρ region. There are two exact solutions (zeros of χ^2), one solution ϕ_1 close to zero according to assumption (B) and a second solution ϕ_2 corresponding to the solution of Ref. 10. In almost all t bins any value between 0 and ϕ_2 is statistically acceptable ($\chi^2 \approx 1$ for a one-constraint fit). Therefore, without imposing further assumptions, the data do not prefer any specific solution between $\phi = 0$ and an upper limit close to ϕ_2. In our method we put $\phi = 0$. The particular value of ϕ influences mainly the magnitude of $\bar{g}_0^0$. This provides a further check since the value of $\bar{g}_0^0$ is restricted by elastic unitarity of T_0, which will be discussed below. It has recently been pointed out[13] that the S-wave in the solution with large ϕ is inconsistent with the broad $\pi^0\pi^0$ mass distribution as observed in $\pi^-p \to \pi^0\pi^0 n$.

4. $\pi\pi$ PHASE-SHIFT ANALYSIS

In order to calculate the $\pi^+\pi^-$ partial-wave amplitudes T_ℓ from the moments via Eq. (3) we apply two methods. First we describe the structure of the moments by parametrizing T_ℓ with functions of $m_{\pi\pi}$ as simply as possible (energy-dependent analysis, "EDA"). In a second step we perform an analysis in each mass bin separately (energy-independent analysis, "EIA").

The T_ℓ are expressed by the phase δ_ℓ^I and elasticities η_ℓ^I for definite isospin I as

$$T_\ell = \begin{cases} \frac{2}{3} T_\ell^0 + \frac{1}{3} T_\ell^2 & \ell \text{ even} \\ T_\ell^1 & \ell \text{ odd} \end{cases} \qquad (9)$$

and

$$T_\ell^I = \frac{1}{2i}\left(\eta_\ell^I e^{2i\delta_\ell^I} - 1\right). \tag{10}$$

In the 1 GeV region the isoscalar S-wave is strongly affected by the opening of the $K\bar{K}$ channel as has been discussed previously in several publications[6,14,15]. To build in the opening of this channel we parametrize our partial-wave amplitude T_ℓ^I by a two-dimensional K matrix[16]. We take this two-channel description also at higher $m_{\pi\pi}$ values. There the second channel will simulate all the other possible inelastic reactions like $\pi\pi \to 4\pi$, K^*K, ... We prefer the K matrix to the $M = K^{-1}$ matrix as the treatment of several resonances in the same partial wave is simpler. Also the usual multi-channel Breit-Wigner formula for T_ℓ^I corresponds to a real pole in the K matrix, but cannot be derived from an M matrix. The relation between the three independent $K^{\ell,I}$ matrix elements and the elastic $\pi\pi$ amplitudes is given by

$$T^I = \frac{K_{\pi\pi}^{\ell,I} q^{2\ell+1} - i \det(K^{\ell,I})(qk)^{2\ell+1}}{1 - \det(K^{\ell,I})(qk)^{2\ell+1} - i(K_{\pi\pi}^{\ell,I} q^{2\ell+1} + K_{K\bar{K}}^{\ell,I} k^{2\ell+1})}, \tag{11}$$

where q (k) denotes the $\pi\pi$ ($K\bar{K}$) centre-of-mass momentum.

We parametrize the $K^{\ell,I}$ matrix for $\ell \leq 2$ by the sum of a real pole term and constant background. After some attempts it turned out that one would not get a satisfactory fit without adding a second pole term in the K matrix for S- and P-waves. For the F-wave (g meson) we use a Breit-Wigner formula with a mass-dependent width. This leads to the following partial-wave amplitudes in I = 0, 1:

$$S_0: \quad K_{ij}^{0,0} = \frac{\alpha_i^{(0)}\alpha_j^{(0)}}{s_1^0 - s} + \frac{\beta_i^{(0)}\beta_j^{(0)}}{s_2^0 - s} + \gamma_{ij}^{(0)} \tag{12a}$$

$$P_1: \quad K_{ij}^{1,1} = \rho \text{ meson} + \frac{\beta_i^{(1)}\beta_j^{(1)}}{s_2^1 - s} + \gamma_{ij}^{(1)} \tag{12b}$$

D_0 : $K_{ij}^{2,0} = f \text{ meson} + \gamma_{ij}^{(2)}$ (12c)

F_1 : $T_3^1 = \dfrac{x_g m_g \Gamma}{s_g - s - im_g \Gamma}$, $\Gamma = \Gamma_g \left(\dfrac{q}{q_0}\right)^7 \dfrac{D_3(q_0 R_g)}{D_3(qR_g)}$. (12d)

In order to obtain Breit Wigner formulae with the kinematical form factors D_ℓ*) for the ρ and f meson, we chose

$K_{\pi\pi}^{1,1}(\rho) = \dfrac{m_\rho \Gamma_\rho}{s_\rho - s} \dfrac{1}{q_0^3} \dfrac{D_1(q_0 R_\rho)}{D_1(q_0 R_\rho)}$, $K_{\pi K}^{1,1}(\rho) = K_{KK}^{1,1}(\rho) = 0$ (13a)

$K_{ij}^{2,0}(f) = \dfrac{\gamma_i \gamma_j}{s_f - s}$, $\gamma_\pi^2 = \dfrac{m_f x_f \Gamma_f}{q_0^3} \dfrac{D_2(q_0 R_f)}{D_2(q R_f)}$, $\gamma_K^2 = \dfrac{m_f \Gamma_f}{k_0^5}$ (13b)

(m, Γ, x denote mass, width and elasticity, q_0, k_0 the momenta at the resonance, R the interaction radius). The ρ meson in Eqs. (13a) has been assumed to be purely elastic. Note that for broad resonances the position of the pole in K need not agree with Re g_0 of the pole s_0 in T.

For the I = 2 S-wave we use a scattering length formula[4]:

$$T_0^2 = \dfrac{a_0 q}{1 - ia_0 q} , \quad a_0 = -0.1 \, m_\pi^{-1} .$$ (14)

This leads to phases of $-15°$ ($-24°$, $-30°$) at the ρ (f,g) mass. The I = 2 D-wave and higher waves were neglected. So far these waves are believed to be small[17]. With the $\pi^+\pi^-$ amplitudes T_ℓ specified by Eqs. (9)-(14) we calculate from Eqs. (3) and (6) the helicity zero density matrix elements. For the remaining density matrix elements we use the energy-dependent parametrization of Eqs. (7) and (8). If we fit the moments with this parametrization we have 36 parameters, i.e. 9 (9, 7, 4) for the S (P, D, F)-waves, the over-all normalization variable c_N and the six parameters of Eq. (8) to determine the non-OPE effects. The χ^2 of this fit was 1865 for 705 data points in Fig. 1.

---

*) See also Ref. 16. We used $D_1(x) = 1 + x^2$, $D_2(x) = 9 + 3x^2 + x^4$, $D_3(x) = 225 + 45x^2 + 6x^4 + x^6$.

The parameters at the minimum are listed in Table 1. The curves in Fig. 1 show how the fit describes the mass dependence of the moments. The main structure of the moments is well represented by the fit even though some systematic deviations are present, especially in the $L = 2$ moments around 1.1 GeV and in the $L = 3$ moments near 1 GeV. Also the sharp drop in the even moments at 1.46 GeV is not reproduced by our amplitudes. The simultaneous description of the $M = 0$ and $M = 1$ moments works well. If there are deviations of the fit from the data they occur usually in both distributions in corresponding directions. This supports the simple $m_{\pi\pi}$ dependence assumed in Eq. (8). As regards the numbers c_ℓ, the Williams model would predict $c_2 = 1/\sqrt{3} = 0.58$ and $c_3 = 1/\sqrt{6} = 0.41$, which may be compared with $c_2 = 0.61 \pm 0.02$ and $c_3 = 0.31 \pm 0.02$ from our fit.

We did not try to improve the EDA by introducing more parameters in Eq. (13), for we think that a real improvement of our parametrization requires a more rigorous treatment of analyticity and crossing symmetry in the parametrization of our $\pi\pi$ amplitudes and a detailed study of the combined $m_{\pi\pi}$ and t dependence of our amplitudes $g_{0,-}^\ell$. A further discussion of the results will be postponed to Section 5.

From meson-baryon phase-shift analysis it is well known that an EDA over a large energy range may lead to χ^2 probabilities far beyond any reasonable acceptance level. This feature is also shown by our fit. In order to check whether the structure found in an EDA is due to the parametrization or required by the data, a reasonable procedure is to perform an EIA and to study in which direction the data want to move the EDA results.

For each mass bin the amplitudes T_ℓ can be calculated from Eqs. (3), (6) and (7). The normalization constant in Eq. (6) is taken from the EDA. In this calculation an over-all phase of the amplitudes remains undetermined. There occurs in addition a discrete ambiguity as discussed below. Out of this ambiguity we want to select those amplitudes T_ℓ which do not only give a good fit to the data, but also are near to the amplitudes T_ℓ^{EDA} of the EDA. This is achieved by minimizing

$$\chi^2 = \chi^2_{exp} + \sum_{\ell} \frac{|T_\ell - T_\ell^{EDA}|^2}{\Delta T_\ell^2}. \tag{15}$$

The tolerances ΔT_ℓ are taken sufficiently large to have an acceptable χ^2_{exp} from the data alone. This procedure has already been shown to be useful in the CERN I πN phase-shift analysis[18]. In these fits we have 11 free parameters η_ℓ^I, δ_ℓ^I and $\bar{r}_\ell$. At lower energies some of these parameters have been fixed to their EDA values (for details see Table 2). The χ^2 distribution of these single energy fits is satisfactory (Fig. 4) i.e. χ^2 is mostly of the order of the number of degrees of freedom (ND).

5. DISCUSSION OF THE PHASE-SHIFT RESULTS

The parameters η_ℓ^I and δ_ℓ^I for various partial waves are shown as functions of $m_{\pi\pi}$ in Fig. 5, for both the EIA and EDA. The errors of the EDA are also given. The Argand plots for the amplitudes from the EDA can be seen in Fig. 6. The agreement between both fits is quite good and the main structures of the EDA are reproduced by the EIA. Deviations occur in the energy region where already the EDA fit to the moments was not satisfactory, e.g. in the D-wave between 800 and 1100 MeV and in η_1^1 between 900 and 1500 MeV.

Figure 7 shows the prediction for $\sigma_{tot}(\pi^+\pi^-)$ and $\sigma_{el}(\pi^+\pi^-)$ as a function of $m_{\pi\pi}$ from our EDA together with the data of Robertson et al.[19] (Note that due to the non-vanishing charge-exchange cross-section σ_{tot} and σ_{el} are not equal below 1 GeV). The agreement between the data is good except for σ_{tot} near the upper end of our mass scale. Whereas σ_{el} corresponds to an integral over the angular distribution, σ_{tot} is calculated via the optical theorem from the elastic amplitude in the forward direction (Z = 1) and is therefore less reliable. Robertson et al.[19], in determining σ_{tot}, neglected the real part of the amplitude in the forward direction. Our value for σ_{tot} may be changed by the inclusion of higher partial waves (we omitted $\ell \geq 4$ contributions). Now we want to discuss the individual waves.

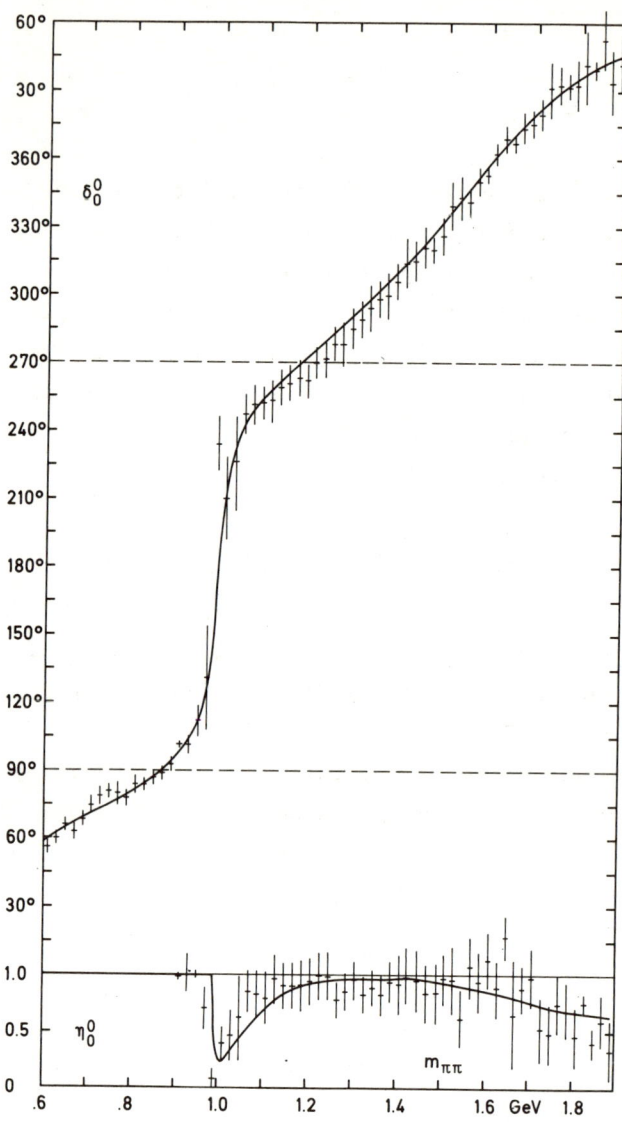

Fig. 5 ππ phase shifts δ_ℓ^I as determined in the energy range $600 \leq m_{\pi\pi} \leq 1900$ MeV. The curves depict the result of the energy-dependent fit to the moments, whereas the points with the errors represent the energy-independent fits. (a) Isospin 0 S-wave, (b) isospin 1 P- and F-waves and isospin 0 D-wave.

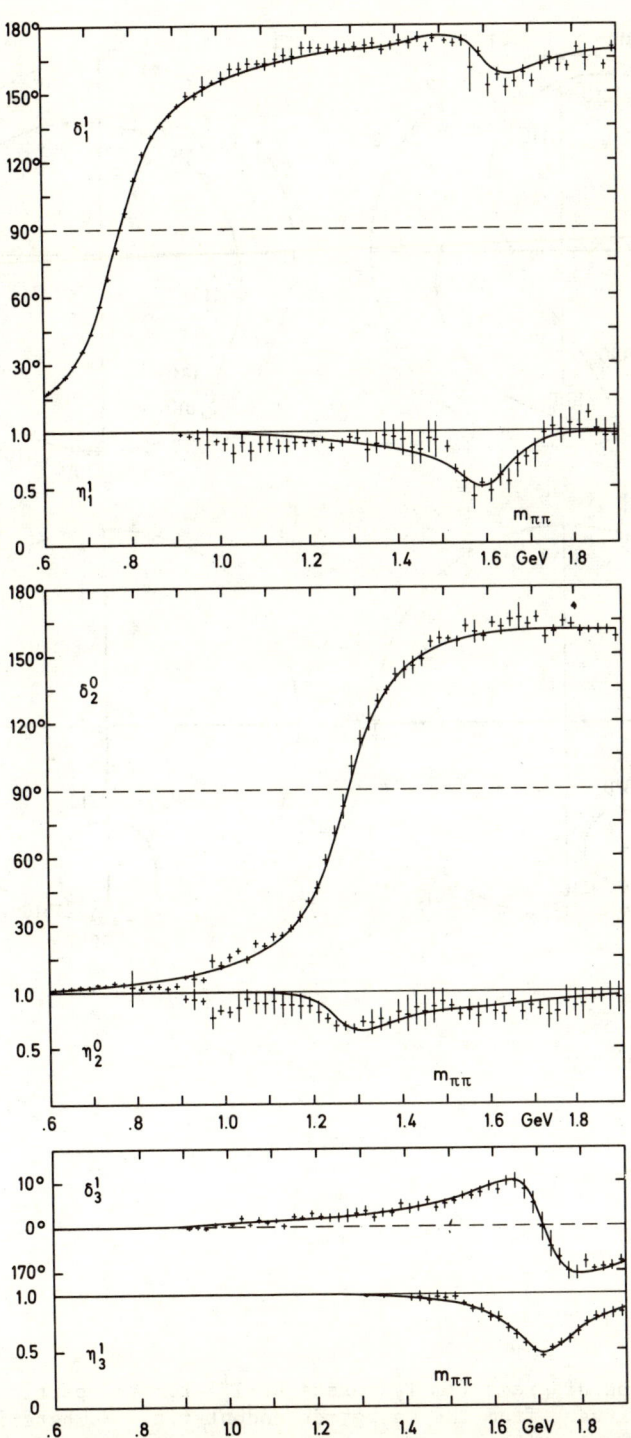

5 b)

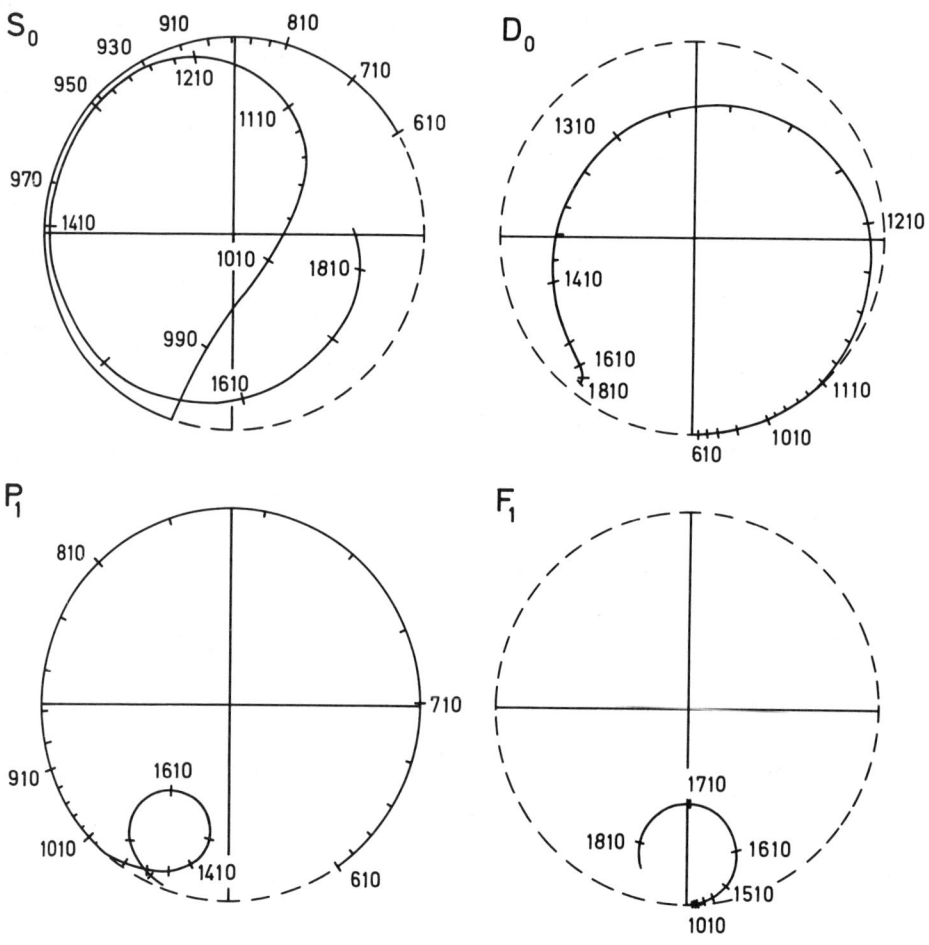

Fig. 6 Argand diagrams (Im T_ℓ^I versus Re T_ℓ^I) for the partial wave amplitudes from the energy-dependent fit. Numbers indicate the $\pi\pi$ energy.

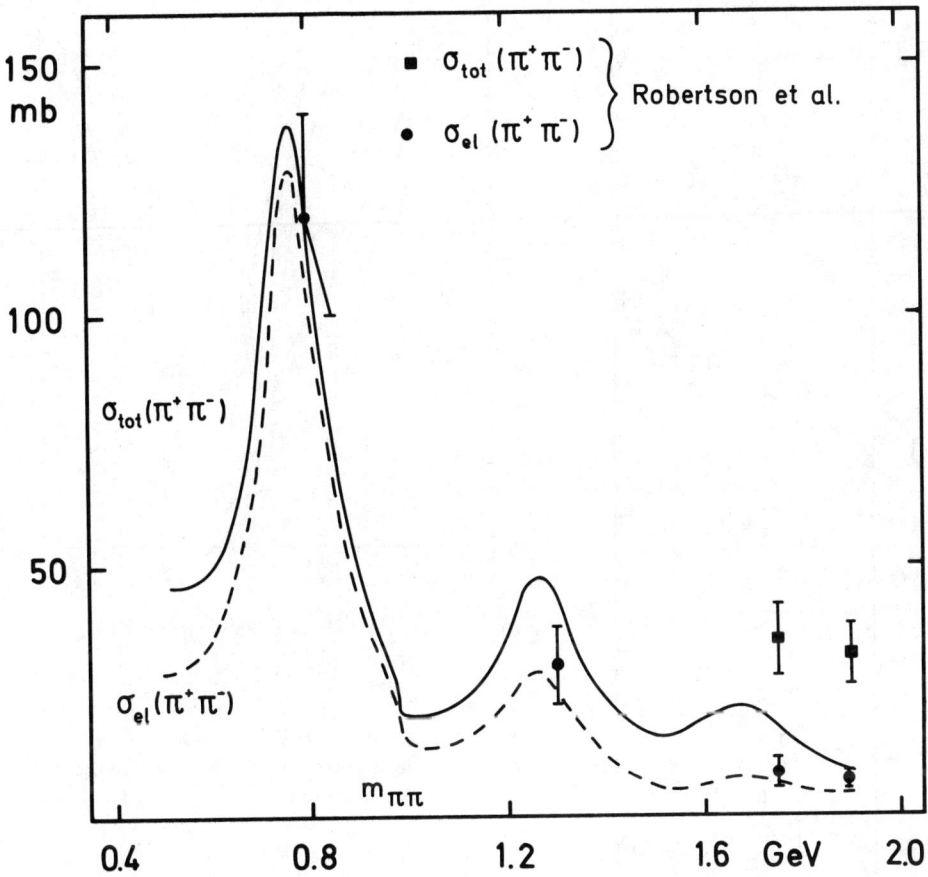

Fig. 7 Total and elastic ($\pi^+\pi^-$) cross-section calculated from the energy-dependent fit.

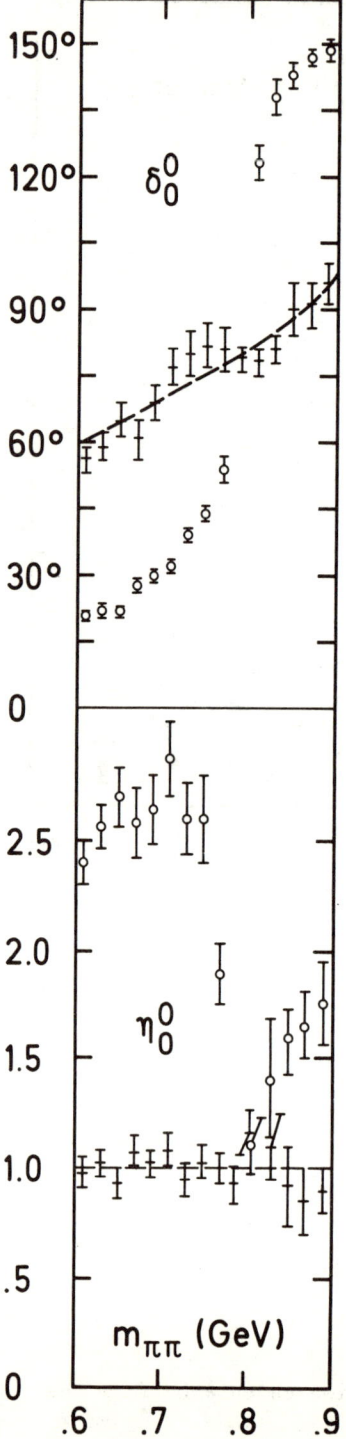

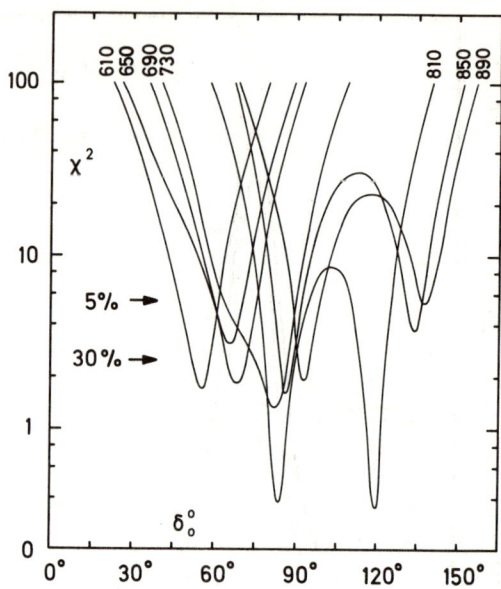

Fig. 9 χ^2 profiles for the S-wave phase shift with all $\eta_\ell^I = 1$ imposed, indicating that below the ρ mass the solution is unique and above the ρ mass one solution is favoured on account of a χ^2 criterion.

Fig. 8 Results of a fit with $\eta_1^1 = \eta_2^0 = 1$ with unconstrained η_0^0 in the ρ region, showing that one of the two solutions yields $\eta_0^0 = 1$.

5.1 The S-wave

The S-wave (Fig. 5a) increases slowly from $\sim 60°$ at 600 MeV to $90°$ at ~ 860 MeV. The sharp rise of this phase at ~ 1000 MeV is connected with a drop in η_0^0, which is also clearly visible in our energy-independent fit (note that we did not use $K\bar{K}$ data in this analysis). This S-wave solution is compatible with the results obtained by Protopopescu et al.[15] with an M matrix fit from data of the reaction $\pi^+ p \to \pi^+ \pi^- \Delta^{++}$ and also with the "down" solution of Grayer et al.[6] based on an analysis of the same data but with an extrapolation method.

At higher energies the S-wave again moves along a circle in the Argand diagram in an anticlockwise direction. The maximal intensity is reached at ~ 1.2 GeV, i.e. close to the f meson mass. A strong S-wave in the f region with a slowly rising phase was shown by Carroll et al.[20] and also in Ref. 6 with a somewhat smaller η_0^0.

If we examine the analytic structure of our S-wave amplitude T_0^0 we find various poles in the complex energy plane. Out of the 12 poles in T_0^0 the positions of the ones which are near to the physical region are given in Table 3. Only one pole with Im E = -15 MeV in sheet II is close to the real axis and we identify it with the S* virtual bound state. The other poles $(\varepsilon, \varepsilon')$ on sheet III are far away from the real axis, so that a resonance interpretation is doubtful.

In order to check how strongly our S* parameters depend on the parametrization, we fitted a constant K matrix directly to the η_0^0 and δ_0^0 parameters as obtained from the EIA in the mass range from 900-1100 MeV. We got

$$K_{11} = (1.0 \pm 0.4) \text{ GeV}^{-1}, \quad K_{12} = K_{21} = (4.4 \pm 0.3) \text{ GeV}^{-1}$$

$$K_{22} = (-3.7 \pm 0.4) \text{ GeV}^{-1}.$$

This yields an S* pole again in sheet II with

$$E = (989 \pm 5) - i(18 \pm 4) \text{ MeV} . \tag{16}$$

In Table 4 we summarize all our resonance parameters. For the S* we insert the parameters of Table 3, but with an error which includes Eq. (16).

We have not yet solved the full ambiguity problem and therefore we cannot exclude the existence of other physical phase-shift solutions. However, in the energy range below 900 MeV, where we neglect F-waves, the ambiguity problem can be resolved easily. We performed an energy-independent fit as above but without the χ^2 term in Eq. (15), which ties the phases to the values of EDA. We required only $\eta_1^1 = \eta_2^0 = 1$ and varied η_0^0 together with the phase shifts. Then we obtained two different s-wave solutions which are plotted in Fig. 8. (This ambiguity should not be confused with the "up-down" ambiguity, which occurs if one does not measure the S-wave cross-section but imposes $\eta_0^0 = 1$.) In Fig. 8 we see one solution with $\eta_0^0 \approx 1$, the corresponding phases being similar to δ_0^0 of Fig. 5a. We find a second solution with a rapidly varying phase, but with large inelasticities $\eta_0^0 > 1$ which are of course unphysical.

This result shows that our method is in fact sensitive to the relatively small S-wave cross-section and moreover predicts η_0^0 correctly. Remember from Section 2 that in a naïve treatment of the M = 0 moments the S-wave cross-section would be too large by a factor of 5.

One may ask what the acceptance level for the second solution is if we constrain η_0^0 to be one. In Fig. 9 we plot some χ^2 profiles for δ_0^0 in selected mass bins below and above the ρ meson. We see clearly that below the ρ there is only one solution. Above the ρ there are two χ^2 minima, where the one with the larger phase is disfavoured with increasing mass.

At energies above 900 MeV the ambiguity problem is much more difficult; as the D-wave gets important, the F-wave starts already to rise and also we can no longer impose elastic unitarity. A classification of our solution within the various ambiguous solutions is given in the next section. A complete treatment will be referred to further study.

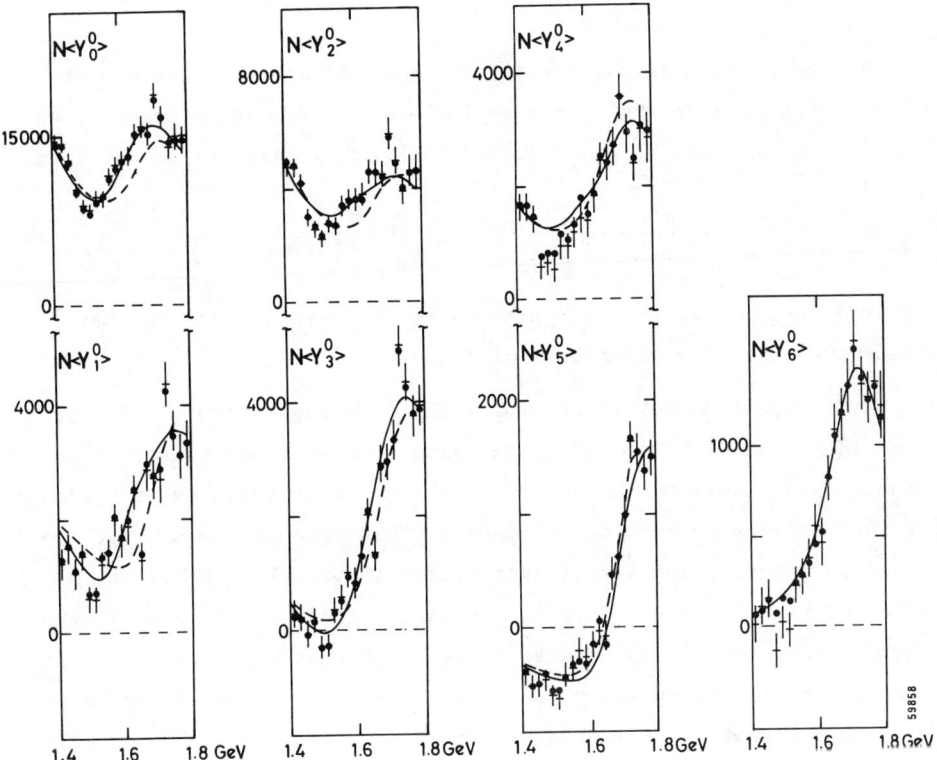

Fig. 10 ππ angular distribution moments in the region between $m_{\pi\pi}$ = 1400 and 1800 MeV, and comparison of two mass-dependent fits. The full line represents the 36-parameter fit of Fig. 1 including a second resonance in the P-wave. The broken line is the result of a 33-parameter fit not including a second resonance in the P-wave; this fit is seen not to reproduce well the Y_0^0, Y_1^0, and Y_2^0 moments. The dots are reconstructed from the EIA and follow closely the drop at 1450 MeV.

5.2 The P-wave

At lower energies the P-wave is dominated by the ρ meson. At 900 MeV this wave starts to become inelastic. We may relate this effect to the $\pi\omega$ decay of the ρ meson. A Breit-Wigner fit of the form

$$T_\ell = \frac{xm\Gamma}{s_0 - s - im\Gamma} , \quad \Gamma = \Gamma_0 \left(\frac{q}{q_0}\right)^{2\ell+1} \frac{D_\ell(q_0 R)}{D_\ell(qR)} , \qquad (17)$$

with $\ell = 1$, $x = 1$ to the EIA phase shifts between 600 and 900 MeV yields the resonance parameters of Table 4.

At around 1500 MeV we observe a rapid decrease of η_1^1, followed by a minimum of $\eta_1^1 \sim 0.5$ and in the same energy region a drop of the phase. This resonant-like behaviour can be interpreted as an evidence for the $\pi\pi$ decay mode of the ρ' meson. This resonance was first required in order to get a satisfactory fit in our EDA. Since the ρ' is found even more pronounced in the EIA, a rapid although small variation of this type in the P-wave is really required by the data. If there exist other physical phase-shift solutions, which cannot be excluded at present as stated above, this effect may change in details, but it seems to be very implausible to have another solution without any resonance structure in this region. In order to show the effect of the ρ' in our parametrization more clearly, we plot in Fig. 10 the M = 0 moments together with the EDA curve as in Fig. 1 and a broken line which results from a second EDA with one pole (ρ) in the P-wave K matrix. From Fig. 10 it is evident that the latter fit is not able to represent the mass dependence of the angular distribution moments in the energy range between 1400 and 1750 MeV. In particular, there is a deficiency in the $\langle Y_0^0 \rangle$ and $\langle Y_2^0 \rangle$ moments. Both EDA's do not have the rapid change of the even moments near 1410 MeV; this, however, is not evidence against the ρ' effect we observe, for the EIA (dots in Fig. 10) follows closely the drop near 1450 MeV and shows the ρ' effect even stronger than the EDA (Fig. 5).

We estimated the mass m, width Γ, and elasticity x of the ρ' from the energy dependence of η_1^1 and δ_1^1 as obtained in the EIA to be as follows:

$$m = 1590 \pm 20 \text{ MeV}, \quad \Gamma = 180 \pm 50 \text{ MeV}, \quad x = 0.25 \pm 0.05.$$

The background amplitude under the ρ' has a phase of $\delta_B \sim 165°$.

So far the ρ' has been observed only in its (4π) decay mode in photoproduction[21,22] and e^+e^- annihilation[23] experiments. In these experiments the width of the ρ' appears to be larger. No evidence for ρ' production was found in the purely hadronic π^+p interaction[24], however the beam momentum of the latter experiment (5 GeV/c) is rather small and not even the g meson is observed. Our result on the small coupling of the ρ' to $\pi\pi$ is compatible with the upper limit of 20% given by the SLAC experiment[22].

5.3 The D-wave

This wave is dominated by the f meson resonance in our energy range. We determined the resonance parameters from a Breit-Wigner fit according to formula (17) with $\ell = 2$ to η_2^0 and δ_2^0 of the EIA in the interval from 1100 to 1500 MeV. The results are listed in Table 4. The width is found to be larger than the world average value 156 ± 25 MeV [25]. However, most of the determinations of this value are based on Breit-Wigner + incoherent background fits to the mass distribution and not on a phase-shift analysis.

At the low-energy tail of the f meson (800-1100 MeV) the agreement between EIA and EDA is rather poor. The EIA indicates a small inelasticity at ~ 1000 MeV. Unlike the P-wave, the D-wave cannot couple to the $\pi\omega$ channel. Above the ρ the D-wave phase is smaller than expected from the f meson Breit-Wigner formula. The bad agreement between EIA and EDA can be traced back to the bad fit of the $L = 3$ moments as seen in Fig. 1. The main difficulty comes from the fact that the zero of $\langle Y_3^0, 1 \rangle$ is found to occur at a higher mass value experimentally as expected from the interference of the ρ meson with an almost real D-wave. A shift of this zero could be achieved by allowing the D-wave to be inelastic already at the ρ mass, which does not seem to be very plausible.

We do not see a signal from the $f'(1514)$ meson. A resonant behaviour of the D-wave in that energy region should be visible directly from the $\langle Y_5^0 \rangle$ and $\langle Y_5^1 \rangle$ moments which are sensitive to the interference of the D- with the F-wave.

5.4 The F-wave

This amplitude finally is well described by the g meson resonance, which is built into our EDA. The elasticity of the g meson is found to be 26%.

6. THE COMPLEX ZEROS OF THE $\pi\pi$ SCATTERING AMPLITUDE

The study of the amplitude zeros is of twofold interest. First they allow a simple classification of the ambiguities which occur in the determination of the partial wave amplitudes. Furthermore, the energy dependence of these amplitude zeros may have a simple dynamical interpretation.

6.1 Some remarks on the ambiguity problem

In a partial-wave analysis the scattering amplitude $F(z)$ (z denotes the cosine of the scattering angle) is assumed to be a polynomial of degree ℓ_{max}, if $\ell_{max} + 1$ partial wave amplitudes are included. Therefore $F(z)$ may also be written in terms of its complex zeros z_k as

$$F(z) = F(1) \prod_{\ell=1}^{\ell_{max}} \frac{z - z_K}{1 - z_K} . \qquad (18)$$

A measurement of $d\sigma/dt \sim |F|^2$ determines the real parts and absolute values of Im z_K, but not the signs of Im z_K [26]. Therefore for a given $d\sigma/dz$ one has $2^{\ell_{max}}$ different phase-shift solutions which can be classified according to the signs of Im z_K. Some of these solutions could be unphysical ($\eta_\ell > 1$). If the total cross-section is not measured as in our case also the phase of $F(1)$ remains undetermined. This simple scheme for the ambiguities holds in our analysis only approximately for two reasons:

i) Apart from the OPE contribution, which measures $|F(z)|^2$, we use also information from the absorptive correction. As shown before the value of $\bar{r}_\ell$ suggests a Williams-type parametrization. For such a model the absorptive corrections are proportional to $|d/dz\ F(z)|^2$ as shown elsewhere[27], and $|d/dz\ F|^2$ depends explicitly on the signs of Im z_K.

ii) An unphysical solution with $\eta_\ell^I > 1$ for a particular wave could still get an acceptable χ^2 after η_ℓ^I has been constrained to be smaller than one. These constraints together with finite errors may enlarge the number of possible solutions considerably.

However, we see that Re z_K and $|\text{Im } z_K|$ are more directly related to the data than the phase shifts. Their calculation needs in addition a particular choice of the signs of Im z_K and of the over-all phase. The real parts of z_K are essentially determined by the dips in $d\sigma/dz$, whereas $|\text{Im } z_K|$ is related to the size of the cross-section at the dip positions. These dips are clearly visible in the data[6].

Our results for Re z_K and Im z_K, as obtained from the EDA and EIA, are shown in Figs. 11 and 12. We have plotted Re z_K in an s-u diagram with

$$s = m_{\pi\pi}^2$$
$$t = -2q^2(1 - z) \tag{19}$$
$$u = -2q^2(1 + z).$$

The dotted lines $t = 0$ and $u = 0$ indicate the boundaries of the physical region. Corresponding to $\ell_{max} = 3$ there are three zero trajectories z_K. Figure 12 shows the particular sign of Im z_K chosen by our energy-dependent fit.

We established a unique solution below the ρ meson which corresponds to Im $z_1 < 0$. The solution with the other sign leads to an unphysical $\eta_0^0 > 1$ (see Fig. 8). Above the ρ meson we have Im $z_1 > 0$. Again the solution with Im $z_1 < 0$ is unphysical. With the constraint $\eta_0^0 = 1$ we still find a solution (see Fig. 9), but with a larger χ^2. This solution is indicated by dots in Fig. 12.

In the energy region above 900 MeV where we allow for an F-wave and inelastic S-, P-, and D-waves we performed in addition to the above-described EDA and EIA a random search for $\pi\pi$ amplitudes, where

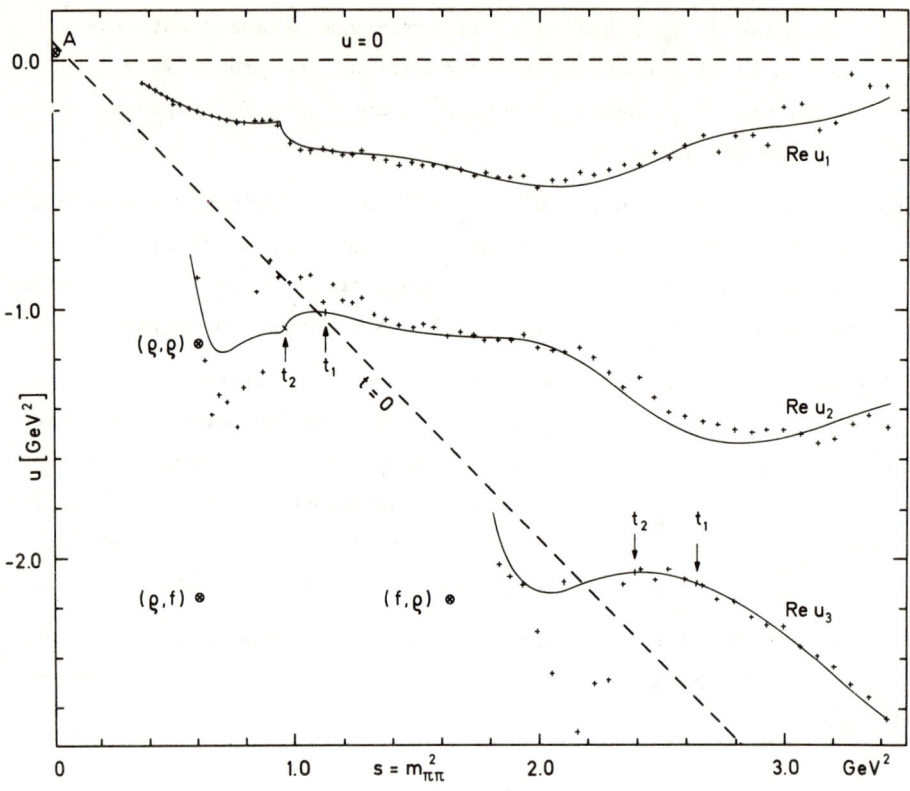

Fig. 11 Real parts of the amplitude zeros plotted in the s-u plane as calculated from the energy-dependent and energy-independent fits.

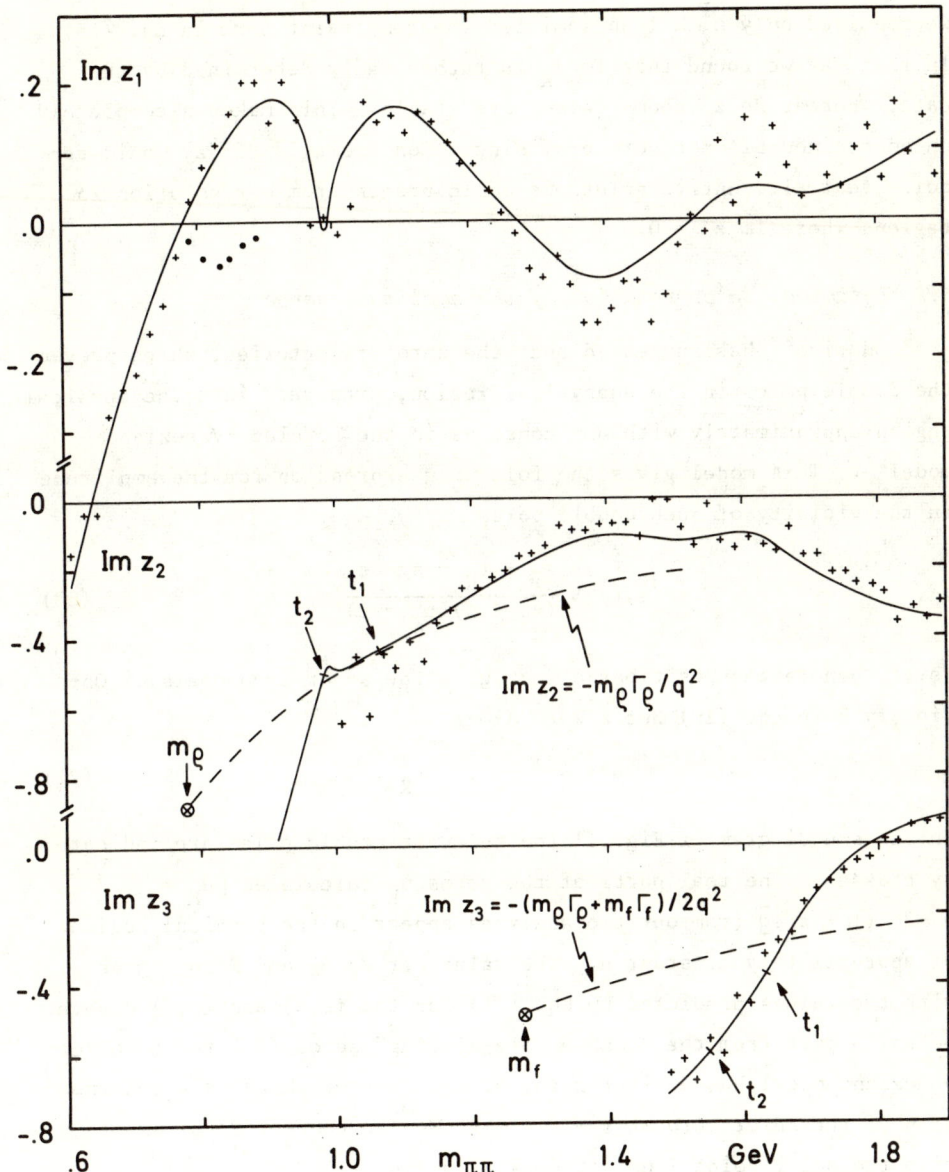

Fig. 12 Imaginary parts of the amplitude zeros as a function of the ($\pi\pi$) mass, calculated from the energy-dependent and energy-independent fits. The dotted curves represent zero trajectories for fixed u passing through the (ρ,ρ) or (ρ,f) double poles.

we required only $\eta_\ell^I \leq 1$ and omitted the constraint term in Eq. (15). In that way we found that Im z_K is rather badly determined by the data, whereas Re z_K seems to be more stable. This makes a completely unconstrained EIA not very promising. Continuity of Im z_K would require that alternative solutions could branch from our solution in regions where Im z_K = 0.

6.2 Dynamical interpretation of the amplitude zeros

Odorico[28] has suggested that the zero trajectories, which prevent the double poles in the unphysical region, propagate into the physical region approximately with u = const as in the Lovelace-Veneziano model[29]. This model gives the following expression for the amplitude in the vicinity of such double poles

$$F(s,t) \sim \frac{s_i + t_K - s - t}{(s_i - s)(t_K - t)} \qquad (20)$$

(s_i, t_K denote the positions of ρ, f, g in the s- or t-channels). Obviously F in Eq. (20) has a zero along

$$u = 4m_\pi^2 - s_i - t_K . \qquad (21)$$

In the s-u diagram of Fig. 11 the relevant double poles are indicated by crosses. The real parts of the zeros u_K calculated [u_K = $-2q^2(1 + z_K)$] from our two analyses appear in the physical region at approximately constant u. The values of Re u_2 and Re u_3 agree with the values predicted by Eq. (21) for the (ρ,ρ) and (ρ,f) double poles. Apart from the "double pole killing" zeros (21) the Lovelace-Veneziano model has a first zero, at constant u, which is a continuation of the Adler zero at $s = t = m_\pi^2$. We can connect Re u_1 smoothly with the Adler point (indicated by A in Fig. 11).

The zeros u_K calculated from a phase-shift analysis can only be regarded as zeros of the amplitude inside the domain of convergence of the partial-wave expansion. This is an ellipse in the complex z plane with foci at $z = \pm 1$ and a major semi-axis of $t_c/2q^2$, where t_c is the nearest important branch point in the crossed channel. The effective t_c may be larger than $4m_\pi^2$ because of the generally-accepted small values of the $\pi\pi$ scattering length[4]. Arrows in Figs. 11 and 12 indicate where the zero trajectories leave an ellipse with $t_c = t_1 = 0.15$ and $t_c = t_2 = 0.35$ GeV^2. Figure 11 suggests a smooth continuation with approximately constant u_2 and u_3 from the physical region to the double poles.

In the Lovelace-Veneziano model Eq. (20) holds only for real s_i, t_K. We can go a step further and assume the relation (20) to be true also for complex s_i and t_K (calculated from the resonance parameters). This gives the following imaginary part for the zeros z_ℓ

$$\text{Im } z_\ell = -(m_i \Gamma_i + m_K \Gamma_K)/2q^2 \ . \tag{22}$$

The trajectories (22) are drawn as dotted lines in Fig. 12. The qualitative agreement with the zero trajectories calculated from our phase-shift analysis at smaller energies suggests that the double pole-killing mechanism holds also for the imaginary parts.

Table 1

Parameters obtained from the energy-dependent fit. The numbers are given in units of the appropriate power of GeV.

S_0-wave	$\sqrt{s_1^0} = 0.11 \pm 0.15$ $\sqrt{s_2^0} = 1.19 \pm 0.01$ $\gamma_{\pi\pi}^{(0)} = 2.86 \pm 0.15$	$\alpha_\pi^{(0)} = 2.28 \pm 0.08$ $\beta_\pi^{(0)} = -1.00 \pm 0.03$ $\gamma_{\pi K}^{(0)} = 1.85 \pm 0.18$	$\alpha_K^{(0)} = 2.02 \pm 0.11$ $\beta_K^{(0)} = 0.47 \pm 0.05$ $\gamma_{KK}^{(0)} = 1.00 \pm 0.53$
P_1-wave	$\sqrt{s_\rho} = 0.777 \pm 0.001$ $\sqrt{s_2^1} = 1.394 \pm 0.015$ $\gamma_{\pi\pi}^{(1)} = 0.14 \pm 0.03$	$\Gamma_\rho = 0.155 \pm 0.001$ $\beta_\pi^{(1)} = 0.476 \pm 0.022$ $\gamma_{\pi K}^{(1)} = 0.37 \pm 0.11$	$R_1 = 3.09 \pm 0.16$ $\beta_K^{(1)} = 2.72 \pm 0.07$ $\gamma_{KK}^{(1)} = 11.18 \pm 0.28$
D_0-wave	$\sqrt{s_f} = 1.281 \pm 0.001$ $R_2 = 4.94 \pm 0.26$ $\gamma_{\pi\pi}^{(2)} = -0.22 \pm 0.04$	$\Gamma_f = 0.205 \pm 0.005$ $\gamma_{\pi K}^{(2)} = 0.45 \pm 0.11$	$x_f = 0.84 \pm 0.01$ $\gamma_{KK}^{(2)} = 1.0 \pm 2.4$
F_1-wave	$m_g = 1.713 \pm 0.004$ $R_3 = 6.38 \pm 0.44$	$\Gamma_g = 0.228 \pm 0.010$	$x_g = 0.26 \pm 0.02$
Non-OPE background	$c_2 = 0.61 \pm 0.02$ $a_1 = 2.64 \pm 0.10$	$c_3 = 0.31 \pm 0.02$ $a_2 = 0.38 \pm 0.10$	$a_0 = 0.06 \pm 0.08$ $a_3 = 2.32 \pm 0.08$
Normalization	$c_N^2 = 4130 \pm 25$		

Table 2

Parameters in the energy-independent fits

Mass range (MeV)	L_{max} for $\langle Y_L^M \rangle$	Fixed parameters	ND	ΔT_0	ΔT_1	ΔT_2	ΔT_3
600-900	3	$\eta_\ell^I = 1;\ \delta_3^1,\ g_-^3$	2	0.15	0.04	0.05	0
900-1400	5	η_3^1	1	0.20	0.10	0.10	0.03
1400-1900	6	—	2	0.20	0.10	0.20	0.03

Table 3

Poles of the S-wave amplitude T_0^0 from our energy-dependent fit. Sheet II is defined by Im q_π < 0, Im q_K > 0, sheet III by Im q_π < 0, Im q_K < 0.

Name	Re E (MeV)	Im E (MeV)	Sheet
S*	1007	15	II
ε	1049	250	III
ε'	1537	233	III

Table 4

Resonance parameter mass m, width Γ, elasticity x and radius R. Note that $\Gamma = 2 \, \text{Im} \, E_0$.

Name	m (MeV)	Γ (MeV)	x	R (GeV^{-1})
ρ	778 ± 2	152 ± 2	1.0 (input)	4.5 ± 0.4
f	1279 ± 3	202 ± 6	0.84 ± 0.02	5.3 ± 1.2
g	1713 ± 4	228 ± 10	0.26 ± 0.02	6.4 ± 0.4
S*	1007 ± 20	30 ± 10	–	–
ρ'	1590 ± 20	180 ± 50	0.25 ± 0.05	–

APPENDIX

CONSTRAINTS FOR THE AVERAGED DENSITY MATRIX ELEMENTS IN THE WILLIAMS MODEL

The parametrization (3) in Section 3.1 of the density matrix is equivalent to the following conditions[7]:

$$\rho_{00}^{\ell\ell'} \rho_{11}^{\ell\ell'} = 2 \rho_{01}^{\ell\ell'} \rho_{01}^{\ell'\ell} . \qquad (A.1)$$

Equation (A.1) will in general not also hold for density matrix elements $\bar{\rho}_{mm'}^{\ell\ell'}$, averaged over an interval of t, if the individual elements have a different t-dependence. In the following we want to show that Eq. (A.1) holds also for the average values if we assume a t dependence as in the Williams model.

These average values are obtained by an integration over t or $\Delta^2 = -t$

$$\bar{\rho}_{mm'}^{\ell\ell'} = \int_0^{a^2 m_\pi^2} d\Delta^2 \, \rho_{mm'}^{\ell\ell'} (\Delta, m_{\pi\pi}) , \qquad (A.2)$$

where a denotes the upper limit of the integration in units of m_π. From the Williams model in the version of Ref. 27 we get the following expressions (apart from a normalization factor) for the amplitudes at $\Delta^2 \ll m_{\pi\pi}^2$:

$$g_0^\ell = F(\Delta) \frac{\Delta}{\Delta^2 + m_\pi^2} f_\ell$$

$$g_-^\ell = F(\Delta) \frac{c_A (m_{\pi\pi})}{2 m_{\pi\pi}} \sqrt{\ell(\ell+1)} f_\ell \qquad (A.3)$$

with

$$f_\ell = \frac{m_{\pi\pi}}{\sqrt{q}} \sqrt{2\ell+1} \, T_\ell .$$

$F(\Delta)$ describes a form factor, c_A the amount of absorption, which depends[11] on $m_{\pi\pi}$. For simplicity we take the same form factor for the Born term and for the absorptive part. The average density matrix elements can be calculated from Eqs. (A.2), (3) and (A.3) to be:

$$\overline{\rho_{mm'}^{\ell\ell'}} = I_{mm'}\, f_\ell f_{\ell'}^* \,. \qquad (A.4)$$

The integrals $I_{mm'}$ are given by

$$I_{00} = \int_0^{a^2 m_\pi^2} d\Delta^2 \, \frac{\Delta^2}{(\Delta^2 + m_\pi^2)^2} \, F^2(\Delta)$$

$$I_{10} = \frac{c_A}{2\, m_{\pi\pi}} \sqrt{\ell(\ell+1)} \int_0^{a^2 m_\pi^2} d\Delta^2 \, \frac{\Delta}{\Delta^2 + m_\pi^2} \, F^2(\Delta) \qquad (A.5)$$

$$I_{11} = \frac{c_A}{2\, m_{\pi\pi}^2} \ell(\ell+1) \int_0^{a^2 m_\pi^2} d\Delta^2 \, F^2(\Delta) \,.$$

In order to get Eq. (A.1) for the average values we have to require

$$\sqrt{2}\, I_{01} = \sqrt{I_{00}\, I_{11}} \,. \qquad (A.6)$$

Evaluating the integrals with a reasonable choice of form factors, it turns out that this relation is extremely well satisfied for a large range of the upper limit. Taking $F^2(\Delta) = 1 - \gamma(\Delta^2 + m_\pi^2)/m_\pi^2$ with $\gamma = 0.1$, and $a = 3$ we find:

$$\sqrt{2}\, I_{01} = 1.70 \, m_\pi^{-1} \, \frac{c_A}{m_{\pi\pi}} \sqrt{\frac{\ell(\ell+1)}{2}}$$

$$\sqrt{I_{00}\, I_{11}} = 1.72 \, m_\pi^{-1} \, \frac{c_A}{m_{\pi\pi}} \sqrt{\frac{\ell(\ell+1)}{2}} \,.$$

This means that relation (A.6) is satisfied within 1% after integration up to $|t| = 0.18$ GeV^2. (It begins to be violated for a > 5). Therefore the averaged moments $\int d\Delta^2 \ N(Y_L^M)$ satisfy the same constraints as $N(Y_L^M)$ and we can perform the same analysis. For the average values $\bar{g}_0^\ell$ and $\bar{r}_\ell$ we get in this model

$$\bar{g}_0^\ell = \sqrt{I_{00}} \ \frac{m_{\pi\pi}}{\sqrt{q}} \ \sqrt{2\ell+1} \ T_\ell \qquad (A.7)$$

$$\bar{r}_\ell = \sqrt{\frac{I_{11}}{I_{00}}} \ \frac{2 \ m_{\pi\pi}}{c_A \ \sqrt{\ell(\ell+1)}} \ . \qquad (A.8)$$

Therefore the constant c_N in Eq. (6) can be calculated from $c_N = \sqrt{I_{00}}$.

The fact that the normalized moments are almost independent of Δ is, of course, closely connected to the particular Δ dependence of the amplitudes g_0^ℓ,- in Eq. (A.3). An example which is equivalent to the Δ^2 dependence of $\rho_{01}^{11}/(\rho_{10}^{11} + 2 \ \rho_{11}^{11})$ has been demonstrated in Ref. 27.

REFERENCES

1. P. Baillon, F. Bulos, R.K. Carnegie, G.E. Fischer, E.E. Kluge, D.W.G.S. Leith, H.L. Lynch, B. Ratcliff, B. Richter, H.H. Williams and S.H. Williams, Phys. Letters $\underline{35}$ B, 453 (1971).

2. G. Grayer, B. Hyams, C. Jones, P. Schlein, W. Blum, H. Dietl, W. Koch, E. Lorenz, G. Lütjens, W. Männer, J. Meissburger, W. Ochs, U. Stierlin and P. Weilhammer, Proc. Fourth Int. Conf. on High-Energy Collisions, Oxford, 1972 (Rutherford High-Energy Laboratory, Chilton, Didcot, Berks., 1972), Vol. 2, p. 26.

3. P.K. Williams, Phys. Rev. $\underline{D1}$, 1312 (1970).

4. For a recent review see: D. Morgan, Questions in $\pi\pi$ scattering and related processes, Rutherford High Energy Laboratory preprint RPP/T/27, 1972.

5. G. Grayer, B. Hyams, C. Jones, P. Schlein and P. Weilhammer, W. Blum, H. Dietl, W. Koch, E. Lorenz, G. Lütjens, W. Männer, J. Meissburger, W. Ochs and U. Stierlin, High statistics study of the reaction $\pi^-p \to \pi^+\pi^-n$ -- Apparatus, method of analysis, and general features of results at 17 GeV/c, to be published.

6. G. Grayer, B. Hyams, C. Jones, P. Schlein, W. Blum, H. Dietl, W. Koch, E. Lorenz, G. Lütjens, W. Männer, J. Meissburger, W. Ochs, U. Stierlin and P. Weilhammer, Proc. 3rd Philadelphia, Conf. on Experimental meson spectroscopy, Philadelphia, 1972 (American Institute of Physics, New York, 1972), p.5.

7. W. Ochs, Nuovo Cimento $\underline{12}$ A, 724 (1972).

8. P. Schlein, Phys. Rev. Letters $\underline{19}$, 1052 (1967).

9. C.D. Froggatt and D. Morgan, Phys. Letters $\underline{40}$ B, 655 (1972).

10. P. Estabrooks and A.D. Martin, Phys. Letters $\underline{41}$ B, 350 (1972).

11. W. Ochs and F. Wagner, On the strength of absorption in single pion production reactions, MPI Munich preprint, to be published in Phys. Letters.

12. G. Grayer, B. Hyams, C. Jones, P. Weilhammer, W. Blum, H. Dietl, W. Koch, E. Lorenz, G. Lütjens, W. Männer, J. Meissburger, W. Ochs and U. Stierlin, Nuclear Phys. $\underline{B50}$, 29 (1972).

13. P. Estabrooks, A.D. Martin, G. Grayer, B. Hyams, C. Jones, P. Weilhammer, W. Blum, H. Dietl, W. Koch, E. Lorenz, G. Lütjens, W. Männer, J. Meissburger and U. Stierlin, Invited talk by A.D. Martin at the Int. Conf. on $\pi\pi$ Scattering and Associated Topics, Tallahassee, Florida, 1973.

14. M. Alston-Garnjost, A. Barbaro-Galtieri, S.M. Flatté,
    J.H. Friedman, G.R. Lynch, S.D. Protopopescu, M.S. Rabin and
    F.T. Solmitz, Phys. Letters 36 B, 152 (1971).
    B.D. Hyams, W. Koch, E. Lorenz, G. Lütjens, W. Ochs, P. Schlein,
    U. Stierlin, P. Weilhammer, W. Beusch, W. Wetzel,
    D. Johnson, V. Stenger and P. Wohlmut, Experimental meson
    spectroscopy (Eds. C. Baltay and A.H. Rosenfeld)
    (Columbia Univ. Press, New York and London, 1970), p. 41.

15. S.D. Protopopescu, M. Alston-Garnjost, A. Barbaro-Galtieri,
    S.M. Flatté, J.H. Friedman, T.A. Lasinski, G.R. Lynch,
    M.S. Rabin and F.T. Solmitz, Proc. 3rd Philadelphia Conf. on
    Experimental Meson Spectroscopy, Philadelphia, 1972
    (American Institute of Physics, New York, 1972), p. 17.

16. For a description of the K matrix formalism and further
    references see: A. Barbaro-Galtieri, *in* Advances in
    Particle Physics (Wiley, New York, 1968), Vol. 2, p. 175.

17. A new measurement of the I = 2 S- and D-wave has been presented
    by W. Hoogland at the Int. Conf. on $\pi\pi$ Scattering and
    Associated Topics, Tallahassee, Florida, 1973.

18. S. Almehed and C. Lovelace, Nuclear Phys. B40, 157 (1972).

19. W.J. Robertson, W.D. Walker and L. Davis, Duke University
    preprint, Durham, N.C., May 1972.

20. J.T. Carroll, R.N. Diamond, M.W. Firebaugh, W.D. Walker,
    J.A.J. Matthews, J.D. Prentice and T.S. Yoon, Phys. Rev.
    Letters 28, 318 (1972).

21. M. Davier, I. Derado, D.C. Fries, F.F. Liu, R.F. Mozley,
    A. Odian, J. Park, W.P. Swanson, F. Villa and D. Yount,
    SLAC-PUB-666, 1969.

22. H.H. Bingham, W.B. Fretter, W.J. Podolsky, M.S. Rabin,
    A.H. Rosenfeld, G. Smadja, G.P. Yost, J. Ballam, G.B. Chadwick,
    Y. Eisenberg, E. Kogan, K.C. Moffeit, P. Seyboth,
    I.O. Skillicorn, H. Spitzer and G. Wolf, Phys. Letters 41 B,
    635 (1972).

23. G. Barbarino, M. Grilli, E. Iarocci, P. Spillantini, V. Valente,
    R. Visentin, F. Ceradini, M. Conversi, L. Paoluzi, R. Santonico,
    M. Nigro, L. Trasatti and G.T. Zorn, Lettere Nuovo Cimento
    3, 689 (1972).
    C. Bacci, G. Penso, G. Salvini, B. Stella, R. Baldini-Celio,
    G. Capon, C. Mencuccini, G.P. Murtas, A. Reale and M. Spinetti,
    Phys. Letters 38 B, 551 (1972).

24. Y. Eisenberg, U. Karshon, J. Mikenberg, S. Pitluck, E.E. Ronat,
    A. Shapira and G. Yekutieli, Phys. Letters 43 B, 149 (1973).

25. Particle Data Group, Phys. Letters 39 B, 1 (April 1972).

26. A. Gersten, Nuclear Phys. B12, 537 (1969).
    E. Barrelet, Nuovo Cimento 8 A, 331 (1972).

27. F. Wagner, On the calculation of spin dependence in the absorption model for π-exchange, MPI preprint, Munich (Feb. 1973), to be published in Nuclear Phys.

28. R. Odorico, Phys. Letters 38 B, 411 (1972).

29. C. Lovelace, Phys. Letters 28 B, 264 (1968).

LEARNING ABOUT MESON STATES
FROM THEIR PRODUCTION PROPERTIES*

G.L. Kane
Randall Laboratory of Physics
University of Michigan
Ann Arbor, Michigan 48104

ABSTRACT

An attempt is made to increase our awareness of
the value of including information on production
mechanisms in our study of hadron resonances and
meson-meson scattering.

My purpose in this talk is to try to convince you to
include information from the production data when you are
studying meson states. Many of you already are doing
that, and others (particularly G. Fox) have urged you to
do it, but I think we still take little advantage of the
production data compared to what is possible.

At this conference we are trying to learn about
meson states and meson-meson scattering, particularly $\pi\pi$
and $K\pi$ scattering. In the past 3-4 years there has been
a lot of work done in this field. It is interesting to
ask how much we have learned. Two things come to mind:
(1) there has been fairly detailed verification that we
have some understanding of small t exchange. Most people
probably agree now that $d\sigma(\pi N \to \pi\pi N)/dt$ has a zero not at t=0
but at a t value about $2m_\pi^4/m_{\pi\pi}^2$ below t_{min}. Even this is
a point some theorists were confident of several years
ago, but it needed experimental verification, most extensively provided by the SLAC experiment. (2) The rapid
variation of the $\pi\pi$ scattering moments at the $K\bar{K}$ threshold has essentially given us the s-wave $\pi\pi$ phase below
the f.

Apart from these results most work has just added
details to what we already understood (in view of some
comments since my talk, it may be worth adding that one
should not count as things we have learned things which
some individual may have learned or worked out but which
were well known to other people or in review talks much
earlier).

To learn more, in this field, basically there are
two possibilities. First, one could find dramatic effects such as the $K\bar{K}$ threshold; generally we will learn
a lot from them because they are dramatic and dominant

---
*Research supported in part by the U.S. Atomic Energy
Commission.

(not because they exist -- we always knew the threshold was there, but the size of the effect is what taught us a lot). Second, one could learn to do better phenomenology with the data and extract more physics information.

To elaborate on the second point, note for example that even in the case of the $\overline{K}K$ threshold one must make a coupled channel model and a fairly sophisticated analysis to get physics out. One is still not able to decide whether there is an s-wave pole in the amplitude below the S^* in mass.

One can make a stronger statement: <u>without a model one cannot in practice learn physics from data</u>. In principle this is not true, but in the real world it is almost completely true. For example, you have heard today several examples of clever amplitude analyses. Even in the best of these, for πN scattering, one only learns the amplitudes up to an overall phase, and one must have additional theoretical input to learn interesting physics about the amplitudes. Similarly, the useful analysis of Estebrooks and Martin[1] assumes that three of the six amplitudes for production of a ρ are zero; this is a good approximation and constitutes a sensible model at this stage. It will be changed if data can be obtained on a polarized target.

Basically, the situation is as in politics. All behavior, even no action, is political. Similarly, there are bad models and there are good models, but never model independence. If you choose to use a bad model to get information from your data, then ...

EXAMPLES OF WAYS TO LEARN FROM PRODUCTION MECHANISMS

(A) Almost model independent

There are a number of ways one can put information about production into the analyses, either to learn more or to gain confidence in what one has already extracted. Here I will mention a few examples. The reader can easily think of more that are relevant to his particular data or interest. (1) In charge exchange production of a pion pair one expects to exchange at high energies only π and A_2. These couple to nucleons mainly by flipping helicity. Thus as long as one is summing over nucleon helicities it is probably a good approximation to neglect all amplitudes involving nucleon non-flip. Then there are four complex amplitudes left if s- and p-wave pairs are produced. One can measure six quantities in this case, so with one further assumption one can determine the remaining amplitudes up to an overall phase. This set of assumptions constitutes a well defined model.

This has been done in most detail by Estebrooks and Martin, as has been discussed at this conference. Whether one can make enough safe assumptions in a given problem to get at important information has to be considered in each case. In the present case of pion pair production the situation will be different when the reaction is measured on a polarized target, as then the nucleon non-flip will show up (hopefully due to A_2 coupling) and we will learn about the relative phases of the amplitudes with flip and non-flip couplings.

For any reaction one can make the necessary number of assumptions and then determine the remaining quantities.

(2) The $\pi\pi$ phase shifts (or anything) must be the same whatever production reaction it is measured in, at whatever energy, if we are in fact correctly extracting them from the production reaction. Near the rho mass that is of course true for the dominant p-wave, but for several small quantities of interest such as the I=2 phase shifts, the s-wave at the kaon mass, the scattering lengths, one should assign a systematic error given by the range in values from different experiments in addition to the usual errors. As an example, the SLAC group[11] has compared the phase shifts from the reactions

$$\pi N \to (\pi\pi) N$$
$$\to (\pi\pi) \Delta^{++}$$
$$\to (\pi\pi) \Delta^{0}$$

They find they can use the same $\pi\pi$ phases if they have some nonflip amplitude at the nucleon vertex and some flip amplitude at the $N\Delta$ vertex. For the latter case for the π exchange contribution alone one can work out that the ratio of the sum of flip couplings to the nonflip coupling is given approximately by

$$-t\left[\sqrt{3}\left((m_\Delta+m_N)/m_N\right)^2 + \left((m_\Delta+m_N)/m_N\right)^2\right] \approx -20t$$

so that at $-t=.05$ GeV^2 one might expect about equal flip and nonflip contributions, which is about what they find. A more careful analysis needs to be done, but qualitatively this gives us some confidence in the resulting $\pi\pi$ phases.

Similar studies should be made for the $K\pi$ phase shifts in KN reactions with N and Δ recoil.

(3) There are some physical region effects which should go away at the π exchange pole after an extrapolation, such as $\rho-\omega$ interference in the mass spectrum, or a φ peak in $m(K\bar{K})$ in $\pi N \to K\bar{K}N$ since $\varphi \not\to 2\pi$.

(4) Production data can provide a very important check on "daughter" states, i.e. on situations where it appears one is observing one state under another, such as the familiar s-wave under the f. If we are producing two resonances of spin S and S-2 (I will concentrate on the f and s-wave example, where S=2) the production amplitude has the general form

$$M \sim M_f(s,t) P_S(\cos\theta) + M_\epsilon(s,t) P_{S-2}(\cos\theta)$$

where the $M_x(s,t)$ describe the production and I have only written the amplitude for helicity zero for simplicity; the symbol "ϵ" represents the state of spin S-2. If it should happen that the two production amplitudes $M_f(s,t)$ and $M_\epsilon(s,t)$ (which in general are unrelated) occur in the ratio (S-1)/S at all s and t values, then the production amplitude would be proportional to $\cos\theta P_{S-1}(\cos\theta)$ by the Legendre function recursion relation, and one would get a pure $\cos^4\theta$ angular distribution in the f region as has sometimes been observed. If one is really observing an s-wave under the f then by going varying s or t so natural parity exchange gets important (natural parity exchange cannot produce an s-wave pion pair) one should see the P_2 angular distribution come back.

An experiment[2] at 13 GeV/c has observed somewhat more bump in the middle than most lower energy experiments, which is encouraging, since the A_2 should get more important relative to the π as the energy increases. To be sure one is seeing an s-wave under the f, one should systematically vary s and t and watch the distribution change from $\cos^4\theta$ to $P_2^2(\cos\theta)$.

In general when one has interfering resonances one can separate them out by varying the production conditions. This is probably the main place where production information should be extensively used to study resonance properties. It has been neglected, partially because people don't often go outside their own data; even in one experiment, however, the t dependence of the decay angular distributions can be used to separate interfering resonances.

(b) More model dependent examples

We have heard extensive discussions from Field and Matthews in this session about the nature of the scattering amplitudes for various reactions. The point of all this is that the reaction you are thinking about at the moment is not the only one in the world, and moreover the amplitudes being measured in any given reaction have usually been at least partially measured in some other reaction already. Thus information can be used from the other experiments to learn more from those currently under study.

For example, there are many places where π and A_2 exchange occur. From np→pn one can see that even at t=0 the A_2 exchange is important; the same thing should be true for $\pi N \to \rho N$ in the appropriate amplitudes, which are those with ρ helicity ±1.

For a long time to come, however, all physics we

learn by comparing reactions will be somewhat model dependent. If you want your results to be correct, you must use a model which is sufficiently correct to avoid misleading results.

The situation is somewhat like the case with phase shift analyses. Beyond a certain partial wave one must set all phase shifts somehow. It used to be that all higher partial waves were set to zero; that is a model although not a very good one. For ten years people have been trying to do better by using our knowledge of the long range forces to fix the higher partial waves, and now I think it is clear that the results for the lower phase shifts are better physically than they were when the higher partial waves were put to zero.

Similarly here we do not yet have long established models for calculating all amplitudes. But we have learned a lot in the past five years and now I think it is possible to be confident of a number of aspects of the situation. The basic useful assumption is the old conjecture of the Michigan group, which seems to be approximately correct, that

FOR A GIVEN EXCHANGE (e.g. $\pi, \rho, \ldots$) AND A GIVEN n,x (these are helicity flip quantum numbers, defined below) AN s-CHANNEL HELICITY AMPLITUDE IS APPROXIMATELY THE SAME FUNCTION OF s AND t IN EVERY REACTION WHERE IT OCCURS.

For a reaction a+b→c+d, with particle a having helicity λ_a, etc., one can label all the amplitudes by the helicities. Then n and x measure the amount of helicity flip and are defined by

$$n = |(\lambda_c - \lambda_a) - (\lambda_d - \lambda_b)|, \quad n+x = |\lambda_c - \lambda_a| + |\lambda_d - \lambda_b|.$$

The above conjecture is that instead of a new amplitude for every set of helicities, all the amplitudes with a given n,x in a given reaction for a given exchange are the same, and even those for different exchanges or different reactions are approximately or qualitatively the same. At a detailed level this is now known not to be true (e.g. vector exchange amplitudes are more peripheral than tensor ones) but in the small t region where most data are the conjecture is approximately true, and it holds well for magnitudes over a larger range.

One important confirmation of this hypothesis is the apparent validity of our prediction[3] that the n=1 π exchange amplitude has a (complex) zero near $-t=0.6$ GeV^2. As you have heard in Matthews' talk, the prediction is basically satisfied. (The situation is not completely clear yet, however, since the higher energy CERN-Munich experiment does not see the dip (but in a detailed model the dip will move out[4] in -t with energy and calculations[5] may be consistent with the CERN-Munich results), and since the effect appears to be cleaner in the non-charge exchange reactions rather than the charge exchange ones, contrary to naive expectations.) Thus all known n=1 amplitudes are consistent with having a dip as in the conjecture. However, the position depends on the exchange and the reaction somewhat, with the short range tensor exchanges (e.g. A_2) having the dip further out in -t with a range about 0.7 times that for the vector exchanges.

We have heard about another result at this meeting which means our conjecture above can only be approximate. Namely, both Martin and Ochs have discovered that in the CERN-Munich data the difference between the full n=0, x=2 amplitude as t→0 and the pion pole is a decreasing function of the pion pair mass. Writing $M(n=0,x=2) = t/(t-m_\pi^2) - C$, they find that C decreases from near one at the rho mass to near $\frac{1}{2}$ at the f mass. With that definition, C is made up of about 1/3 A_2 exchange and about 2/3 absorption of the π in $\pi N \to \rho N$. It seems likely that part of the effect they have found is a decrease in the strength of the coupling of the A_2 to pion pairs of increasing mass; indeed, such an effect has been predicted by Hoyer, Roberts and Roy[6]. The rest of the effect will mainly arise from a decrease in the total cross section of the pion pair-nucleon system. To get precise numbers a detailed calculation is needed, but simple estimates suggest that one should predict that $\sigma_T(fN) \approx 15$ mb. It will be very interesting to see if measurements on nuclei can give such a result. A small part of the decrease should come from a decrease in the sum over non-elastic intermediate states because of the increased change in mass, but this should not be more than about 10%.

At this point I could take the amplitudes from our detailed analysis[7] of np→pn and give detailed predictions for the polarization measurements in ρ and K^* production, because the same s-channel helicity amplitudes are involved. Since the polarization measurements will not be done for some time and there will be detailed predictions available[5,8] I will restrict myself to only using the NN analysis as a guide to make some remarks on two topics of interest here.

PHASE COHERENCE

It has often been assumed that the three amplitudes for producing a ρ with nucleon helicity flip (four counting the s-wave production) have zero relative phase. The analysis[1] of Estebrooks and Martin disagrees with this, and so does the absorption model or the lessons of the np→pn analysis assuming the s-channel helicity amplitudes have a common structure. The basic point is very simple.

At small t the amplitudes with net helicity flip n>0 feel little absorption whatever the model, so if the pion pole is mainly real they are mainly real. But the amplitude with n=0, x=2 has a pion pole that is mainly real and which is cancelled at some t value (exactly where is model dependent, with any -t value in the range 0.02-0.05 GeV^2 being reasonable) so that that amplitude is purely imaginary at that t value. Thus at a point near $-t = m_\pi^2$ one amplitude is purely imaginary and the others are mainly real, with almost complete phase <u>incoherence</u>. If it should turn out that phase coherence held for $\pi N \rightarrow \rho N$ and not for np→pn it would have important implications not only for ρ production but for our entire view of particle reactions.

THE WILLIAMS' MODEL

In the past few years the Williams' model[9] has been exceedingly valuable in the study of pion pair production. It has been very effective in increasing the utilization of proper extrapolation techniques and in increasing our insight into the details of pion pair production.

However, there are situations where continued use of the Williams' model will get us into trouble. I suggest that it is time to go beyond the approximations of the Williams' model to a more realistic treatment. To repeat what I said above, the results one gets out of the data will only be as valid as the model used to get them.

Some of the shortcomings of the Williams' model are

-- Its amplitudes are coherent in phase; see the previous section.

-- It allows one to fit pion pair production data out to -t=0.15 or so with no other contributions. But there are many indications[10] that considerable A_2 exchange must be present there. Indeed, if the np→pn analysis[7] is a good guide the A_2 is important even at t=0 in the amplitudes with $\lambda_\rho = \pm 1$, perhaps as much as 1/3 of the full contribution.

-- More theoretically, it is not really an absorption model (in spite of what it is called) because it does not remove partial waves in a smooth way but instead artificially simulates the effects of absorption. When one has reached the level of looking at extensive data in detail and of needing to consider interferences with other exchanges then it may be very important to be as realistic as possible.

SUMMARY

I would like to emphasize the following points.

(1) If you want to stay in the business of meson-meson scattering, and you want to learn something, then

either (a) find new effects of unexpected importance, or (b) use production information and theoretical models much more.

(2) Your experiment or model is not the only one in the world.

(3) Always publish normalized $d\sigma/dt$ and <u>s-channel</u> density matrices, as well as anything else you want. Then other physicists can utilize the production data in trying to understand what is going on.

(4) All assumptions made to extract physics from raw data are models. Some are more correct and more useful than others.

(5) There is a good possibility that s-channel helicity amplitudes are simple and approximately common to many reactions. There are only a few kinds of s-channel helicity amplitudes in the world, and much may be learned about your reaction by studying the amplitudes it has in common with data from other sources.

ACKNOWLEDGMENTS

I would like to thank R. Field, S. Kramer, J. Gaidos and S. Barish for useful information.

REFERENCES

1. P. Estebrooks and A.D. Martin, Phys. Lett. <u>41B</u>, 350 (1972).
2. J.A. Gaidos, et al., Nucl. Phys. <u>B46</u>, 449 (1972).
3. M. Ross, F. Henyey, and G.L. Kane, Nucl. Phys. <u>B23</u>, 269 (1970).
4. It is probably a general effect that the strength of absorption decreases with energy over the energy range where non-diffractive experiments can be done, so that dips and crossovers move out with increasing energy until Regge shrinkage takes over and moves them in.
5. R. Field, private communication.
6. P. Hoyer, R.G. Roberts, and D.P. Roy, Rutherford preprint RPP/T/35.
7. M Vaughn and G.L. Kane, in preparation.
8. J.D. Kimel and E. Reya, to be published.
9. P.K. Williams, Phys. Rev. <u>D1</u>, 1312 (1970).
10. J.T. Carroll, et al., Phys. Rev. Lett. <u>27</u>, 1025 (1971).
11. R.K. Carnegie, et al., to be published.

Chapter 4.   Extrapolation Panel   255

RELATIVE PHASES OF SINGLE PION PRODUCTION AMPLITUDES

L. J. Gutay[*]
Purdue University, W. Lafayette, Indiana 47907

K. V. Vasavada
Indiana-Purdue University, Indianapolis, Indiana 46205

ABSTRACT

Recently it has been claimed that the coherent phase hypothesis of the P wave amplitudes in ρ^0 production is wrong. We show that this conclusion is not valid.

From the inception of absorptive amplitude description of charged[1,2] and neutral[3] ρ production, it has been implicitly assumed that the phase of the helicity amplitudes for a given dipion spin is independent of both the nucleon and the vector meson helicity states. Hereafter we refer to the former as the nucleon helicity coherence (NC) and to the latter as the vector meson helicity coherence (VC) hypothesis. These two distinct and independent assumptions combined with the factorization hypothesis[3] (FH) served the basis of earlier S-P-wave phase shift analyses.[3,4,5,6,7,8] In ρ^0 production there are six independent experimentally measurable quantities.[3] On the other hand the measured quantities depend on eight unknown quantities[5] even after the NC, VC, and FH. In view of these facts it would be of great importance if with fewer assumptions, one could determine all the parameters on which the density matrix elements depend.

In a recent letter Estabrooks and Martin[9] (EM) performed an analysis for single pion production process ($\pi^-p \to \pi^+\pi^-n$) in the ρ^0 region by assuming some restrictive relations between certain helicity amplitudes. Their results contradict the VC hypothesis for the helicity one and helicity zero P-wave amplitudes and they question the results of those analyses which were based on this assumption.[3,4,5,6,7,8] We will show in the following that their conclusion does not necessarily follow from the data and it arises from a specific choice of coordinate system and identification of a change of sign of an amplitude with the change of phase.

For the sake of clarity we give in the following, explicit expressions for the un-normalized s-channel density matrix elements $R_{\lambda\lambda'}^{\ell\ell'}$ in terms of the s channel helicity amplitudes $H_{\lambda_n,\lambda_p}^{\lambda}$ (P wave) and $H_{\lambda_n,\lambda_p}^s$ (S wave). Here the symbols ℓ, λ, λ_n and λ_p denote the di-pion angular momentum, its helicity and the neutron, proton helicities respectively. For convenience, we introduce[5,6,7] the two dimensional vectors, representing the helicity amplitudes:

---

[*]Work supported in part by the U.S. Atomic Energy Commission.

$$S_\pm = H^s_{+,\pm} \tag{1}$$

$$T_\pm = H^o_{+,\pm} \tag{2}$$

$$L_\pm = H^1_{+,\pm} + H^{-1}_{+,\pm} \tag{3}$$

$$D_\pm = H^1_{+,\pm} - H^{-1}_{+,\pm} \tag{4}$$

In terms of these vectors the density matrix elements and the differential cross-section can be expressed as

$$R^{00}_{00} = |S_+|^2 + |S_-|^2 = |S|^2 \tag{5}$$

$$3R^{11}_{00} = |T_+|^2 + |T_-|^2 = |T|^2 \tag{6}$$

$$6(R^{11}_{11} - R^{11}_{1-1}) = |D_+|^2 + |D_-|^2 = |D|^2 \tag{7}$$

$$6(R^{11}_{11} + R^{11}_{1-1}) = |L_+|^2 + |L_-|^2 = |L|^2 \tag{8}$$

$$6\, R^{11}_{10} = D_+ T^*_+ + D_- T^*_- = D \cdot T^* \tag{9}$$

$$2\sqrt{3}\, R^{10}_{10} = D_+ S^*_+ + D_- S^*_- = D \cdot S^* \tag{10}$$

$$\sqrt{3}\, R^{10}_{00} = T_+ S^*_+ + T_- S^*_- = T \cdot S^* \tag{11}$$

$$\frac{d\sigma}{dt} = N\left[|S|^2 + |T|^2/3 + (|D|^2 + |L|^2)/6\right] \tag{12}$$

where N is a kinematic factor. These density matrix elements are related to those EM by certain normalization factors which are not relevant to the present discussion.

Next we introduce the NC hypothesis for the relative phases of the nucleon non flip and flip amplitudes.

$$S_+/\widetilde{S}_+ = S_-/\widetilde{S}_- = e^{i\,{}^o\alpha_o} \tag{13}$$

$$T_+/\widetilde{T}_+ = T_-/\widetilde{T}_- = e^{i\,{}^o\alpha^1_1} \tag{14}$$

$$D_+/\widetilde{D}_+ = D_-/\widetilde{D}_- = e^{i\,{}^1\alpha^1_1} \tag{15}$$

At the pion pole ${}^o\alpha^I_\ell \to \delta^I_\ell$, the on mass shell isotopic spin I and angular momentum ℓ pion-pion phase shift.

EM set $S_+ = D_+ = 0$ and take $T_+ = \sqrt{\frac{t_{min}}{t-t_{min}}}\ T_-$ where t denotes the four momentum transfer squared to the nucleon and $t=t_{min}$ in the forward direction. Thus clearly EM hypothesis is much more restrictive than ours. If EM hypothesis is true, ours will be true but the converse does not follow. For example, as we have pointed out recently,[10] the proportionality hypothesis of EM for T_+ and T_- does lead to a disagreement with the available experimental data at $|t| \sim 0.6$ GeV^2.

Combining the phase relations in Equations (13), (14), (15) with Eqs. (9), (10), (11) we have

$$6\ \text{Re} R_{10}^{11} = \vec{D} \cdot \vec{T} \cos({}^1\alpha_1^1 - {}^0\alpha_1^1) = |D|\ |T|\cos\gamma \cos\varphi \tag{16}$$

$$2\sqrt{3}\ \text{Re} R_{10}^{10} = \vec{D} \cdot \vec{S} \cos({}^1\alpha_1^1 - {}^0\alpha_0^0) = |D|\ |S|\cos(\gamma-\tau)\cos(\varphi+\Delta) \tag{17}$$

$$\sqrt{3}\ \text{Re} R_{10}^{10} = \vec{T} \cdot \vec{S} \cos({}^0\alpha_1^1 - {}^0\alpha_0^0) = |T|\ |S|\cos\tau \cos\Delta \tag{18}$$

Here γ and τ are the angles between the relevant two dimensional helicity vectors and φ and Δ are the relative phases. For convenience we have interchanged γ and $\gamma-\tau$ as defined in reference 8. Our expressions would reduce to those of EM by setting $\tau=\gamma=0$ and making the above assumptions.

So far we have confined our discussion to the s-channel helicity frame. But is well known that all the equations given above remain the same in form if the density matrix elements and the amplitudes are defined in the Gottfried-Jackson (GJ) frame.[8] The amplitudes in GJ frame are linear combinations of the s-channel helicity amplitudes with real coefficients. Consequently if there is a phase coherence of all the amplitudes describing a dipion state with a definite angular momentum in the GJ frame the same will be true in any other frame related by rotation e.g. the s-channel frame.

In order to reduce the number of unknowns, EM[9] assumed that the A_1 exchange amplitude is negligible and hence $D_+ = 0$. Note that under their approximation we would have for the real parts

$$6\ \text{Re} R_{10}^{11} = \widetilde{D}_- \widetilde{T}_- \cos\varphi \tag{19}$$

$$2\sqrt{3}\ \text{Re} R_{10}^{10} = \widetilde{D}_- \widetilde{S}_- \cos(\varphi+\Delta) \tag{20}$$

and for the imaginary parts

$$6\ \text{Im}\ R_{10}^{11} = \widetilde{D}_- \widetilde{T}_- \sin\varphi \tag{21}$$

$$2\sqrt{3}\ \text{Im}\ R_{10}^{10} = \widetilde{D}_- \widetilde{S}_- \sin(\varphi+\Delta) \tag{22}$$

This would imply that zero of $\widetilde{D}_-$ would give rise to zeros in all the four density matrix elements. On the other hand zeros of $\cos\varphi$ would give rise to maximum at least in Eq. (21). Zeros of $\text{Re} R_{10}^{11}$ and

ReR_{10}^{10} are observed in the s channel frame.[9] Absorptive one pion exchange models[1,2,3,11] do predict a change of sign of $\tilde{D}_-$ in the s-channel. Also the presently available experimental data indicates that the equality sign in the positivity condition

$$|R_{10}^{11}|^2 \le \tfrac{1}{2} R_{00}^{11} (R_{11}^{11} - R_{1-1}^{11})$$

is satisfied[12,13] for $-t < 0.2$ GeV^2 both at 6 GeV/c and 17 GeV/c when only ReR_{10}^{11} is used for R_{10}^{11} and hence $\sin\varphi$ does not exhibit violent variation in the relevant t-region and is consistent with zero. This shows that the zero in ReR_{10}^{11} and ReR_{10}^{10} are due to the zero in D and supports the VC hypothesis. Unfortunately EM set $\tilde{D}_-$ as $|D_-|$ in Eq. (19) and thus the change of sign of ReR_{10}^{11} results in a sudden variation of $\cos\varphi$ essentially from -1 to 1. This is then interpreted by them as a phase incoherence between the helicity zero and the helicity one P wave amplitudes.

Note that the zeros are not seen in the GJ frame where these density matrix elements are relatively flat in this momentum transfer region.[12] Thus an approach in the GJ frame formally identical to that of EM would not cause this problem. In fact in reference 8, all the density matrix elements were fitted well in the 2.7 - 11 GeV momentum range without requiring any phase incoherence for $|t| < 0.2$ GeV^2.

Finally we would like to present a phase independent argument to show that $D_- \approx 0$ at $t \approx 0.02$. It follows from our equations 7 and 8 without any approximation that

$$A = \frac{R_{1-1}^{11}}{R_{11}^{11}} = \frac{|L|^2 - |D|^2}{|L|^2 + |D|^2} \tag{24}$$

we claim that if $D_+ = 0$, the asymmetry parameter $A \approx 1$ at $t \approx -0.02$ since $|D_-| = |D| = 0$ at this value of t at 17 GeV/c incident beam momentum. The deviation of A from unity at its maximum value is the measure of the magnitude of $|D_+|$, and thus the sum of A_1 exchange and the cut contribution to D_+. In an earlier paper we showed[10] that indeed the asymmetry parameter is unity at $t \approx -0.02$. This shows that not only A_1 exchange is negligible as assumed by EM, but also the cut contribution to D_+, however their solution for D_- ($|M_-|$) and $\cos\varphi$ is incorrect.

Thus we conclude that the phase incoherence found by EM is essentially due to their identification of a change of sign of a real amplitude with the change of phase by π. Thus their results do not invalidate earlier phase shift analyses based on the vector meson helicity coherence hypothesis.

We would like to thank M. Block, R. Field, D. Kimel and F. T. Meiere for valuable discussions; A. D. Martin for clarifying his assumptions and W. Manner for information about his experimental results presented at the conference on $\pi\pi$ and $K\pi$ interactions at Tallahassee. Further we would like to thank F. J. Loeffler for encouragement.

REFERENCES

1. K. Gottfried and J. D. Jackson, Nuovo Cimento $\underline{34}$, (1964) 735.
2. L. Durand and Y. T. Chiu, Phys. Rev. $\underline{139}$, (1965) 646.
3. L. J. Gutay et al., Phys. Rev. Letters $\underline{18}$, (1967) 142.
4. W. D. Walker et al., Phys. Rev. Letters $\underline{18}$, (1967) 630.
5. E. Malamud and P. E. Schlein, Phys. Rev. Letters $\underline{19}$, (1967) 1056.
6. J. H. Scharenguivel et al., Phys. Rev. $\underline{186}$, (1969) 1387.
7. L. J. Gutay, et al., Nucl. Phys. $\underline{B12}$, (1969) 31.
8. J. H. Scharenguival et al., Phys. Rev. Letters $\underline{24}$, (1970) 332.
9. P. Estabrooks and A. D. Martin, Phys. Letters $\underline{41B}$, (1972) 350.
10. L. J. Gutay, et al., Phys. Rev. Letters $\underline{30}$, (1973) 465.
11. J. D. Kimel and E. Reya, Florida State Univ. Preprint 73-2-20.
12. G. Grayer et al., Nucl. Phys. $\underline{B50}$, (1972) 29.
13. H. A. Gordon et al., Phys. Review, to be published.

EXTRAPOLATION TECHNIQUE USED BY THE S.L.A.C. GROUP IN THEIR MEASUREMENT OF THE $\pi^+\pi^-$ CROSS SECTION

Paul Baillon
CERN-Geneva

ABSTRACT

We describe the extrapolation technique used by the S.L.A.C. Boson spectrometer group to measure the $\pi^+\pi^-$ cross section and phase shift. The errors introduced by a limited polynomial expansion of the extrapolated curve are discussed.

This talk will be devoted to the technique we used to find via extrapolation the $\pi^+\pi^-$ cross section. I shall recall first the experiment, then the formulae used for the extrapolation. Next I shall say some words about the conformal mapping used and I shall evaluate approximatively the errors due to limited expansion and I shall recall the results we found.

The detail and the authors of the works appear in Ref. (1). Let us here say simply that the experiment was done at S.L.A.C. using a boson spectrometer (Fig. 1). We recorded the reaction $\pi^-p \to \pi^+\pi^-X^0$ which was divided into

(1) $\pi^-p \to \pi^+\pi^-n$ 17121 events
(2) $\pi^-p \to \pi^+\pi^-X^0$ 48516 events

both with $|t| < 11\mu^2$ (X^0 stands for all possible states at the nucleon vertex).

In order to keep only events obtained with a good detection efficiency we eliminated those for which $|\cos\theta| > 0.7$, θ being the angle between the incident and outgoing π^-'s in the ($\pi^+\pi^-$) rest system. With the events of type (1) we define a first extrapolation

$$\int_{-0.7}^{0.7} \frac{d\sigma_{\pi\pi}}{d\cos\theta} d\cos\theta = \lim_{t \to \mu^2} \frac{2\pi k_{lab}^2 (t-\mu^2)^2}{f^2 M_{\pi\pi} \sqrt{0.25 M_{\pi\pi}^2 - \mu^2}} \cdot \int_{-0.7}^{0.7} \frac{d^3\sigma_n}{dt\, dM_{\pi\pi}\, d\cos\theta} d\cos\theta$$

where k_{lab} is the incident pion momentum in the laboratory, μ is the pion mass, $M_{\pi\pi}$ is the mass of the outgoing π-π system, $f^2 = 0.081$ the πN coupling constant, $d^2\sigma_n/dt\, dM_{\pi\pi}$ the cross section for the reaction $\pi^-p \to \pi^+\pi^-n$ in the intervals dt and $dM_{\pi\pi}$, and $\sigma_{\pi\pi}$ the $\pi^+\pi^- \to \pi^+\pi^-$ the cross section. With events of type (2) we define a second extrapolation formula which gives the differential cross section.

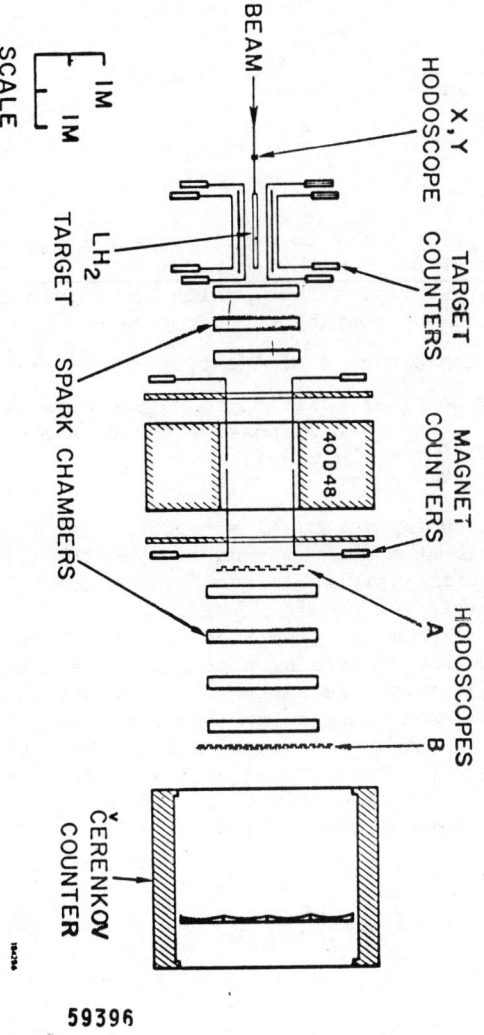

fig. 1

$$\frac{d\sigma_{\pi\pi}}{d\cos\theta} = \left\{ \int_{-0.7}^{0.7} \frac{d\sigma_{\pi\pi}}{d\cos\theta} \cdot d\cos\theta \right\}_{\pi^+\pi^- \to \pi\pi} \cdot \lim_{t\to\mu^2} \frac{\dfrac{d^3\sigma_{x^0}}{dt\, dM_{\pi\pi}\, d\cos\theta}}{\displaystyle\int_{-0.7}^{0.7} \dfrac{d^3\sigma_{x^0}}{dt\, dM_{\pi\pi}\, d\cos\theta} \cdot d\cos\theta} \cdot d\cos\theta$$

In order to perform the extrapolation we used a conformal mapping technique similar to Ref. (2). It consists first to look at the singularities in t, which are a pole at $t = \mu^2$ and a cut from $t = 9\mu^2$ to $t = \infty$ corresponding to the opening of $\pi \to \pi\pi\pi$. Then we made a transformation of the type $x = \dfrac{at + b}{t + d}$ in order to bring the cut to a symmetrical position round the physical region, Fig. (2).

For this purpose we choose a,b,d such that:
$x(\infty) = -x(9\mu^2)$
$x(-0.275) = -x(-0.002) = 1$.

which imposes $x(\mu^2) = -1.26$, $x(\infty) = -x(9\mu^2) = 4.36$. The double pole at $t = \mu^2$ is removed by multiplying by $(t - \mu^2)^2$, and in the x plane only the singularities coming from the cut $[-\infty, x(9\mu^2)]$, $[x(\infty), +\infty]$ remain.

From the general property of analytical functions we know that the Taylor expansion of our extrapolating function converges in the biggest circle centered at the origin of the expansion such as it does not contain any singularities. We know also that the expansion diverges outside this circle. Here the circle is centered at the origin and has a radius $R = |x(9\mu^2)| = 4.36$.

Let us call g(t) the function to extrapolate at $t = \mu^2$ and f(x) its transformed. We have:

$f(x) = g(g(x)) = g(\dfrac{b - dx}{x - y})$

$g(t) = f(\dfrac{at + b}{t + d})$.

The Taylor series of f(x) at the origin is written

$f(x) = \sum\limits_{0}^{\infty} A_i\, x^i$, $A_i = \dfrac{1}{n!} \dfrac{d^n f(o)}{dx^n}$

Considering now the fact that the radius of convergence is $R = 4.36$ we have a conservative method of evaluating the errors. Roughly speaking we have

$f(x) \sim B \sum\limits_{0}^{\infty} (\dfrac{x}{R})^i$

B being the average value of $A_i\, R^i$ which is finite in order to insure the convergence for $|x| = R - \varepsilon$ and the divergence for $|x| = R + \varepsilon$.

When we limit the development to n we make a relative error on the value at x:

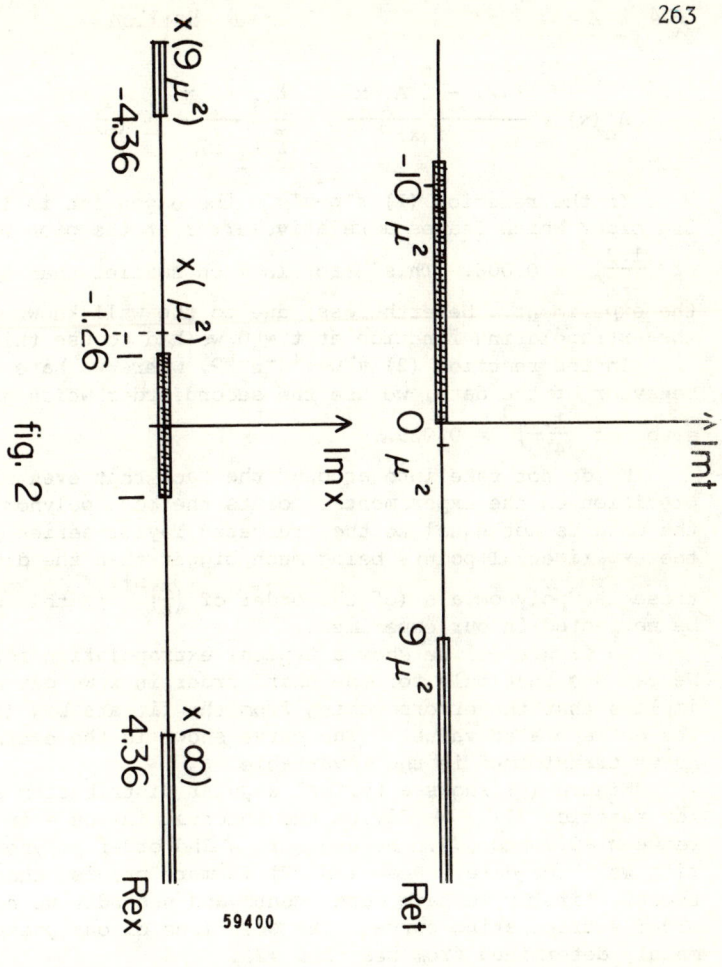

fig. 2

$$\Delta_n(x) = \frac{f(x) - \sum_0^n A_i x^i}{f(x)} = \frac{\sum_{n+1}^\infty A_i x^i}{\sum_0^\infty A_i x^i} \sim \left(\frac{x}{R}\right)^{n+1}$$

In the reaction (1) $\pi^-p \to \pi^+\pi^-n$ the expansion is limited to the 3rd order which leaves a relative error at the pion pole of the order of $\left(\frac{1.2}{4.3}\right)^4 = 0.006$. This error is much smaller than the precision of the experiment. Nevertheless, due to the well known rapid fall of the extrapolating function at t = 0 we had to use this high order.

In the reaction (2) $\pi^-p \to \pi^+\pi^-n X^0$, where we have a much smoother behavior of the data, we use the second order which gives a relative error of $\left(\frac{1.2}{4.3}\right)^3 = 0.025$.

We do not take into account the fact that even with an infinite precision on the experimental points the best polynomial fitted on the data is not equal to the truncated Taylor series. The errors on the experimental points being much bigger than the differences between these two polynomials (of the order of $\left(\frac{x}{R}\right)^{n+1}$), this difference can be neglected in our experiment.

In figure (3) we show a typical extrapolation for reaction (1). We can see that only for the third order in x we get a good χ^2. It implies that the errors coming from the fit are big (50 ± 8 mb was the extrapolated value). The curve shown is the extrapolating curve transformed in the t variable.

Figure (4) shows a typical angular distribution extrapolation for reaction (2). We divide the interval in cos θ in four parts between -0.7 and 0.7. We use here a 2nd order polynomial in x which fits well the data. Reaction (2) is more precise than reaction (1) because firstly we have more events and secondly we need a lower order extrapolating curve. The solutions of our phase shift are mainly determined from reaction (2).

Figure (5) displays the angular distribution obtained combining results from reaction 1 and 2. The points are the result of the phase shift analysis done by fitting separately results from reaction 1 and 2.

Figure (6) shows the I = 0 S wave phase shift. The bottom solution is in perfect agreement with the bottom solution found later by the experiments of Ref. (3) and (4).

In Ref. (3) only one solution was obtained for the S wave phase shift. Their extrapolation was a linear expansion in t multiplied by a Dü rr-Pilkuhn form factor. Clearly this form factor does not eliminate the cut in t and they are left with the same singularities as in our case. From the same considerations as before we can estimate at least an extra relative error of 10% at the π pole due to the limited expansion. The fact that they obtain a good χ^2 in the extrapolation does not prove that the remaining term is negligible. It proves only that it is small. Then it would have been much safer to use a second order polynomial. In fact doing this

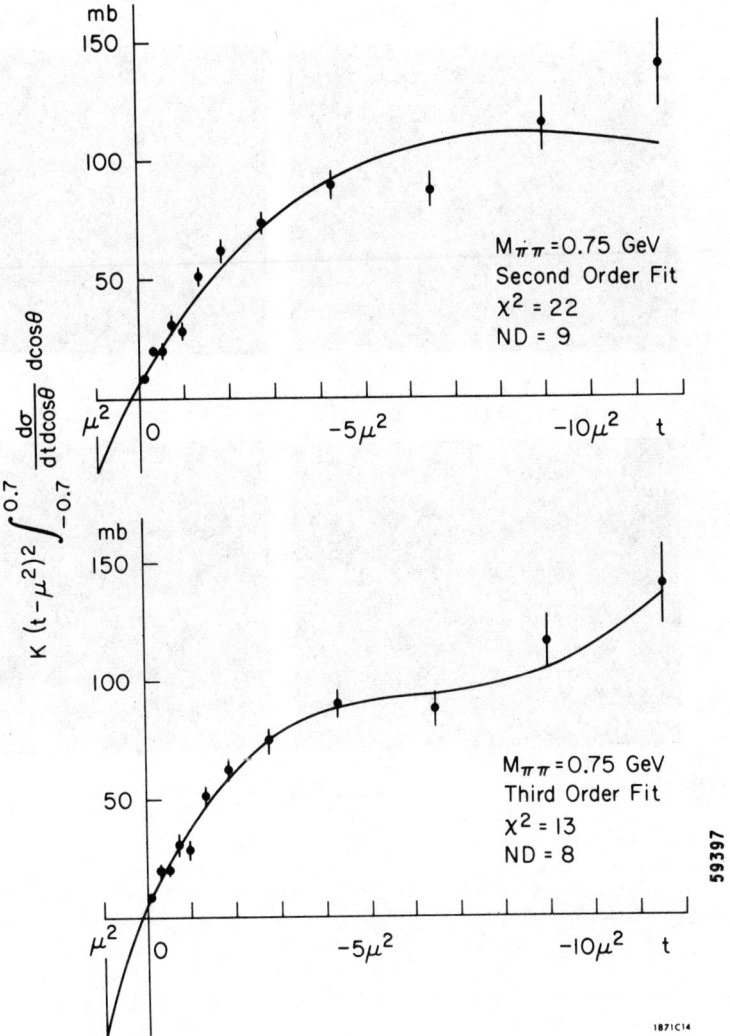

fig. 3

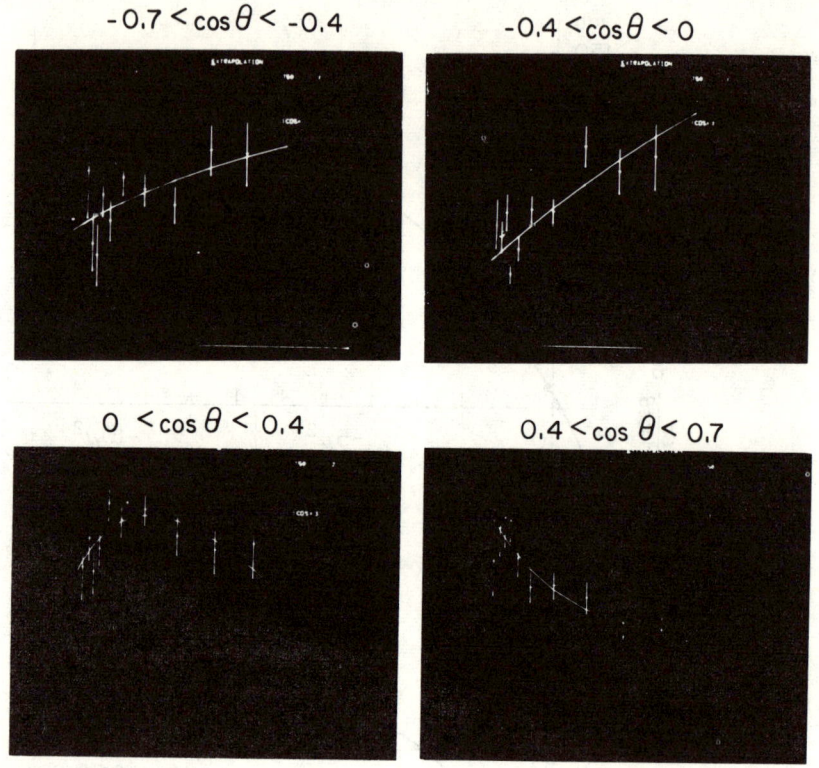

fig. 4

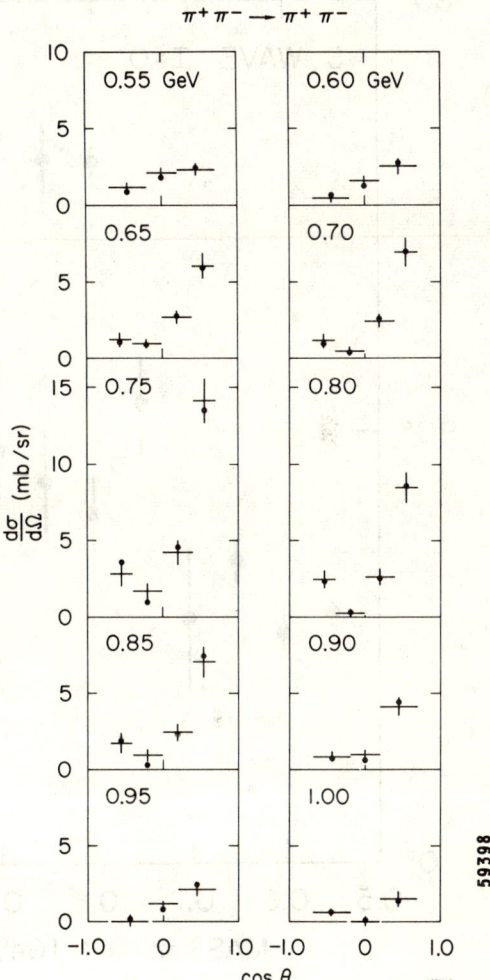

fig. 5

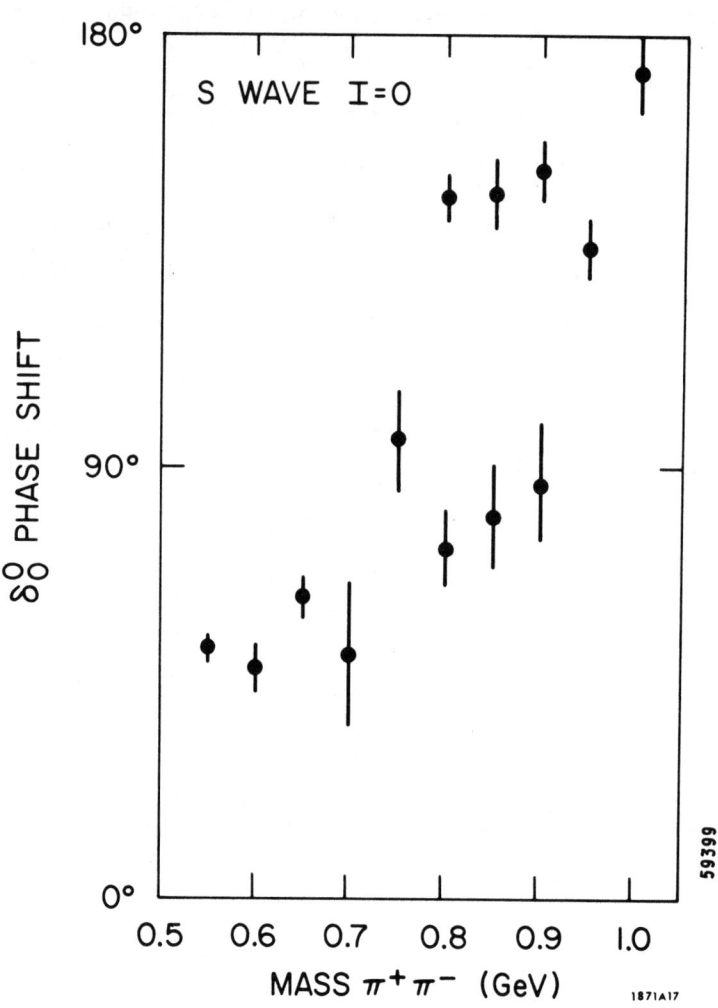

fig. 6

they would have increased the error coming from the fit and perhaps would not have excluded the up solution so clearly.

REFERENCES

1. P. Baillon, R. K. Carnegie, E. E. Kluge, D.W.G.S. Leith, H. L. Lynch, B. Ratcliff, B. Ritcher, H. H. Williams and S. H. Williams, Phys. Letters 38B (1972) 555.
2. J. P. Baton, G. Laurens and J. Reignier, Phys. Letter 33B (1970) 525 and 528.
3. S. D. Protopopescu, M. Alston-Garnjost, A. Barbaro-Galtieri, S. M. Flatté, J. H. Friedman, T. A. Lasinskj, G. R. Lynch, M. S. Rabin and F. T. Solmitz, 3rd International Conference on Experimental Meson Spectroscopy Philadelphia 28-29 April 1972. Submitted to Phys. Rev.
4. G. Grayer, B. Hyams, C. Jones, P. Schlein, W. Blum, H. Dietl, W. Koch, E. Lorenz, G. Lütjens, W. Männer, J. Meissburger, W. Ochs, U. Stierlin and P. Weilhammer, 3rd International Conference on Experimental Meson Spectroscopy Philadelphia 28-29 April 1972.

EXTRAPOLATION WITH COMPLEX TRANSFORMATIONS

G. Laurens
Département de Physique des Particules Elémentaires
CEN-Saclay, B.P.2, 91190 Gif sur Yvette, France

ABSTRACT

Some complex transformations used to obtain the ππ cross section by Chew and Low extrapolation are succintly described.

The problem I have to talk about is a very large one. This problem is in fact: how to use the information we have accumulated in the physical region, in order to obtain the residue of an analytic function at a point outside the physical region.

You know that this is a mathematicaly improperly posed problem, that is to say a problem without an unique solution and even unstable with respect to weak fluctuations of the data.

First of all one needs high statistics data in order to minimize the influence of the bining on the final result. We know[1] that the meaningful maximum degree for the fitting polynomial grows as the logarithm of the number of events: high statistics are needed to reduce the truncation errors done in limiting the expansion of our extrapolation function.

After that, the problem becomes a purely mathematical one, and several authors have worked on this question. As we know the analytic structure of the function to be extrapolated, we will use this very important information to improve the convergence of the process and to force ourselves in the best mathematical conditions, by doing suitable mappings.

For these mappings I restrict myself to the first papers of Ciulli[2] and Cutkosky and Deo[3]. The mathematical transformations are summarized below:

Initial situation in the complex Δ^2 plane

$$\begin{array}{c} \text{-}\infty \quad \text{cut} \qquad \text{-c} \qquad \text{pole} \quad \text{Physical region} \\ \text{\hrulefill} \qquad \qquad \ast \qquad \cdot \quad \text{\vrule}\text{------}\text{\vrule} \longrightarrow \quad \Delta^2 = -t \\ \qquad \text{-}9\mu^2 \qquad \text{-}\mu^2 \; 0 \; A \qquad \qquad B \end{array}$$

Homographic transformation

$$x(\Delta) = \frac{q\Delta^2 + \beta}{\Delta^2 + \delta}$$

(diagram: complex X plane with pole at A, physical region from A to B, hatched regions from $-\infty$ to $-c'$ and c' to $+\infty$)

$k^2 = \frac{1}{q}$ example $A = 0$ $B = 12\rho^2$ $k^2 = 0.044$

Elliptic transformation

$$\mathfrak{z}(x) = \int_0^x \frac{du}{\sqrt{(1-u^2)(1-k^2 u^2)}}$$

(diagram: complex $\mathfrak{z}$ plane, rectangular physical region with pole at A, B at right, corners $-c'$ and c' at top)

Circular transformation

$$Z(\mathfrak{z}) = i e^{-i\frac{\pi}{2}\frac{\mathfrak{z}}{K}} \qquad K = \mathfrak{z}(1)$$

(diagram: complex Z plane, semicircular physical region between radii $1/R$ and R, pole at A)

Approximations

k^2 small, at the first order we have:

$$\mathfrak{z}(x) = (\text{Arcsin } x)\left[1 + \frac{k^2}{4} - \frac{k^2}{4} x\sqrt{(1-x^2)}\right]$$

$$Z(\mathfrak{z}) = i e^{-i\frac{\pi}{2}\frac{\mathfrak{z}(x)}{K}} = \rho e^{i\varphi} \quad ; \quad K = \mathfrak{z}(1)$$

$$\rho = \frac{\pi}{2}\left(1 - \frac{\text{Re }\mathfrak{z}}{K}\right) \qquad \varphi = e^{\frac{\pi}{2}\frac{\text{Im}(\mathfrak{z})}{K}}$$

$F(\varphi)$ the function to be extrapolated, is expanded as:

$$F(\varphi) = \sum_{n=0}^{\infty} A_n (z + z^{-1})^n = \sum_{n=0}^{\infty} A_n' \cos n\varphi$$

$$\cos\varphi = \sin\left[\text{Arcsin}\, X - \frac{k^2}{4} X \sqrt{(1-X^2)}\,\right]$$

$$\cos\varphi \simeq X\left[1 - \frac{k^2}{4}(1-X^2)\right]$$

$$F(\varphi) \simeq \sum_{n=0}^{N} B_n X^n \left(1 - \frac{k^2}{4}(1-X^2)\right)^n$$

$$F(\varphi) \simeq \sum_{n=0}^{N} \beta_n X^n$$

The expansion in a Laurent serie or in a Fourier serie with the φ variable is the optimal convergent approximation of our analytic function.

Applying these considerations to a $\pi^-p \to \pi\pi N$ experiment at 2.77 GeV/c we have found[4] that, at this energy, for our particular and simple reaction, the X variable was sufficient, because the parameter of the elliptic transformation is small. This is certainly not the case for the reaction K p → K π N where the cut is closer to the physical region.

The improvement we have obtained by taking into account in such a way, the analyticity of the function was:
- a better stability of the results according to various binings in t and ππ mass.
- a lower degree for the extrapolating polynomial; i.e. for $\pi^-\pi^0$ in the X variable we need a degree one instead of a degree two in the t variable. For $\pi^+\pi^-$ it is a degree two instead of a degree three.

This implies lower errors on the extrapolated values.

In addition I would point out that it is not convenient to perform mappings undiscriminately: for example, if we use the Durr-Pilkhun form factors they introduce complex singularities (outside the physical region) which come near the interesting region after the mapping is done.

In conclusion and to illustrate this, I would say that a Saclay group is working on a 4 GeV/c experiment $\pi^-p \to \pi\pi N$. Their preliminary results on the unconstrained extrapolated cross sections are shown here.

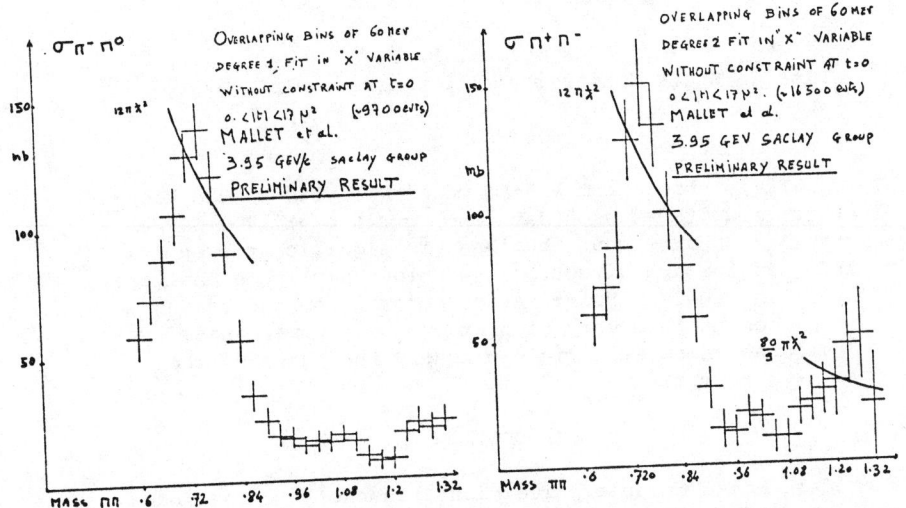

I must point out that mappings are not a magic trick to extract ππ phase shifts; since the first papers mentioned elsewhere, other papers have been published[5] proposing other mathematical tools, they must be used and compared.

REFERENCES

1. M. Froissart, Physique des hautes énergies, Ecole d'été des Houches 1965 (Dunod, Paris, 1965) p.170.

2. S. Ciulli, Nuovo Cimento 61 A, 787 (1969), 62 A, 301 (1969).

3. R.E. Cutkosky and B.B. Deo, Phys. Rev. 174, 1859 (1969).

4. J.P. Baton, G. Laurens, J. Reignier, Physics Letters 33 B, 528 (1970).

5. J. Pisůt, Springer Tracts in modern physics, 55, 43 (1969),
R.E. Cutkosky, Ann. Phys. 54, 350 (1969),
Chia Chang Shin, Carnegie-Mellon University preprint (1970).

A_2-EXCHANGE AND POLARIZATION EFFECTS IN THE ANALYSIS OF $\pi\pi$ SCATTERING*

J. D. Kimel and E. Reya
Florida State University, Tallahassee, Fla. 32306

ABSTRACT

It is shown from a $\pi + A_2$ model analysis of the high-statistics CERN-Munich data on $\pi^- p \to \pi^- \pi^+ n$ at 17.2 GeV/c that A_2 exchange contributions are significant even in the near forward direction. A unique amplitude analysis, utilizing nucleon polarization information and taking into account this more complex production mechanism, is developed which will yield uniquely the phase of the $\pi\pi$ s-wave amplitude.

INTRODUCTION

Models based on absorptive pion exchange[1,2,3] have been remarkably successful in predicting the highly structured production and decay distributions for $\pi^- p \to \pi^- \pi^+ n$ when the dipion effective mass is in the ρ region. Although analyses of pion photoproduction[4] suggest the presence of significant A_2-exchange contributions in the transverse ρ production amplitudes, their effects in $\pi^- p \to \pi^- \pi^+ n$ have not been readily apparent in the region of small $|t|$. The high statistics CERN-Munich experiment,[5] however, over the broad t-range up to $-1(GeV/c)^2$, now makes possible a strongly constrained study of A_2-exchange in this reaction. Here we show that the A_2-exchange contribution is <u>substantial</u> in the transverse ρ helicity amplitudes, even at $t=t_{min}$, and that the presence of A_2-exchange should be evident through predicted energy-dependent shifts in the structure of the data.

Because of the complexity of the production mechanism, analyses of $\pi\pi$ scattering based solely on absorptive π-exchange are inadequate and possibly misleading. More model-independent methods are required, e.g. of the kind suggested by Estabrooks and Martin.[6] But since their partial amplitude analysis does not include nucleon polarization information, their resulting $\pi\pi$ s-wave amplitude suffers an up-down ambiguity as in previous work. This ambiguity can, however, be eliminated on the basis of $\pi^- p \to \pi^- \pi^+ n$ data alone if polarization information is included. We delineate here the new information which can be obtained from target (or recoil) nucleon polarization measurements. Neglecting the contributions of A_1-exchange, we develop a unique amplitude analysis for the reaction $\pi^- p \to \pi^- \pi^+ n$ which yields the phase of the

---

*In part supported by the U. S. Atomic Energy Commission and by the National Science Foundation.

ππ s-wave amplitude uniquely. We also discuss the generalization
of the method to include the possibility of A_1-exchange effects.

$$\pi + A_2 \text{ MODEL FOR } \pi^- p \rightarrow \pi^- \pi^+ n$$

In the reaction

$$\pi^- + p(\beta) \rightarrow \sum_{\ell,m} \pi^- \pi^+(\ell,m) + n(c), \tag{1}$$

where the symbols in parentheses label the helicities of the
nucleons and the spin and helicity of the dipion states, the
amplitudes $(f^\ell_{mc\beta})^\pm$ corresponding to natural (+) and unnatural (−)
parity exchanges in the t-channel can, at high energies, be
expressed as a simple linear combination of s-channel helicity
amplitudes,

$$(f^\ell_{mc\beta})^\pm = \frac{1}{\sqrt{2}} [f^\ell_{mc\beta} \mp (-)^m f^\ell_{-mc\beta}]. \tag{2}$$

For ππ effective masses in the ρ region, where only s- and p-wave
dipion states strongly mediate reaction (1), eight independent
complex amplitudes contribute. If, as in this model, the contribution of A_1-exchange is neglected,[7] three of these amplitudes,
f^0_{0++}, f^1_{0++}, and $(f^1_{1++})^-$, vanish. The five which survive are

$$\begin{aligned}
M_s &= f^0_{0+-} \, [\pi, \, \pi\text{-cut}] \\
M_0 &= f^1_{0+-} \, [\pi, \, \pi\text{-cut}] \\
M_- &= (f^1_{1+-})^- \, [\pi, \, \pi\text{-cut}, \, A_2\text{-cut}] \\
M^+_{+-} &= (f^1_{1+-})^+ \, [A_2, \, A_2\text{-cut}, \, \pi\text{-cut}] \\
M^+_{++} &= (f^1_{1++})^+ \, [A_2, \, A_2\text{-cut}] \\
|M_+|^2 &= |M^+_{+-}|^2 + |M^+_{++}|^2
\end{aligned} \tag{3}$$

where the first three of these amplitudes correspond to unnatural
parity exchange while the remaining ones correspond to natural (in
brackets are given the possible pole and cut contributions).

The high statistics CERN-Munich data on reaction (1)
over the broad t-range up to $t = -1 (GeV/c)^2$ at 17.2 GeV/c makes
possible for the first time an analysis of the A_2 contribution to
this reaction strongly constrained by accurate data. It is
important in such an analysis not to bias the results by using
a rigid model for the absorbed π-exchange contribution. Thus we
apply the same general absorption model for π-exchange which
successfully describes[2,3] the SLAC and CERN-Munich data in the
near forward direction.

<u>Absorbed π amplitude:</u> $f^\ell_{m+-} = (-t)^{n/2} \dfrac{P^\ell_n(\mu^2,s')}{\mu^2-t} F_n(t) [(2\ell+1)^{1/2} b_\ell]$ (4)

where μ is the pion mass, s' the square of the dipion mass, n the net helicity flip, $P^\ell_n(\mu^2,s')$ is the well-known[2] residue from the π-exchange Born term, and b_ℓ is the ππ partial wave amplitude. The absorption-induced collimating functions $F_n(t)$ are generally expected to <u>differ</u> for each net helicity n. We take this into account by

$$F_n(t) = F_1(t)[1 + a_n(t-\mu^2)], \quad n = 0, 2;$$ (5)

this form was motivated[2] by specific calculations based on the "canonical" absorption model of Gottfried and Jackson.[8] The generality of (5) is shown by its being able to represent rather disparate models of absorptive corrections, e.g., the model of Gottfried and Jackson (or strong cut model) corresponds to

$$a_0 \simeq 10(\text{GeV/c})^{-2}, \quad a_2 \simeq -3(\text{GeV/c})^{-2}$$ (6)

whereas the OPE-δ model of Williams[9] is represented by $a_0=a_2=0$. The collimating function $F_1(t)$ is determined by the effective rescattering slope A, as usual in <u>absorption models</u>. Being suppressed relative to (4) by a factor $\sqrt{t_{min}/t}$, the π-exchange nucleon non-flip amplitudes are neglected. The A_2 Regge poles for the n = 0, 2 amplitudes are written in the form

<u>A_2 pole amplitude:</u> $f^1_{-1+-} = (-t)\gamma e^{-i\pi\alpha/2} s^\alpha [(2\ell+1)^{1/2} b_\ell]$ (7)

where parity conservation requires (for the A_2 pole contribution) $f^1_{1+-} = f^1_{-1+-}$. In Eq. (7) the scale factor is taken to be $1(\text{GeV})^2$ and the trajectory to be $\alpha(t) = 0.5 + 0.9t$. The ππ scattering amplitude b_ℓ is included in (7) in accordance with Watson's theorem which states[10] that the phase variation of (7) with ππ mass is given by the ππ phase shifts. For A_2-exchange poles the nucleon non-flip are related to the flip by

$$f^1_{m++} = rf^1_{m+-}/\sqrt{-t}$$ (8)

with r expected[11] from ρ-A_2 exchange degeneracy (EXD) and pion photoproduction to be $r \simeq \frac{1}{4}$ GeV, so that the nucleon non-flip amplitudes are rather small compared to the flip. The absorptive cut corrections are calculated by the conventional convolution integral, which of course introduces the cut parameters λ_n ($\lambda_n = 1$ corresponds to the weak cut model).

RESULTS

Figure 1 shows our model description of the CERN-Munich data. Since in fitting the absorptive pion parameters a_0 and a_2, we always obtained results consistent with the canonical absorption model, we used the values of Eq. (6). The t-dependence of the cross-section requires $A = 12.5$ $(GeV/c)^{-2}$ and the A_2 residue we obtain is

$$\gamma = -0.85 \text{ GeV}^{-3}, \tag{9}$$

in approximate agreement[11] with ρ-A_2 EXD. Although the model description of the data is not strongly dependent on the A_2-cut strength, our best fit corresponds to $\lambda_0 \approx \lambda_2 \approx 1.8$.

The evidence for A_2 exchange is most apparent in the density matrix elements $\rho_{00}^{11} - \rho_{11}^{11}$ and ρ_{1-1}^{11} where the data show for $|t| > 0.4 (GeV/c)^2$ that these density matrix elements are approaching their limiting values (for negligible absorptive π and A_2-cut contributions) of -0.5 and 0.5 respectively. In Fig. 2 we compare our model amplitudes directly with the amplitude analysis of Estabrooks and Martin,[6] with excellent agreement. Although their amplitude analysis corresponds to "solution 2", rather than the now favored[12] "solution 1", we understand that the main difference in the results is for the $\cos\phi$ distribution of "solution 1" to tend more toward a step function, more in agreement with our model analysis as seen in Fig. 2.

Of particular importance to analyses of $\pi\pi$ phase shifts is the magnitude of the absorptive A_2 contribution to $|M_+|$ which from Fig. 2 is about 30% that of the absorbed π at $t = t_{min}$. The same holds for $|M_-|$. This means that $\text{Re}\rho_{10}^{+0}$ is badly contaminated by A_2-exchange but that $\text{Re}\rho_{00}^{10}$ is still dominated by absorptive π exchange in the near forward direction.

The presence of considerable A_2 exchange is not apparent if one analyzes the data at small $|t|$ at one energy, as is well known. Models based on absorptive π exchange alone can describe[3] the data quite well for $|t| < 0.2 (GeV/c)^2$. The point is that only by considering the whole data over the complete t-range can the considerable A_2 contamination of the data in the near forward direction be discerned. Thus in model analyses of $\pi\pi$ phase shifts, it is dangerous not to consider the behavior of the data for large $|t|$. Further, with A_2 exchange present, the π-exchange amplitudes are in good agreement with the "canonical" absorption model of Gottfried and Jackson; this is <u>not</u> true when absorptive π-exchange alone is used to fit the data.

Otherwise the presence of A_2 exchange is expected to show up in <u>energy</u> dependent effects relative to the 17.2 GeV/c data. At lower beam momenta, as the A_2 contribution becomes smaller relative to π at small $|t|$, we expect from Fig. 1 for the data to appear more like the prediction of pure absorptive pion exchange: (i) ρ_{1-1}^{11} should become <u>negative</u> in the region around $t = -0.15$ $(GeV/c)^2$, and (ii) the zeros in $\text{Re}\rho_{10}^{11}$ and $\text{Re}\rho_{10}^{10}$, occurring near $t = -\mu^2$ at 17.2 GeV/c, should shift to somewhat <u>smaller</u> values of

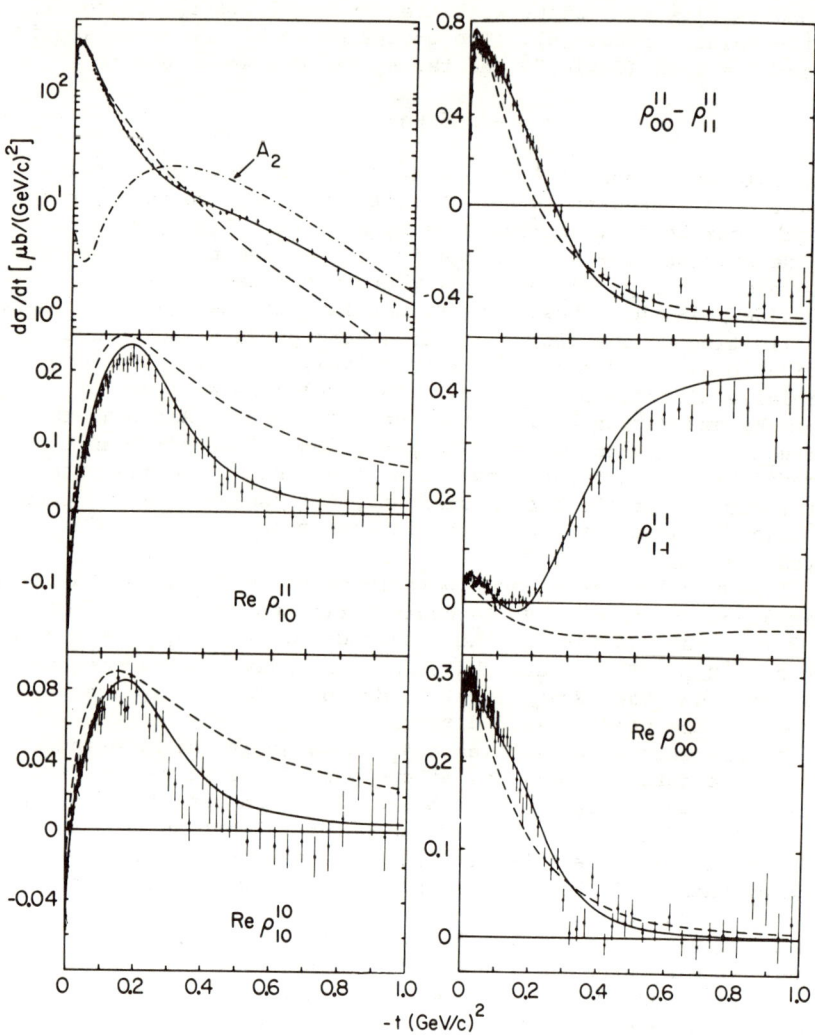

Fig. 1. Differential cross section and density matrix elements in the helicity frame at 17.2 GeV/c. The dashed curves correspond to the pion contribution.

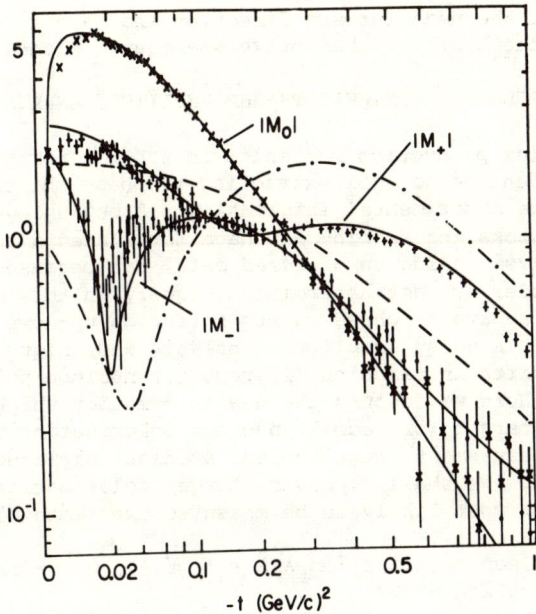

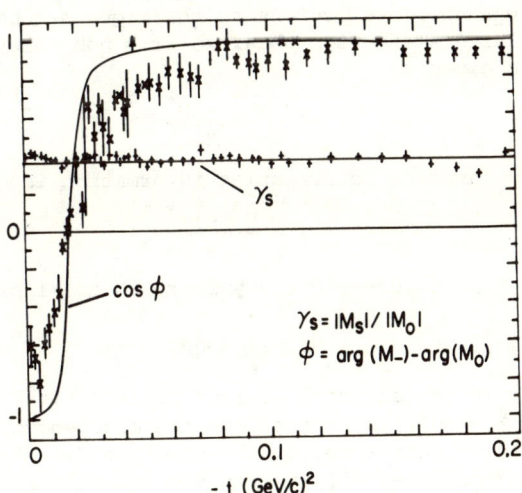

Fig. 2. Absorption model predictions for the Estabrooks-Martin[6] amplitude analysis. The dashed curve shows the π-cut contribution to $|M_+|$ and the dashed dotted one that of the A_2.

$|t|$ at lower momenta. The opposite trend should be seen at higher beam momenta. These effects, recently reported[13], are the imprint of A_2 exchange in the near forward direction and cannot reasonably be explained on the basis of absorptive π-exchange alone.

POLARIZATION EFFECTS AND AMPLITUDE ANALYSES

The complex production mechanism in $\pi^-p \to \pi^-\pi^+n$ requires more model-independent methods of extracting $\pi\pi$ phase shifts, which still utilize all the experimental information. A beginning has been made by Estabrooks and Martin[6] who have formulated a partial amplitude analysis using unpolarized data, but because the phases of the amplitudes are not determined uniquely in this analysis, the resulting s-wave $\pi\pi$ phase shifts suffer an up-down ambiguity as previously. A unique amplitude analysis and unique determination of $\pi\pi$ phase shifts in reaction (1) requires nucleon polarization information. Here we discuss the new information which can be obtained from target (or recoil) nucleon polarization measurements, and minimum measurements required for a unique amplitude analysis are determined. We shall emphasize target polarization effects, since these are most likely to be measured experimentally in the near future.

For a nucleon polarization λ ($\lambda = \uparrow$ or $\downarrow$), we define the angular distribution W^λ by

$$W^\lambda = \frac{d^3\sigma(\lambda)}{dt\,ds'\,d\Omega} \Big/ \int d\Omega \, \frac{d^3\sigma}{dt\,ds'\,d\Omega} \qquad (10)$$

where $d\Omega$, in what follows, refers to the s-channel helicity frame. Thus the angular distribution obtained from non-polarization experiments is given by

$$W = \frac{W^\uparrow + W^\downarrow}{2} \qquad (11)$$

whereas the new nucleon polarization information is contained in

$$\Delta W = \frac{W^\uparrow - W^\downarrow}{2}. \qquad (12)$$

The polarized target left-right asymmetry P_T is simply

$$P_T = \int d\Omega \, \Delta W \qquad (13)$$

Each of the three conventional nucleon polarization directions yields different angular distributions and independent observables. These are the two transverse (to the beam) polarizations, perpendicular ($\perp$) and parallel ($\parallel$) to the production plane, and the longitudinal polarization. For these target nucleon polarizations, ΔW can be written as follows (keeping s- and p-waves only):

Transverse Polarization ($\perp$)

$$\Delta W = \frac{1}{4\pi} \{a_0 + 3\cos^2\theta a_1 + 2\sqrt{3}\cos\theta a_2 - 2\sqrt{6}\sin\theta\cos\phi a_3$$
$$- 3\sqrt{2}\sin 2\theta\cos\phi a_4 - 3\sin^2\theta\cos 2\phi a_5\} \tag{14}$$

$$P_T = a_0 + a_1$$

Transverse Polarization ($\parallel$)

$$\Delta W = \frac{1}{4\pi}\{-2\sqrt{6}\sin\theta\sin\phi b_3 - 3\sqrt{2}\sin 2\theta\sin\phi b_4 \tag{15}$$
$$- 3\sin^2\theta\sin 2\phi b_5\}$$

Longitudinal Polarization

$$\Delta W = \frac{1}{4\pi}\{-2\sqrt{6}\sin\theta\sin\phi c_3 - 3\sqrt{2}\sin 2\theta\sin\phi c_4$$
$$-3\sin^2\theta\sin 2\phi c_5\} \tag{16}$$

The a_i, b_i, and c_i are the new observables. The same parametrization (14) - (16) of $\Delta\widetilde{W}$ for recoil polarization measurements yields the observables $\tilde{a}_i$, $\tilde{b}_i$, and $\tilde{c}_i$. However, the a_i are linearly related to the $\tilde{a}_i$, so that altogether 18 new observables are obtained from all possible target and recoil single nucleon polarization measurements (12 from target polarization alone).

If, as we shall assume, A_1-exchange is negligible in this reaction, eleven high energy predictions result for the polarization observables. They are $a_1 = -P_T/2$, $a_2 = 0$, $a_3 = 0$, $a_4 = 0$, $a_5 = P_T/2$, $b_i = \tilde{b}_i$, and $c_i = -\tilde{c}_i$. The failure of any one of these predictions would be evidence for A_1 exchange. It should be noted that polarization measurements are much more sensitive to the presence of A_1 contributions than are the non-polarization results. In the former A_1 exchange contributes linearly, but quadratically in the latter.

If A_1 exchange is negligible,[7] i.e. if the above predictions are verified, then a complete (unique) amplitude analysis on $\pi^- p \to \pi^- \pi^+ n$ is easy to formulate. Let us label our five independent amplitudes by

$$A_1 = f^0_{0+-},\ A_2 = f^1_{0+-},\ A_3 = (f^1_{1+-})^-,\ A_4 = (f^1_{1+-})^+,\ A_5 = (f^1_{1++})^+ \tag{17}$$

and introduce the notation

$$Q_i = |A_i|^2,\ R_{ij} = \text{Re}(A_i A_j^*),\ I_{ij} = \text{Im}(A_i A_j^*) \tag{18}$$

Then the observables from unpolarized and polarized measurements can be succinctly summarized:

$$\sigma \equiv d^2\sigma/dtds': \quad d_1 = \varrho_1 + \varrho_2 + \varrho_3 + \varrho_4 + \varrho_5$$
$$\sigma(\rho_{00}^{11} - \rho_{11}^{11}): \quad d_2 = \varrho_2 - \tfrac{1}{2}(\varrho_3 + \varrho_4 + \varrho_5)$$
$$\sigma \rho_{1-1}^{11}: \quad d_3 = -\varrho_3 + \varrho_4 + \varrho_5$$
$$\sigma \operatorname{Re}\rho_{10}^{11}: \quad d_4 = R_{23}$$
$$\sigma \operatorname{Re}\rho_{00}^{10}: \quad d_5 = R_{12}$$
$$\sigma \operatorname{Re}\rho_{10}^{10}: \quad d_6 = R_{13}$$
$$\sigma P_T: \quad d_7 = I_{45} \qquad (19)$$
$$\sigma b_3: \quad d_8 = I_{15}$$
$$\sigma b_4: \quad d_9 = I_{25}$$
$$\sigma b_5: \quad d_{10} = I_{35}$$
$$\sigma c_3: \quad d_{11} = I_{14}$$
$$\sigma c_4: \quad d_{12} = I_{24}$$
$$\sigma c_5: \quad d_{13} = I_{34}$$

The d_i correspond to the observables on the left times known numerical and kinematic factors. The first six of these equations correspond to the unpolarized data used by Estabrooks and Martin.[6] It is easy to see, using the elegant graphical technique of Moravcsik and Yu[14] that the first nine of these equations are sufficient to determine these five complex amplitudes up to an (unmeasurable) overall phase with no continuum of ambiguity. Solving the equations explicitly shows that measurements of d_{10} and d_{11} are sufficient to eliminate the remaining discrete ambiguities. Thus a unique amplitude analysis requires at least one longitudinal polarization measurement.

This analysis eliminates the "up-down" ambiguity of Estabrooks and Martin,[6] since all of the complex amplitudes contributing are determined uniquely (up to an overall phase). To obtain the phases of the $\pi\pi$ amplitudes, it is convenient to take $\mathrm{Im} A_2 = 0$. Then

$$\cos\Delta = \frac{\mathrm{Re} A_1}{|A_1|} \qquad (20)$$

$$\sin\Delta = \frac{\mathrm{Im} A_1}{|A_1|} \qquad (21)$$

where $\Delta = \arg A_1 - \arg A_2$. Both of these functions should be independent of s and t in the physical region and should be extrapolated smoothly to the pion pole, determining Δ uniquely at $t = \mu^2$ at which point

$$\Delta = \delta_0 - \delta_1 \qquad (22)$$

where δ_0 and δ_1 are the phases of the s- and p-wave scattering amplitudes, respectively.

Finally we comment on the general case in which A_1 exchange turns out not to be negligible. This more complex situation is most conveniently analyzed in terms of transversity amplitudes[15] in which the quantization axis is chosen perpendicular to the reaction plane. The eight complex transversity amplitudes $\hat{f}^\ell_{mc\beta}$ which contribute are ($\hat{f}^0_{0+-}$, $\hat{f}^1_{0++}$, $\hat{f}^1_{1+-}$, $\hat{f}^1_{-1+-}$) and ($\hat{f}^0_{0-+}$, $\hat{f}^1_{0--}$, $\hat{f}^1_{1-+}$, $\hat{f}^1_{-1-+}$). One finds that no experiment with unpolarized targets or with target polarization alone determines the phase of any one of the first group of amplitudes ($c = +\frac{1}{2}$) relative to any one of the second group ($c = -\frac{1}{2}$). Thus if A_1 exchange is not negligible, there will be a continuum of ambiguity in amplitude analyses even when all possible target polarization measurements are carried out. Elimination of this ambiguity requires a recoil polarization experiment. Except for this phase ambiguity between the two groups of transversity amplitudes, transverse target experiments alone determine all eight transversity amplitudes with only discrete ambiguities. As before, when A_1 exchange was neglected, elimination of these ambiguities requires longitudinal target polarization measurements.

REFERENCES

1. P. Baillon et al., Phys. Letters 35B, 453 (1971).
2. J. D. Kimel and E. Reya, Nucl. Phys. B47, 589 (1972).
3. J. D. Kimel and E. Reya, Phys. Letters 42B, 249 (1972).
4. R. Worden, Nucl. Phys. B37, 253 (1972).
5. G. Grayer et al., Proceedings of the IVth International Conference on High Energy Collisions, Oxford, 1972.
6. P. Estabrooks and A. D. Martin, Phys. Letters 41B, 350 (1972).
7. G. Grayer et al., Nucl. Phys. B50, 29 (1972).
8. K. Gottfried and J. D. Jackson, Nuovo Cimento 34, 735 (1964).
9. P. K. Williams, Phys. Rev. D1, 1312 (1970).
10. G. Fox, Proceedings of the Philadelphia Conference on Experimental Meson Spectroscopy, 1972.
11. J. D. Kimel and E. Reya, Florida State University report HEP 73-1-15, 1973 (to be published in Nucl. Phys. B).
12. P. Estabrooks and A. D. Martin, this Conference; and CERN report TH.1647-CERN (1973).
13. R. E. Diebold, this Conference.
14. M. J. Moravcsik and Wing-Yin Yu, J. Math. Phys. 10, 925 (1969).
15. A. Kotanski, Acta Phys. Polonica 29, 699 (1966); 30, 629 (1966).

284 Chapter 5.   Panel on New Directions

ππ AND Kπ SCATTERING STUDIES WITH THE ARGONNE EFFECTIVE MASS SPECTROMETER*

D. S. Ayres, R. Diebold, A. F. Greene,[†]
S. L. Kramer, A. J. Pawlicki and A. B. Wicklund
Argonne National Laboratory
Argonne, IL 60439

ABSTRACT

A total of 580 000 events of the type $\pi^-N \to \pi^-\pi^+N$ has been obtained in the region 3 to 6 GeV/c. Isolation of $\rho\omega$ interference is made in a straight-forward way by using both π^+ and π^- beams with a deuterium target. An amplitude analysis of ρ^0 production at 6 GeV/c yields results similar to those previously found at 17 GeV/c, except that the natural-parity exchange contribution is relatively less important. Results on $K^+n \to K^+\pi^-p$ and $\pi^-p \to K^-K^+n$ are also discussed.

INTRODUCTION

The CERN-Munich results[1] on the reaction $\pi^-p \to \pi^-\pi^+n$ at 17 GeV/c have shown that experiments with very high statistics can lead to new and exciting physics. We have obtained similar data at Argonne in the range 3 to 6 GeV/c. In addition to reactions off hydrogen, we have also taken data with a deuterium target. Although the results presented here are preliminary, they show very interesting qualitative features and indicate the scope of the physics that can be expected in the near future from more detailed analyses.

THE SPECTROMETER

A sketch of the Argonne Effective Mass Spectrometer[2,3] is shown in Fig. 1. This facility sits in an unseparated secondary beam of maximum momentum 6 GeV/c. A six-counter hodoscope at the first focus tags the momentum to ±0.2% and four threshold Cerenkov counters are used to identify the beam particles. The

---

*Work supported by the U. S. Atomic Energy Commission.
†Present address: National Accelerator Laboratory, Batavia, IL.

Invited paper presented by R. Diebold at the International Conference on π-π Scattering and Associated Topics, The Florida State University, March 28-30, 1973.

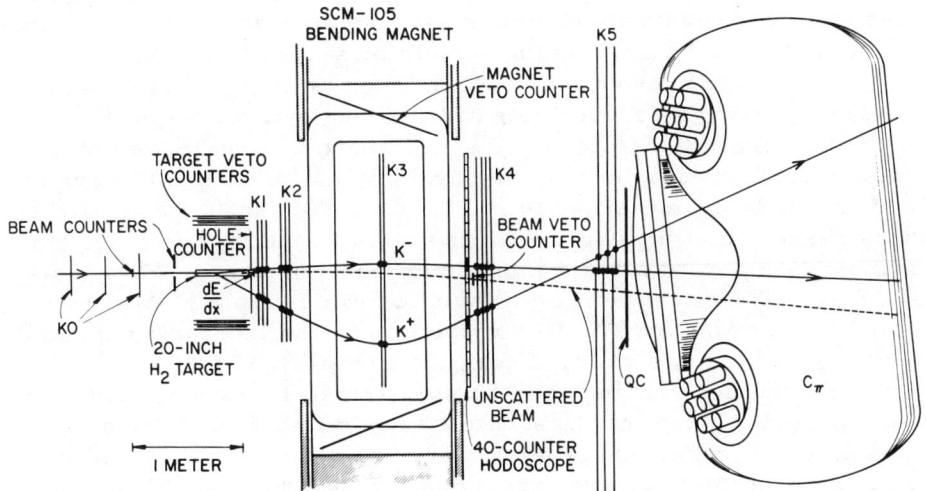

Fig. 1 Sketch of the spectrometer.

angle and position of the beam at the hydrogen target is determined by the magnetostrictive wire spark chambers marked K0. Up to 200 000 beam particles are taken each 0.7 second spill of the ZGS; special gating is used to avoid extra beam tracks in the chambers.

The momenta and angles of forward-going charged particles are measured with a large bending magnet and associated magnetostrictive wire spark chamber planes K1 through K5. The spectrometer magnet has a gap of 66 cm, a width of 2 m, an effective length of 1 m and $\int BdL = 11.4$ kG-m. Special programs are used to achieve high precision for the track momenta and angles with a minimum of computing time.

The data relating to $\pi\pi$ and $K\pi$ scattering were taken with a 20-inch target containing liquid hydrogen or deuterium. Veto counters around the target and lining the magnet aperture were used to suppress unwanted events. The target and nearby counters were mounted on a cart allowing rapid changeover of experiments.

The hole and dE/dx counters were used in the trigger to detect interactions in the hydrogen target; one or more large-angle particles through the hole counter or at least two particles through the dE/dx counter were required. Noninteracting beam triggers were further suppressed by the beam veto counter. The trigger also required at least two charged particles in the 40-counter hodoscope at the magnet exit. During some of the data taking, the large Cerenkov counter behind the last wire chamber plane was used to suppress events with pions in the final state.

The counter and spark chamber information was read into an EMR-6050 computer via SAC electronics; up to four sparks per readout were digitized with 1/4 mm least count. The computer not only logged the raw data onto tape but also analyzed a sample of the events online and displayed the results on an oscilloscope.

The spatial resolution of a single spark chamber is typically ±0.5 mm, giving measurements of track angle at the target to ±0.5 mr. The K3 chambers make an important contribution to the good momentum resolution of the system, the momentum of single fast tracks being measured to ±0.7% at 3 GeV/c and ±1.1% at 6 GeV/c. This results in a missing mass resolution on the recoiling neutron mass of ±13 MeV at a beam momentum of 3 GeV/c and ±47 MeV at 6 GeV/c. The effective mass resolution varies from ±1 MeV for ϕ and Λ to ±4 MeV for f^o. The resolution on the laboratory production angle of the two-charged-particle system is ±2 mr.

Production running with the spectrometer first began in August 1971. Since that time data taking has been completed for seven experiments, and several more are scheduled.[3] Preliminary results from the first five experiments were presented at the XVI International Conference on High Energy Physics in September 1972. The experiments cover a wide range of topics, but only two of them bear on $\pi\pi$ and $K\pi$ scattering and are discussed here.

$\pi^+\pi^-$ EFFECTIVE MASS SPECTRUM

Data were taken on the reactions

$$\pi^- p \rightarrow \pi^- \pi^+ n \qquad (1)$$

and

$$\pi^+ n \rightarrow \pi^+ \pi^- p \qquad (2)$$

at laboratory momenta of 3, 4 and 6 GeV/c as summarized in Table I. A total of approximately 580 000 events were obtained.

The raw $\pi^+\pi^-$ mass spectrum obtained at 6 GeV/c from hydrogen is shown in Fig. 2; the distribution is dominated by the ρ^o. Although there are many events in the f region, the asymmetric decay has a relatively low spectrometer acceptance, leaving only a shoulder at the f.

With high statistics and good resolution, one can search for fine structure in the spectrum. The small peak near 500 MeV is presumably K^o contamination from the reaction $\pi^- p \rightarrow K^o \Lambda$, while the small but significant peak at the ω mass is due to $\rho\omega$ interference.

The fall-off near 970 MeV and the associated sharp change in the forward-backward asymmetry shown in Fig. 3 are related to the onset of inelasticity at $K\bar{K}$ threshold. These features of the $K\bar{K}$

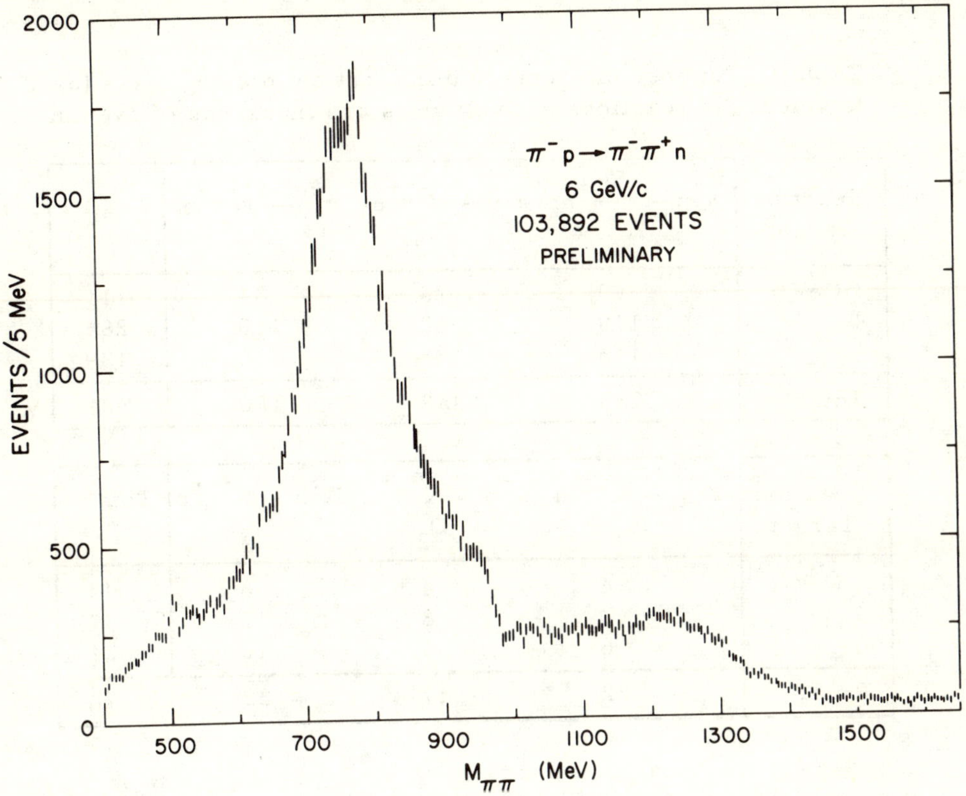

Fig. 2 $\pi^-\pi^+$ mass spectrum, uncorrected for spectrometer acceptance.

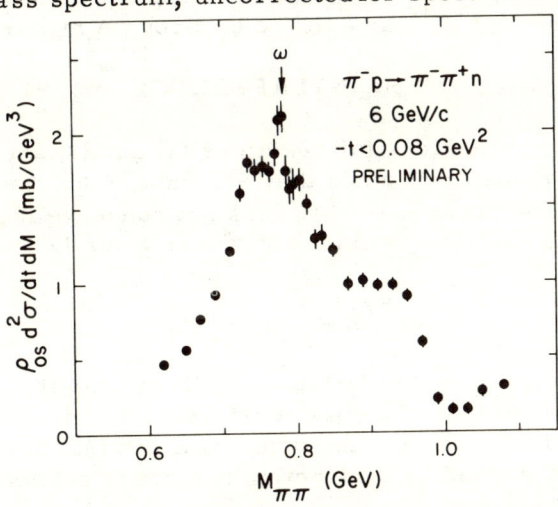

Fig. 3 Interference between s and p-wave amplitudes in the $\pi\pi$ system, as seen in the s-channel (helicity) frame.

Table I. Number of events within a cut on missing mass (used to select the reactions shown); units are thousands of events.

Reaction Target	$\pi^- p \to \pi^- \pi^+ n$ H_2	$\pi^- p \to \pi^- \pi^+ n$ D_2	$\pi^+ n \to \pi^+ \pi^- p$ D_2	Total
3 GeV/c	49	28	34	111
4	119	52	110	281
6	115	38	36	189
Total	283	118	180	581

Reaction Target	$K^- p \to K^- \pi^+ n$ H_2	$K^- p \to K^- \pi^+ n$ D_2	$K^+ n \to K^+ \pi^- p$ D_2	Total
3 GeV/c	4	4	6	14
4	12	8	17	37
6	6	5	10	21
Total	22	17	33	72

threshold region are similar to those observed previously for the same reaction by the CERN-Munich Group[1] at 17 GeV/c and for the reaction $\pi^- p \to \pi^+ \pi^- \Delta^{++}$ at 7 GeV/c by Group A at Berkeley.[4]

$\rho\omega$ INTERFERENCE

The cross sections for reactions (1) and (2) are generally assumed to be identical since they are related by charge symmetry $(I_z \to -I_z)$. The decay $\omega \to \pi^+ \pi^-$ does not conserve isospin, however, and the cross sections for incident $\pi^\pm$ are given by

$$d\sigma^\pm/dt = \sum_{i=1}^{6} |A_i^\omega \mp A_i^\rho|^2 \tag{3}$$

where the sum is taken over the six helicity amplitudes. The important thing to note is that the $\rho\omega$ interference terms change sign when going from one reaction to the other, and the interference effects can be directly found by subtracting one cross section from the other.

The quantity $(\rho_{11} + \rho_{1-1} + \rho_{ss}/3) d^2\sigma/dt\, dM$ is shown in Fig. 4 for both π^- and π^+ incident at 4 GeV/c. This particular combination of density-matrix elements times cross section isolates two of the

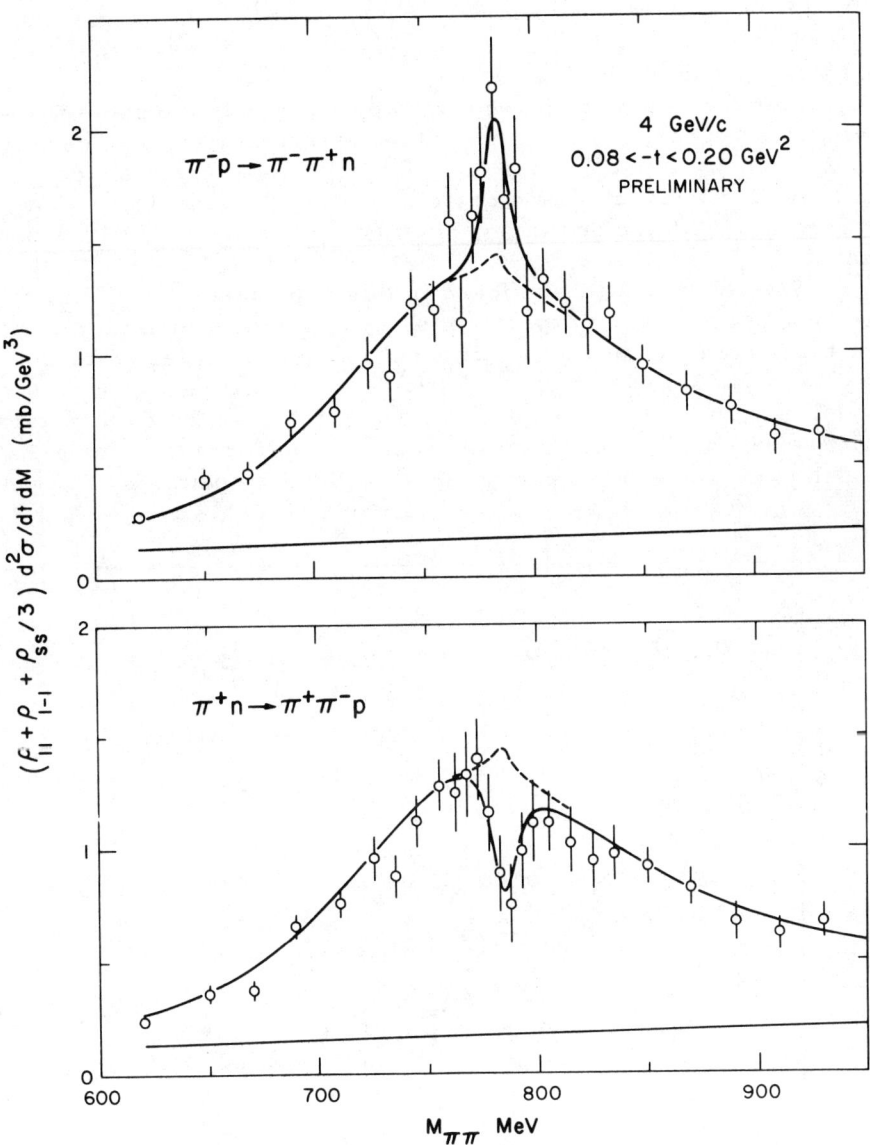

Fig. 4 ρω interference effects for natural-parity exchange.

six p-wave helicity amplitudes, those corresponding asymptotically to natural-parity exchange in the t channel. Sizeable interference patterns can be seen in Fig. 4, with opposite sign for incident π^+ and π^- as predicted by Eq. (3).

Interference patterns showing approximately the same relative phase between ρ and ω amplitudes were also found for unnatural-parity exchange, in all cases the interference being mainly constructive for incident π^- and destructive for π^+. The phase shows no variation with incident momentum but may have a small t dependence.

The relative $\rho\omega$ phase found in this experiment is similar to that observed by other experiments[5] using incident pion beams in our momentum range, but is different from that suggested by the 15-GeV/c SLAC experiment.[6] While a π-B exchange degenerate model[7] has been used to explain the relative phase for the unnatural-parity exchange part of the cross section, the corresponding ρ-A_2 model gives the wrong sign[8] for the interference patterns of Fig. 4.

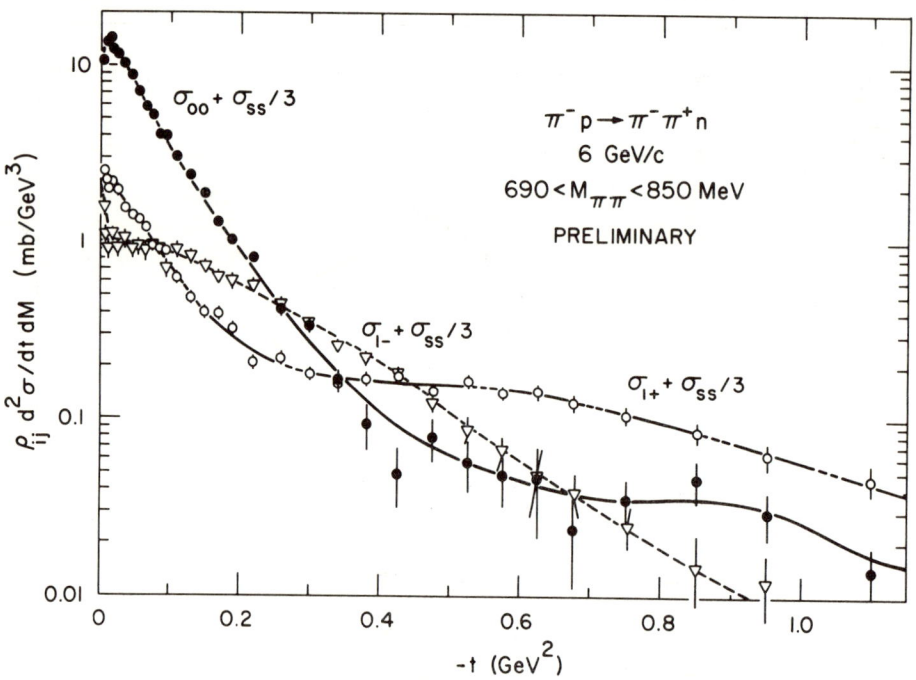

Fig. 5 Momentum transfer distributions for $\pi\pi$ mass in the ρ^0 region. The decay angular distribution in the helicity frame has been used to split the data into three components, $\sigma_{ij} = \rho_{ij}^{hel} d^2\sigma/dt\,dM$.

AMPLITUDE ANALYSIS FOR ρ^0 PRODUCTION

The $\pi\pi$ decay angular distribution can be used to split vector-meson production into three pieces, each corresponding to a pair of helicity amplitudes. Fig. 5 shows the 6-GeV/c π^- hydrogen data split in such a manner. This figure gives an indication of the momentum-transfer range covered by the experiment. Although 98% of the events at 6 GeV/c are at $-t \leq 0.5$ GeV^2, the remaining 2 000 events at large momentum transfers are still more than the entire data sample of many bubble chamber experiments. While σ_{00} is relatively flat from 0.4 to 1.0 GeV^2, there is no real indication for the dip at 0.5 GeV^2 that has recently been suggested.[9]

As indicated in Fig. 5, the s-wave $\pi\pi$ system contributes equally to each of the three terms. In order to separate s-wave from p-wave and to find the relative phases between amplitudes, Estabrooks and Martin[10] assumed that amplitudes with the quantum numbers of A_1 exchanges are negligible. They were then able to perform an amplitude analysis on the 17-GeV/c data of the CERN-Munich collaboration; Kimel and Reya[11] have made similar calculations.

We have applied such an analysis to our 6-GeV/c data; although the quadratic ambiguities need further study, the first results are shown in Fig. 6. For $-t < 0.5$ GeV^2 the amplitude $|M_0|$, presumably dominated by one pion exchange, looks much the same as at 17 GeV/c except for an overall change of normalization. The ratio $\gamma_s = |M_s/M_0|$ remains constant near 0.36 independent of s and t, while $|M_-/M_0|$ appears to be about 20% larger at 6 GeV/c as compared with 17 GeV/c. The unnatural parity amplitudes all seem to be relatively real, or nearly so, M_- going from antiparallel to parallel with M_0 near $-t = m_\pi^2$.

At t_{min}, M_+ is necessarily equal to M_-; at larger t, however, $0.2 < -t < 0.5$ GeV^2, $|M_+/M_0|$ is about 1.6 times larger at 17 GeV/c, than at 6 GeV/c, indicating a difference in effective trajectory for natural and unnatural parity exchange amplitudes of $\Delta\alpha \approx 0.4$. This would seem to confirm the speculation that M_+ has more than just pion cut contributions in this medium-t range and presumably includes considerable A_2 exchange.

Kπ ELASTIC SCATTERING

As shown in Table I, we have a large collection of data bearing on Kπ elastic scattering, a total of about 70 000 events taken with hydrogen and deuterium targets. If one pion exchange were the only important t-channel exchange, one would expect an equality between the reactions

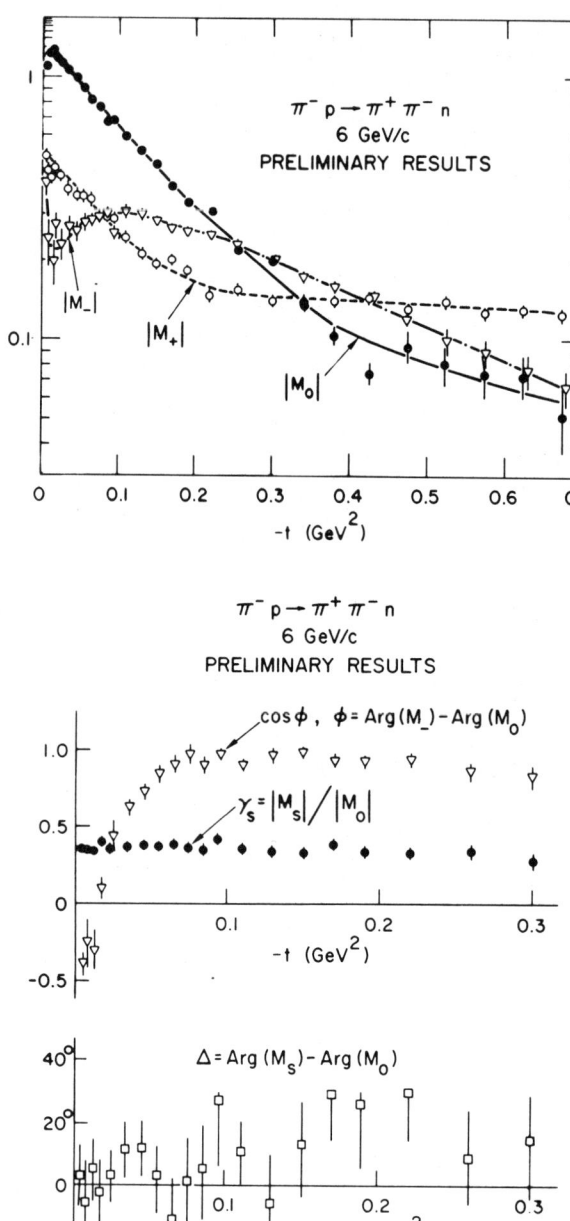

Fig. 6 Results of an Estabrooks-Martin (Ref. 10) analysis. $|M_+|^2$ and $|M_-|^2$ give the cross sections for helicity-one ρ production via natural and unnatural parity exchange respectively. $|M_{s,0}|^2$ give the helicity-zero cross sections for s and p wave $\pi^-\pi^+$ production.

$$K^-p \to K^-\pi^+ n, \tag{4}$$
$$K^+n \to K^+\pi^- p. \tag{5}$$

A change of sign will occur when going from one reaction to the other for interference terms between amplitudes of opposite charge conjugation in the t channel. With data from both reactions, one can check to see if such interference terms are important, and by taking an average of the two reactions can make the interference terms cancel.

Results on reaction (5) at 4 GeV/c are shown in Fig. 7; the spectrum is dominated by the K^*_{890} peak, produced mainly with

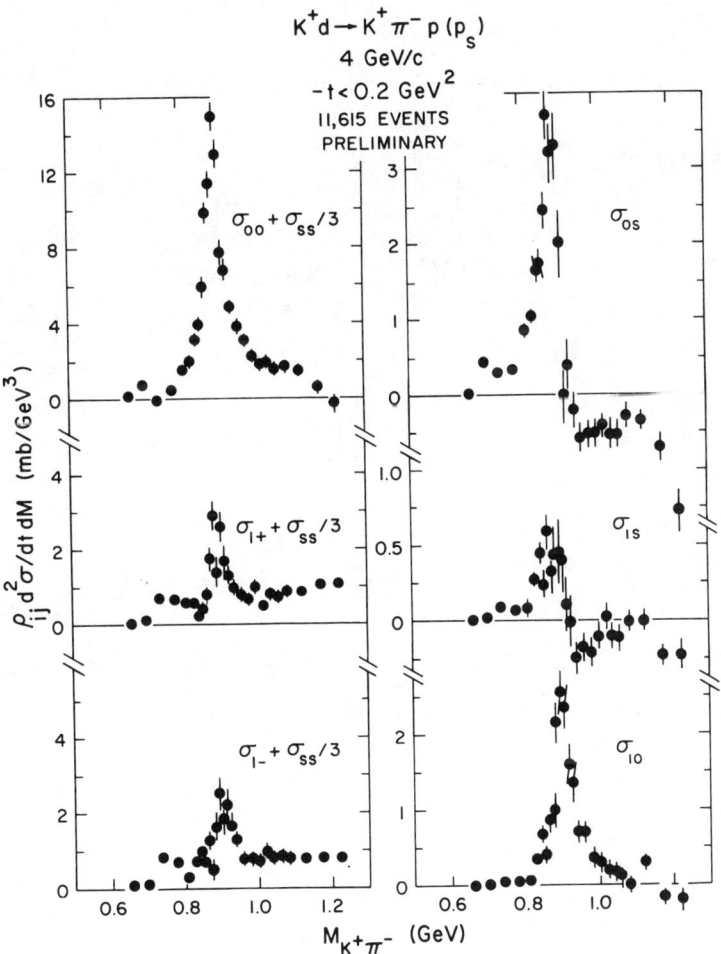

Fig. 7. $\sigma_{ij} = \rho^{hel}_{ij} d^2\sigma/dt\,dM$ for the $K^+\pi^-$ system at low momentum transfer.

helicity zero as expected for one pion exchange. As in the case of ρ^0, the K^*_{890} interferes with a s-wave background which is slightly out of phase, as shown by σ_{0s} and σ_{01}. There is also substantial interference between the helicity 0 and 1 p-wave indicated by σ_{10}. It will clearly be of interest to do an Estabrooks-Martin type of amplitude analysis and to extrapolate to the pion pole using an OPE absorption model as a guide.

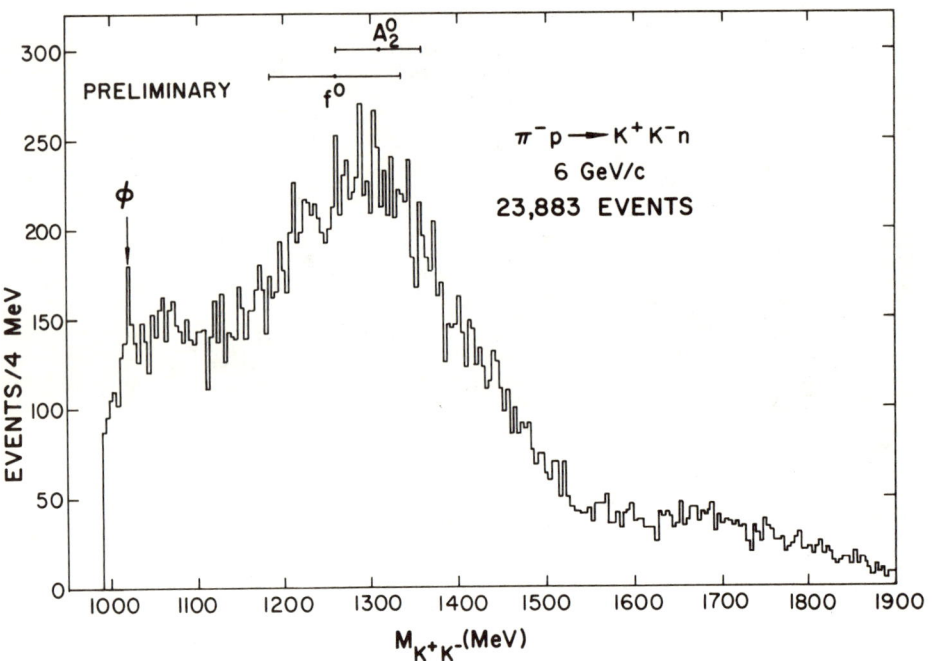

Fig. 8 K^-K^+ mass spectrum, uncorrected for spectrometer acceptance.

ππ INELASTIC SCATTERING

The large Cerenkov counter behind the spectrometer (see Fig. 1) was used during part of the running to suppress events with pions in the final state in order to study the reactions

$$\pi^- p \to K^- K^+ n , \qquad (6)$$

$$\pi^- p \to \bar{p} p n . \qquad (7)$$

Reaction (7) was interpreted by the CERN-Munich Group[12] at 19 GeV/c as being dominated by one pion exchange, i.e., inelastic ππ scattering, $\pi^- \pi^+ \to \bar{p} p$; at our energies t_{min} is so large that OPE may no longer be dominant, however.

The $K^+ K^-$ mass spectrum for reaction (6) is shown in Fig. 8 where the raw 6-GeV/c data are plotted in 4-MeV bins. After a very fast rise at threshold, there is a small ϕ peak which can be enhanced by cutting on large t. Beyond the ϕ the mass spectrum is relatively flat followed by a peak in the $f-A_2$ region.

The combined 5 and 6-GeV/c data below 1200 MeV are shown in greater detail in Fig. 9. In the small-t region ϕ effects are relatively small and the data appear quite smooth as a function of mass. Presumably the reaction is dominated by the s-wave S^* system produced by OPE.

Some p wave is also present, however, as shown by σ_{0s}. Taken at face value there also appears to be considerable σ_{1+} and σ_{1-} present. This interpretation must be viewed cautiously, however, since only s and p waves were taken into account, and we may be seeing the results of some d-wave background.

We see no evidence for narrow structure other than the ϕ in this mass region. In particular, previous experiments[13] have sometimes claimed a bump near 1070 MeV which is not present in our data.

SUMMARY

The data presented here should give a good picture of the Effective Mass Spectrometer data related to ππ and Kπ elastic scattering as well as ππ inelastic scattering. By working at lower energies and with both hydrogen and deuterium targets, our data are complementary to those of the CERN-Munich Group, and together the two experiments can provide valuable energy-dependence information. In addition to providing ππ and Kπ scattering information, the spectrometer has also yielded very interesting data on many other topics not the subject of this conference.[2,3]

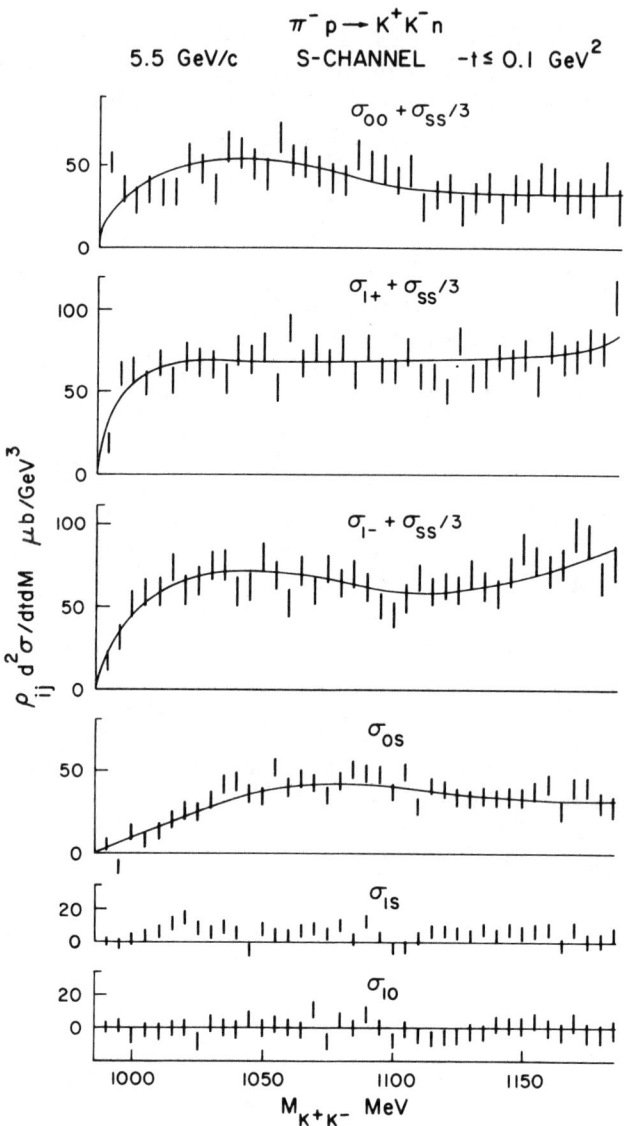

Fig. 9 $\sigma_{ij} = \rho_{ij}^{hel} d^2\sigma/dt\,dM$ for the combined 5 and 6-GeV/c K^-K^+ data.

REFERENCES

1. G. Grayer et al., Experimental Meson Spectroscopy - 1972 (American Institute of Physics, NY, 1972), p. 1; Phys. Letters 35B, 610 (1971).
2. The spectrometer was built as a collaborative effort between the present authors and I. Ambats, D. R. Rust, C. E. W. Ward and D. D. Yovanovitch.
3. For a more detailed description of the apparatus and a brief summary of the experimental program, see R. Diebold, Proc. XVI Int. Conf. on High Energy Physics 2, 447 (1972), and D. S. Ayres, ANL/HEP 7314 (to be published in Proc. of Frascatti Int. Conf. on Instrumentation for High Energy Physics, May 1973).
4. S. D. Protopopescu, Phys. Rev. D7, 1279 (1973).
5. G. Goldhaber et al., Phys. Rev. Letters 23, 1351 (1969); S. Hagopian et al., Phys. Rev. Letters 25, 1050 (1970); M. Abramovich et al., Nucl. Phys. B20, 209 (1970); I. J. Bloodworth, Nucl. Phys. B35, 133 (1971).
6. B. N. Ratcliff et al., Phys. Letters 38B, 345, (1972).
7. A. S. Goldhaber et al., Phys. Letters 30B, 249 (1969).
8. The ρ-A_2 exchange degenerate prediction has the opposite sign to that for π-B because the trajectories have opposite sign, $\alpha_\rho > 0$ and $\alpha_\pi < 0$, in the t range considered.
9. Howard A. Gordon, Kwan-Wu Lai and J. Michael Scarr, BNL preprint submitted to this conference.
10. P. Estabrooks and A. D. Martin, Phys. Letters 41B, 350 (1972).
11. J. D. Kimel and E. Reya, Phys. Letters 42B, 249 (1972).
12. G. Grayer et al., Phys. Letters 39B, 563 (1972).
13. W. Beusch et al., Phys. Letters 25B, 357 (1967); J. Alitti et al., Phys. Rev. Letters 21, 1705 (1968); B. Hyams et al., Nucl Phys. B22, 189 (1970); R. Diamond et al., Phys. Rev. D7, 1977 (1973).

NEW DIRECTIONS IN $\pi\pi$ AND $K\pi$ EXPERIMENTS*

Status of a 13 GeV $K^{\pm}p$ Experiment at SLAC

R. K. Carnegie
Stanford Linear Accelerator Center
Stanford University, Stanford, California 94305

ABSTRACT

A new SLAC experiment to study $K\pi$ scattering using both K^+p and K^-p reactions at 13 GeV is described.

A session on new directions in $\pi\pi$ and $K\pi$ experiments permits a diversity of possible topics for emphasis. My remarks will be confined to a discussion of the tasks immediately ahead in the $\pi\pi$ and $K\pi$ resonance regions using the data from the present generation of experiments, [1-5] and to a brief report on a new experiment in progress at SLAC to study $K\pi$ scattering using the reactions K^+p and K^-p at 13 GeV.

Recent experiments and analyses [1-6] exhibit general agreement on the qualitative results of $\pi\pi$ and $K\pi$ scattering in the lower mass region below the d wave. However, there are still a number of questions to be further studied and confirmed, especially in $K\pi$ scattering. One area for subsequent analyses to explore concerns the features of the higher mass region, such as studying the possible s wave resonances in the f^o and $K^*(1400)$ regions, [6-8] and verifying the existence of and/or determining the parameters of higher mass states such as the spin 3 K^*,[9] or even the ρ'. A second area to focus on concerns the quantitative uncertainties of the present phase shifts in the low mass region. In the case of the I=0 $\pi\pi$ s wave, how are we to interpret and choose between the range of solutions that have been presented,[1,3,6] all qualitatively consistent, but exhibiting systematic differences? In the $K\pi$ system, can we definitely rule out the existence of a narrow K^* s wave resonance near the $K^*(890)$?[4,5]

Future experiments must address themselves to these issues. In this regard, the ultimate limitation of these experiments is in understanding the extrapolation to the pion pole. This in turn implies a need for as detailed an understanding as possible of the $\pi\pi$ and $K\pi$ production amplitudes in the physical region and an attempt to construct experiments that have cross checks on the specific extrapolation procedures employed.

With these remarks in mind, let me now turn to a brief report on an experiment at SLAC[10] which, among other topics, will study $K\pi$ scattering. The experiment involves a study of both K^+p and K^-p reactions at 13 GeV. Figure 1 shows a plan view of the apparatus. The experiment uses a conventional wire spark chamber dipole spectrometer to detect the decay products of the peripherally produced mesons. The magnet aperture is 72 in. wide by

---

*Work supported by the U. S. Atomic Energy Commission.

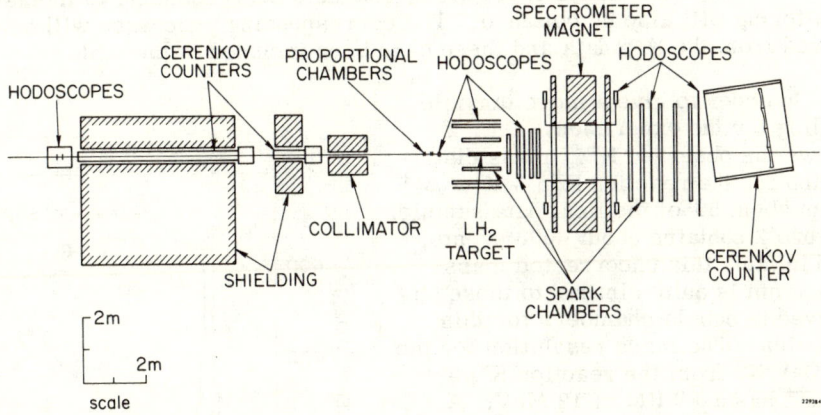

Fig. 1. Plan view of experimental apparatus.

24 in. high. Scintillation counter hodoscopes, Cerenkov counters and proportional chambers are used to detect each incident K in the rf separated beam. The trigger requirement is simply that 2 or more charged particles pass through the spectrometer after originating from an interaction in the 1 meter hydrogen target. The large aperture cellular Cerenkov counter is used to give K, π identification of the meson decay products.

The relatively loose trigger criteria allow us to study a wide range of final states; the dominant processes are listed in Table I. The event numbers

Table I Observed event rates (13 GeV, K^+p, K^-p)

Process	Events	Process	Events
$K^+p \rightarrow K^+\pi^-\Delta^{++}$	115,000	$K^-p \rightarrow K^-\pi^+n$	48,000
$\rightarrow K^*(890)\Delta^{++}$	45,000	$\rightarrow K^*(890)n$	18,000
$\rightarrow K^*(1400)\Delta^{++}$	14,500	$\rightarrow K^*(1400)n$	6,500
$K^+p \rightarrow K^+\pi^+n$	8,500	$K^-p \rightarrow K^-\pi^-\Delta^{++}$	
$K^+p \rightarrow K^0\pi^+p$	14,000	$K^-p \rightarrow \bar{K}^0\pi^-p$	11,000
$\rightarrow K^*(890)p$	9,600	$\rightarrow K^*(890)p$	7,500
$K^+p \rightarrow (K^+\pi^+\pi^-)p$	46,000	$K^-p \rightarrow (K^-\pi^-\pi^+)p$	36,000

quoted are based on the observed reconstructed event rates for the first half of the data, and these are then projected to the full data sample. For the K^+p data, in addition to the dominant $K^+p \rightarrow K^+\pi^-\Delta^{++}$ channel, we also detect $K^+p \rightarrow K^+\pi^+n$ to study the I=3/2 states and $K^+p \rightarrow K^0\pi^+p$. The large sample of $(K^+\pi^+\pi^-)p$ events will be used for a study of the Q meson. Some additional information on the higher multiplicity channels corresponding to inelastic $K\pi$ scattering will also be obtained. The corresponding processes will be obtained from the K^-p data and these have been included in the table.

In order to give a more tangible feeling for the experiment, Fig. 2 shows the observed $K^+\pi^-$ mass distribution for the reaction $K^+p \to K^+\pi^-\Delta^{++}$ from about 5% of the final data sample. Figure 2 contains about 6000 events, and in fact, this uncorrected mass spectrum is quite similar to those observed in bubble chambers for this reaction. The mass resolution for the 13 GeV K^0 from the reaction $K^+p \to K^0\Delta^{++}$ has a FWHM of 12 MeV. A traditional concern with spectrometer data is the angular acceptance, i.e., the acceptance as a function of the K^* decay angles in the s or t channel production frame. For the $K\pi$ case, in contrast with $\pi\pi$ data, the acceptance is good over the entire backward half of the angular distribution, a consequence of the Lorentz transformation properties of the K. There remains an acceptance loss for events with a low momentum backward pion.

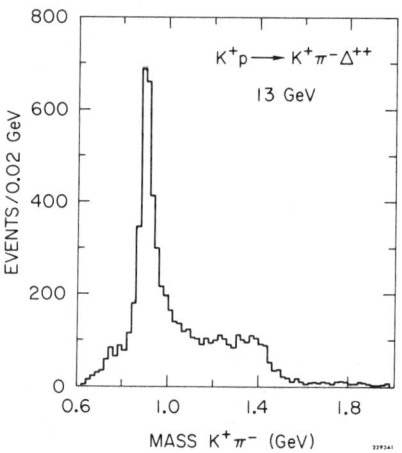

Fig. 2. Uncorrected $K^+\pi^-$ mass distribution for $K^+p \to K^+\pi^-\Delta^{++}$.

Let me summarize the present status of the experiment. About half of the data has been recorded, and the rest will be obtained this summer (1973). The reconstruction programs are in production on the first part of the data, and we are currently working on folding in the apparatus acceptance with the data in order to obtain the corrected $K\pi$ cross section and moments in the physical region.

Finally, I would like to emphasize some of the features of this experiment relevant to a study of $K\pi$ scattering, in addition to the obvious point of high statistics data with good mass resolution.

The first point is that we are studying $K\pi$ scattering for two reactions; first where the target proton dissociates into a π^+n and secondly into $\pi^-\Delta^{++}$. The neutron and Δ^{++} reactions have different t dependence characteristics in the physical region. An important cross check on the results obtained to describe $K\pi$ scattering after extrapolation to the pion pole is that they must be the same for both reactions. The quality of the extrapolated results is related to the ability to understand and extract the pion exchange contribution in the physical region. An important aspect of the π exchange study is that one obtains a unified description of both the neutron and Δ^{++} data, and even further that this phenomenology should describe both $\pi\pi$ and $K\pi$ production. Differences between $\pi\pi$ and $K\pi$ production in the physical region are expected and will be an interesting aspect of our data since additional exchanges can contribute in the Kπ system.

The second point is an extension of the first. This concerns the fact that we will have good data on the pure I=3/2 channel for both the n and Δ^{++} processes mentioned above for the neutral Kπ system. This means one can group the pairs of reactions (K$^+$p $\to$ K$^+\pi^-\Delta^{++}$, K$^-$p $\to$ K$^-\pi^-\Delta^{++}$) and (K$^-$p $\to$ K$^-\pi^+$n, K$^+$p $\to$ K$^+\pi^+$n) and process the data, extrapolate, and analyze each pair in a consistent way. This should lead to a good knowledge of the relative cross sections internal to a single experiment and hence remove some of the uncertainties associated with specifying the small I=3/2 contributions.

The last point concerns the angular acceptance characteristic that we do detect the backward K. As a result, we expect to be able to study questions associated with higher spin partial waves such as the s wave behavior in the K*(1400) region.

REFERENCES

1. S. Protopopescu et al., International Conference on Experimental Meson Spectroscopy, 3rd, April 28-29, 1972, Philadelphia, Pennsylvania (American Institute of Physics, N. Y., 1972).
2. P. Baillon et al., Phys. Letters 38B, 555 (1972).
3. CERN-Munich collaboration: G. Grayer et al., International Conference on Experimental Meson Spectroscopy, 3rd, April 28-29, 1972, Philadelphia, Pennsylvania (American Institute of Physics, N. Y., 1972); E. Lorenz, proceedings of this conference; W. Ochs, proceedings of this conference.
4. H. Bingham et al., Nucl. Phys. B41, 1 (1972).
5. A. Barbaro-Galtieri et al., proceedings of this conference.
6. A. Martin, paper presented at this conference; see also, P. Estabrooks and A. Martin, Phys. Letters 41B, 350 (1972).
7. J. Beaupre et al., Nucl. Phys. B28, 77 (1971); J. Carroll et al., Phys. Rev. Letters 28, 318 (1972).
8. A. Firestone et al., Phys. Rev. Letters 26, 1460 (1971); H. Yuta et al., Nucl. Phys. B52, 70 (1973).
9. D. D. Carmony et al., Phys. Rev. Letters 27, 1160 (1971); A. Firestone et al., Phys. Letters 36B, 573 (1971).
10. G. Brandenburg, R. Carnegie, R. Cashmore, G. Charlton, M. Davier, D. Leith, J. Matthews, P. Walden, S. Williams, and F. Winkelmann, SLAC Proposal E-75, Stanford Linear Accelerator Center.

RHO-OMEGA INTERFERENCE
IN THE REACTION $\pi N \to \pi\pi N$ AT 3, 4 AND 6 GEV/C*

D. S. Ayres, R. Diebold, A. F. Greene,[†] S. L. Kramer
A. J. Pawlicki and A. B. Wicklund
Argonne National Laboratory
Argonne, IL 60439

ABSTRACT

Preliminary results are presented on ρ-ω interference in a high statistics experiment on the reaction $\pi^{\pm} N \to \pi^{+}\pi^{-} N$ at 3, 4 and 6 GeV/c. Significant constructive (destructive) ρ-ω interference is observed in both the unnatural and natural parity exchange cross section for incident $\pi^{-}(\pi^{+})$. This sign of the interference agrees with that predicted from exchange degenerate π-B Regge poles but disagrees with the prediction of the exchange degenerate ρ-A_2 model.

The Argonne Effective Mass Spectrometer[1] has been used to observe the reactions

$$\pi^{-} p \to \pi^{+}\pi^{-} n \quad (1)$$

$$\pi^{+} n \to \pi^{+}\pi^{-} p \quad (2)$$

at 3, 4 and 6 GeV/c. More than 500 K $\pi\pi N$ events were obtained with excellent resolution on the $\pi^{+}\pi^{-}$ effective mass (±2 to 4 MeV).

In Fig. 1(a) the unweighted dipion mass spectrum is presented for reaction (1) at 4 GeV/c. A four-momentum transfer cut of $0.1 < t < 0.3$ GeV^2 has been applied to enhance the ω production. A narrow peak is observed near $m_{\pi\pi} = 780$ MeV indicating a significant constructive ρ-ω interference. If the mass spectrum for reaction (1) is expressed as a coherent sum of ρ and ω amplitudes

$$N(\pi^{-}) = |A_{\rho} + A_{\omega}|^2, \quad (3)$$

then by charge symmetry the mass spectrum for reaction (2) is given by

$$N(\pi^{+}) = |A_{\rho} - A_{\omega}|^2. \quad (4)$$

---
*Work supported by the U. S. Atomic Energy Commission.
[†]Present address: National Accelerator Laboratory, Batavia, IL.

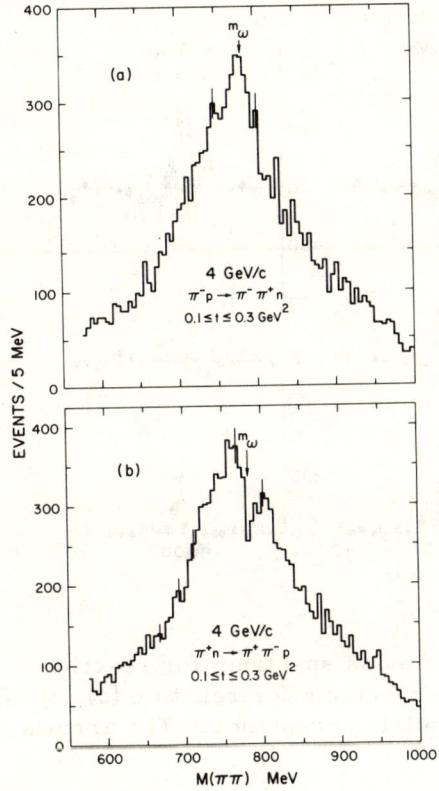

Fig. 1 The unweighted $\pi\pi$ mass spectra for (a) reaction (1) and (b) reaction (2).

Therefore, if a constructive interference is observed for reaction (1), a destructive interference is expected for reaction (2). The corresponding mass spectrum for the latter reaction is presented in Fig. 1(b). Instead of the peak observed in reaction (1), a dip of a similar magnitude, position and width is observed.

The interference term between ρ and ω amplitudes can be isolated by taking the difference between Eqs. (3) and (4)

$$N(\pi^-) - N(\pi^+) = 4 \operatorname{Re}(A_\rho^* A_\omega). \quad (5)$$

The result of this subtraction[2] is presented in Fig. 2. For all energies and t regions a significant peak is observed at the ω mass. At 4 GeV/c the interference term has a statistical significance of greater than six standard deviations (S.D.) and appears to be symmetric about the ω mass, indicating a total phase angle (ϕ) near zero. In Fig. 3 the normalized difference

$$\Delta = \frac{N(\pi^-) - N^*(\pi^+)}{N(\pi^-) + N^*(\pi^+)} \quad (6)$$

for the ω mass region is presented to indicate the significance of these observations as well as the energy dependence of the interference term.

Although the previous data were uncorrected for geometric acceptance, this should have little effect on the shape of the ρ-ω interference. However, in order to project out the density matrix elements, these corrections must be applied to the data. Assuming only $\ell \leq 1$ waves contribute in the ρ mass region and calculating the density matrix elements in the s-channel helicity frame, we define

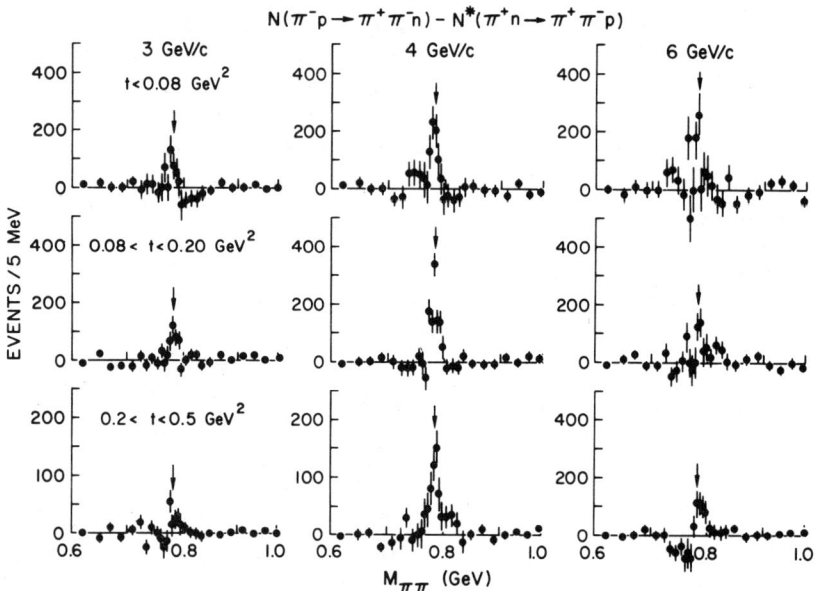

Fig. 2 The difference between the $\pi\pi$ mass spectrum for reaction (1) and the normalized (Ref. 2) mass spectrum for reaction (2). These data are uncorrected for geometric acceptance. The arrows indicate the ω mass value.

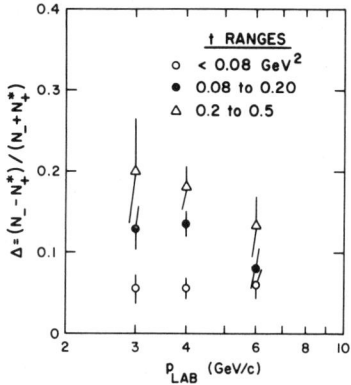

Fig. 3 The normalized difference, Δ, as a function of laboratory momentum and t range, for $775 < m_{\pi\pi} < 790$ MeV.

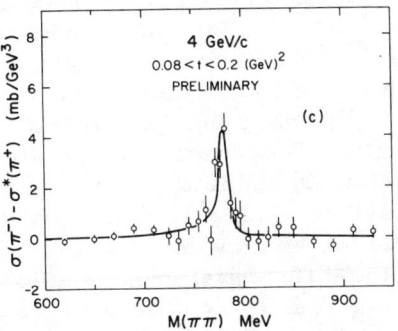

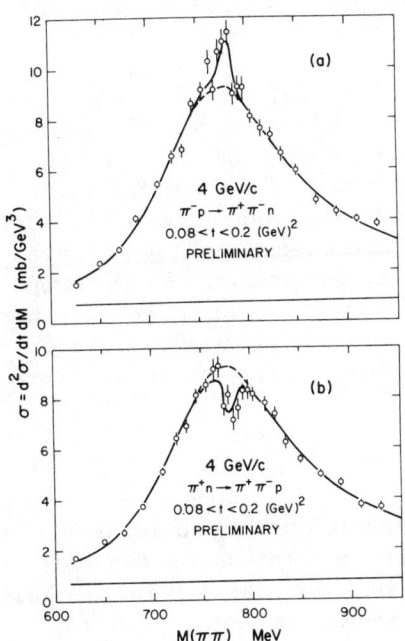

Fig. 4 The differential cross-section (σ) for (a) reaction (1), (b) reaction (2) and (c) their difference, with $\sigma^*(\pi^+)$ including a 1% normalization correction (Ref. 2). The curve through the data is the result of the fit described in the text, the dashed curve is the fitted ρ shape and the thin line the fitted background.

$$\sigma = d^2\sigma/dt\, dm_{\pi\pi} \qquad \sigma_{1+} = (\rho_{11} + \rho_{1-1} + 1/3\, \rho_{ss})\sigma \qquad (7)$$
$$\sigma_{00} = (\rho_{00} + 1/3\, \rho_{ss})\sigma \qquad \sigma_{1-} = (\rho_{11} - \rho_{1-1} + 1/3\, \rho_{ss})\sigma\ .$$

Asymptotically the quantities σ_{00} and σ_{1-} correspond to unnatural parity exchange,[3] and σ_{1+} is related to natural parity exchange. By studying the mass dependence of these cross sections, it is possible to isolate the ρ-ω interference for the different exchange amplitudes. In order to make this study more quantitative, we have attempted a simple phenomenological fit similar to that given by G. Goldhaber in reference (4). For each of the quantities in Eqs. (7), a simultaneous fit to the π^- and π^+ data was performed assuming $\phi(\pi^+)$ = $\phi(\pi^-) + 180°$ and complete coherence.

An example of this fit to the differential cross section (σ) is presented in Fig. 4. The fit yields a χ^2/DF = 59/46, resonance parameters for the ρ and ω in excellent agreement with their accepted values, and an overall phase angle[5] $\phi = 26 \pm 12°$. Fits to σ_{00} and σ_{1+} both required significant[1] ρ-ω interference and gave $\phi \approx 45°$ and $\phi \approx -10°$ respectively. The phases obtained for σ_{00}, σ_{1+} and σ show little or no energy dependence and only slight t dependence; however, the magnitude of the interference varied considerably with t. The helicity one unnatural parity exchange cross section (σ_{1-}) requires very little ρ-ω interference at low energies and small t value, but significant interference is observed at 6 GeV/c and t > .08 GeV^2, with a phase angle near zero.

Previous theoretical treatment of ρ-ω interference considered mainly π-B exchange degenerate models. However, our observation of interference in σ_{1+} indicates that ρ-A_2 exchanges must also be considered. The exchange degenerate π-B model predicts[6] a phase angle for π^- incident of $\phi \approx 16°$, in good agreement with our result for the differential cross section (σ). However the phase observed for σ_{00} at 4 GeV/c differs by ≈ 2 standard deviations. An exchange degenerate ρ-A_2 model predicts[7] ϕ to be 180° out of phase from the π-B model in the t region being considered. This would produce destructive interference in the π^- data for σ_{1+}, instead of the constructive interference observed in our data.[1]

REFERENCES

1. See R. Diebold's invited talk presented to this conference.
2. There may be systematic differences between reactions (1) and (2) due to deuterium effects and veto biases from the final state nucleon. These effects have not yet been completely determined, and we have arbitrarily renormalized the π^+ data to the π^- data in the ρ mass band, outside the ω mass region. For all but the high t data, the normalization corrections were less than 10%.
3. J. P. Ader et al., Nuovo Cimento 56A, 953 (1968).
4. The phenomenological expression is given by Eq. (2) of G. Goldhaber, Experimental Meson Spectroscopy, edited by C. Baltay and A. H. Rosenfeld (Columbia University Press, NY, 1970), p. 59.
   The ρ and ω shapes were taken to be p-wave relativistic Breit-Wigner with $\Gamma_\rho(m_{\pi\pi})$ given by Eq. (30) of J. Pisut and M. Roos, Nucl. Phys. B6, 325 (1968).
5. The overall phase angle ϕ includes the decay phase for the ω which is usually taken to be $\approx 160°$. The quantity ϕ referred to in the text is always referenced to reaction (1).
6. A. S. Goldhaber, G. C. Fox, and C. Quigg, Phys. Letters 30B, 249 (1969).
7. G. V. Dass, Proceedings of the Daresbury Study Weekend No. 1, DNPL/R7, 207 (1970).

A $\pi\pi$ PHASE SHIFT ANALYSIS FROM AN EXPERIMENT $\pi^-p \to \pi^°\pi^°n$
AT 2 GeV/c SUPPLEMENTED BY $\pi^+\pi^-$, $\pi^-\pi^°$, $\pi^\pm\pi^\pm$ FINAL STATES

G.Villet, M.David, R.Ayed, P.Bareyre, P.Borgeaud,
J.Ernwein, J.Feltesse, Y.Lemoigne, P.Marty, A.V.Stirling,
D.Ph.P.E., C.E.N.-Saclay, France.

APRIL, 1973

ABSTRACT

$\pi^+\pi^- \to \pi^°\pi^°$ elastic and differential cross sections from threshold to 1 GeV $\pi\pi$ mass have been obtained by Chew-Low extrapolation methods from the reaction $\pi^-p \to \pi^°\pi^°n$ at 2 GeV/c. We find large J = 0 production approaching the unitarity limit in the 450-750 MeV $\pi\pi$ mass. An energy-independent phase shift analysis up to F waves using both the $\pi^°\pi^°$ final state and the already known $\pi^\pm\pi^\pm$, $\pi^-\pi^°$, $\pi^+\pi^-$ states (giving I = 0,1,2) verifies the overall consistency of the experimental data. Supplemented by the use of forward dispersion relations the phase shift analysis gives a unique solution characterized by the scattering lengths $a_0^0 = 0.67$ and $a_0^2 = 0.021$ and by the presence of an ε (J = I = 0) of mass 940 MeV and width 180 MeV.

INTRODUCTION

We present here results of a $\pi\pi$ energy-independent phase shift analysis from threshold to 1.2 GeV involving 4 channels. For $\pi^+\pi^- \to \pi^°\pi^°$ we have used our data. For the three other final states ($\pi^+\pi^-$, $\pi^-\pi^°$, $\pi^{\pm\pm}$) we have used the most statistically significant data after checking their overall consistency. In our experiment, the $\pi^°$ were detected through pair conversion in an optical cylindrical large gap spark chamber array placed in a magnetic field. The average mass resolution for a $\pi^°$ is found to be around 15% in the 1-c fit of $\pi^-p \to \gamma\gamma n$ events. After subtracting the $N^* \to n\pi^°$ and the $\eta \to 3\pi^°$ (with only 4γ rays converted) events we are left with 2400 statistically good $\pi^°\pi^°$ events with $\Delta^2/\mu^2 \le 15$. The apparatus as well as details of the data analysis are described elsewhere.[1]

DISCUSSION OF THE DATA

The differential cross sections have been reduced to the expansion $d\sigma/d\Omega = \lambda^2 \Sigma_\ell C_\ell P_\ell (\cos\theta)$. The forward backward asymmetries were also used, but only when obtained independently of the C_ℓ coefficients. The total inelastic cross sections, mainly $\pi\pi \to \pi\pi\pi\pi$ and $\pi\pi \to K\bar{K}$ have been taken into account with large errors.

a) $\pi^+\pi^- \to \pi^°\pi^°$

The C_0 and C_2 coefficients were obtained by extrapolating our angular distribution to the pion pole. The Chew-Low function $F(\omega,\Delta^2)$ serves as the basis for this procedure.[2] After verifying that $F(\omega,0)$ is near 0, we have included this constraint in our extrapolation. The results are shown in Figure 1. The C_0 variation up to 720 MeV is in agreement with other experiments [3,4,5], but in the 720-1000 MeV region some results [4,5], stick longer to the unitarity limit and afterwards decrease more slowly. We have used our data because ours is the only experiment giving both coefficients C_0 and C_2.

b) $\pi^\pm\pi^\pm \to \pi^\pm\pi^\pm$

We have been led to eliminate some data [6] because of inconsistency with other channels. The remaining results can be divided into two sets : (A)[7] having a large cross-section (10-15 mb) for $M_{\pi\pi}$ 400 MeV, and (B)[8]. Each set leads to a phase shift solution.

c) $\pi^-\pi^° \to \pi^-\pi^°$

We have used the Saclay results [2] at 2.77 GeV/c.

d) $\pi^+\pi^- \to \pi^+\pi^-$

We have at our disposal 3 high statistic measurements from the following reactions [9,10,2] : $\pi^-p \to \pi^-\pi^+n$ at 17.2 GeV/c, $\pi^+p \to N_{33}^{++}\pi^+\pi^-$ at 7.1 GeV/c, and $\pi^-p \to \pi^+\pi^-n$ at 2.77 GeV/c. Unfortunately, the normalization was not known for the first one.

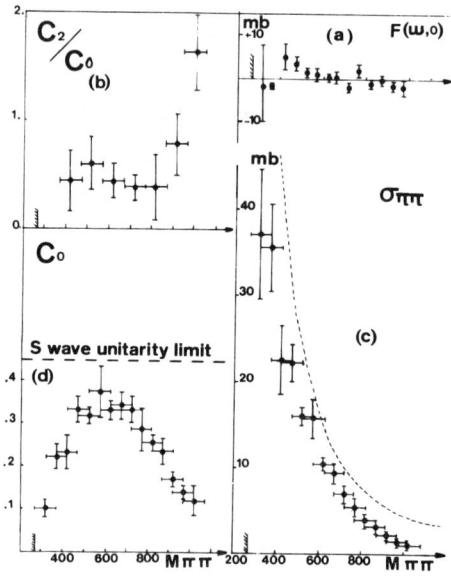

Figure 1 :

a) $F(\omega,\Delta^2)$ for $\dfrac{\Delta^2}{\mu^2} = 0$, versus dipion mass.

b) Ratio C_2/C_0 versus dipion mass.

c) Extrapolated cross section for $\pi^+\pi^- \to \pi^°\pi^°$ with the constraint at $\Delta^2 = 0$, versus dipion mass: Note the two different mass binings.

d) Variation of $C_0 = \dfrac{\sigma_{\pi^+\pi^- \to \pi^°\pi^°}}{2\pi\lambda^2}$ versus dipion mass.

The agreement is good up to 900 MeV $\pi\pi$ mass except for a mass shift ($\simeq$ 30 MeV) mainly visible on C_0 and C_2 coefficients. To keep the data consistent with the $\pi^-\pi^0$ channel [essentially for $C_2(+0)$ and $C_2(+-)$ in which the $|P_1\text{ wave}|^2$ appears with the same high weight] we were led to use the Saclay data,[2] supplemented by the C_5 and C_6 coefficients from Berkeley.[10]

DETERMINATION OF THE PHASE SHIFTS

The analysis has been carried out at 18 energies up to 1200 MeV $\pi\pi$ mass. The method is exactly the same as for π-N scattering.[11] By considering partial waves up to F_1, 12 parameters are to be determined at each $M_{\pi\pi}$. The use of 4 channels for 3 isospin states gives in theory a maximum of 20 constraints. All the coefficient structures are well fitted : this shows a good consistency of the data with regards to charge independence. Figure 2 displays the results. Below 500 MeV we find two solutions (A) and (B) (visible in S_2 partial wave) corresponding to the two sets of data (A) and (B) for the (++) channel. By considering the speed of the partial amplitude in the Argand plot, we have found that the resonant behaviour of S_0 around 900 MeV corresponds to a pole below the $K\bar{K}$ threshold : m = 940 MeV, Γ = 180 MeV.

CONCLUSION

In order to choose between solutions (A) and (B), we have calculated at each energy the scattering length a_0^2 using a forward dispersion relation for $\pi^-\pi^0$ scattering as done in reference 2. With this point by point determination, solution (B) gives clearly a more stable a_0^2 value. Finally, using the parametrization of the phase shifts near threshold

$$\delta^I(\nu) = \left(\frac{\nu}{\nu+1}\right)^{\ell+\frac{1}{2}} \cdot \text{(polynomial in } \nu\text{)}$$

where   I : isospin, $\nu = q^2 = \frac{s-4}{4}$, ℓ : angular momentum,

the scattering lengths were calculated. We find for solution (B) :

$a_0^0 = 0.67 \pm 0.06$; $a_1^1 = 0.027 \pm 0.002$; $a_0^2 = 0.021 \pm 0.004$.

Note the reasonable agreement with solution 3 (requiring crossing, unitarity and analyticity) proposed by Basdevant,[12] with the results of a Ke_4 experiment [13] and with the analysis of $K \to \pi\pi$ decay.[15]

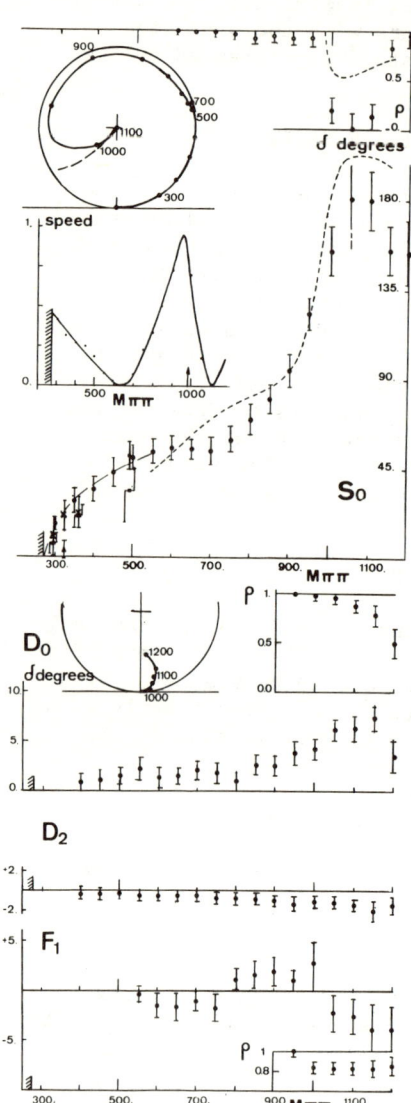
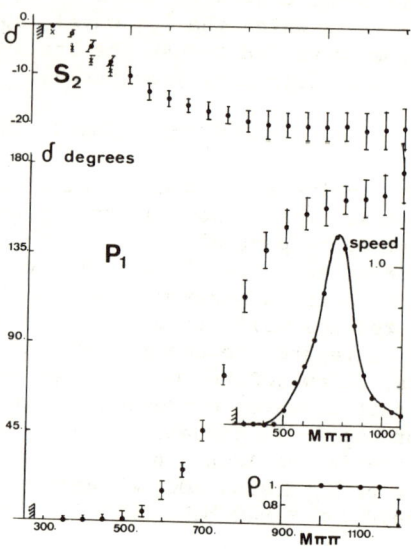

Figure 2 :

Plot of the S_0, S_2, P_1, D_0, D_2 and F_1 phase shifts and inelasticity parameters.

The speed of the partial wave amplitudes S_0 and P_1 is displayed.

S_0 wave : the solid line represents the fit giving the a_0^0 scattering length. Values of S_0 coming from K decays are also given :
- Black squares are from $K \to 2\pi$.[15]
- Crosses and triangles are from K_{e4}.[13,14]

The dashed line is the phase shift from Protopopescu et al.[10]

S_2 wave : crosses refer to solution A.

REFERENCES

1. M.David, G.Villet, R.Ayed, P.Bareyre, P.Borgeaud, J.Ernwein, J.Feltesse, P.Marty, B.Ollivier, J.Poinsignon, J.Rousseau, A Wide Magnetic Analyser for Pion-Proton $\rightarrow$ Multi γ's$(\rightarrow e^+e^-)$ + Neutron Reactions. (To be published).
2. G.Laurens, Thesis, Paris Report CEA-N-1497 (1971); J.P.Baton, G.Laurens, J.Reignier, Phys.Letters $\underline{33B}$, 528 (1970).
3. P.Sonderegger, P.Bonamy, Lund Intern. Conf. on High Energy Physics (1969).
4. A.Skuja, M.A.Wahlig, T.B.Risser, M.Pripstein, J.E.Nelson, I.R.Linscott, R.W.Kenney, O.I.Dahl, R.B.Chaffee, XVIth Intern. Conf. on High Energy Physics, Batavia (1972).
5. J.R.Bensinger, Ph.D. Thesis, University of Wisconsin (1970).
6. O.R.Sander, J.P.Prukop, J.A.Poirier, C.A.Rey, A.J.Lennox, N.N.Biswas, N.M.Cason, W.D.Shephard, V.P.Kenney, XVIth Intern. Conf. on High Energy Physics, Batavia (1972).
7. J.Alitti, Thesis, Report CEA-R-3035 (1966); N.Schmitz, Nuovo Cimento $\underline{31}$, 255 (1964); Aachen-Berlin-Birmingham-Bonn-Hambourg-London(I.C.)-Munchen Collaboration, Phys. Rev. $\underline{138}$, B897 (1965); B.Gandois, Thesis 3° cycle, Paris (1967).
8. E.Colton, E.Malamud, P.E.Schlein, A.D.Johnson, V.J.Stenger, P.G.Wohlmut, Report UCRL 20053 (1970); D.Cohen, T.Ferbel, P.Slattery, B.Werner, Phys.Rev. $\underline{D7}$,73(1973); M.Baubillier, B.Burusoy, R.George, M.Goldberg, A.M.Touchard, N.Armenise, M.T.Fogli-Muciaccia, A.Silvestri, XVIth Intern. Conf. on High Energy Physics, Batavia (1972).
9. G.Grayer, B.Hyams, C.Jones, P.Schlein, W.Blum, H.Dietl, W.Koch, E.Lorenz, G.Lutjens, W.Manner, J.Meissburger, W.Ochs, U.Stierlin, P.Wielhammer, Proc. of Experimental Meson Spectroscopy Conference (Third Philadelphia Conf.) (1972) p.5.
10. S.D.Protopopescu, M.Alston-Garnjost, A.Barbaro-Galtieri, S.M.Flatte, J.H.Friedman, T.A.Lasinski, G.R.Lynch, M.S.Rabin, F.Solmitz, Report LBL 970 (1972).
11. R.Ayed, P.Bareyre, G.Villet, Phys. Letters $\underline{31B}$, 598 (1970).
12. J.L.Basdevant, C.D.Froggatt, J.L.Petersen, P.L. $\underline{41B}$, 457 (1972).
13. A.Zylbersztejn, P.Basile, M.Bourquin, J.P.Boymond, A.Diamant-Berger, P.Extermann, P.Kunz, R.Mermod, H.Suter, R.Turlay, Phys. Letters $\underline{38B}$, 457 (1972).
14. E.W.Beier, D.A.Buckholz, A.T.Mann, S.H.Parker, J.B.Roberts, Phys.Rev.Letters $\underline{30}$, 399 (1973).
15. A.Q.Sarker, Phys.Letters $\underline{41B}$, 157 (1972).

π-π PHASE SHIFTS IN THE ENERGY REGION 0.6 to 1.42 GeV

S. Toaff, J. C. Anderson, A. Engler,
R. W. Kraemer, and F. Weisser
Carnegie-Mellon University, Pittsburgh, Pennsylvania 15213[*]

J. Diaz, F. A. DiBianca, W. Fickinger,
D. K. Robinson, and C. A. Sullivan
Case-Western Reserve University, Cleveland, Ohio 44106[**]

ABSTRACT

We present results on π-π scattering via the reaction $\pi^+ n \rightarrow \pi^+ \pi^- p$ at 6 GeV/c. Using a one-pion exchange model modified by absorption and I=2 phase shifts from other experiments we have extracted S, P, and D wave π-π phase shifts under certain energy dependent assumptions. The data sample consists of approximately 13000 events in the energy region 0.6 to 1.42 GeV.

We present results of a ππ phase shift analysis in the region 0.6 GeV $\leq$ M(ππ) < 1.42 GeV. Using the reaction $\pi^+ d \rightarrow \pi^+ \pi^- p p_s$, our experiment is a 910,000 picture exposure of the ANL deuterium filled 30" bubble chamber to a 6 GeV/c separated π^+ beam. The analysis is based on approximately 75% of the total available data. After cutting the spectator momentum at 250 MeV/c, $|MM^2| \leq .06$ GeV^2, and $P(\chi^2) \geq .01$, our total sample for the reaction $\pi^+ n \rightarrow \pi^+ \pi^- p$ consists of 13,024 events. For the phase shift analysis we use only events with four-momentum transfer between the outgoing dipion system and the incident π^+ less than 0.3 (GeV/c)^2, and we are left with 9500 events.

We have used a one pion exchange model modified by absorption, as described by Oh[1] et al. The total amplitude is given by 2 parts: the first is the regular one pion exchange amplitude and the second is a "Nuclear Pole Term". The total amplitude is

$$\langle \mu \lambda' | T | \lambda \rangle = \langle \mu \lambda' | T_{\pi\,ex.} | \lambda \rangle + \langle \lambda' | T_{NPT} | \lambda \rangle$$

where λ, λ' are the helicities of the incident and outgoing nucleon and μ is the helicity of the dipion system. The Nuclear Pole Term as parameterized by Oh et al.[1] yields a strongly peaked forward cross section for the diffraction dissociation process. The π exchange term is given by the following form:

$$\langle \mu \lambda' | T_{\pi\,ex.} | \lambda \rangle = \sum_{L,I} \frac{m_{\pi\pi}}{K} C^N C^I \left[\eta_L^I \exp(2i\delta_L^I) - 1 \right]$$
$$\times f_{\mu\lambda\lambda'}(t)\, d_{\mu 0}^L(-\psi)\, Y_L^\mu(\theta, \phi)$$

where: L, I indicate the partial wave and I spin of the ππ system,

---

[*] Supported by the U. S. Atomic Energy Commission.
[**] Supported by the National Science Foundation.

c^N, c^I are the Clebsch-Gordan I-spin coefficients for the nucleon and $\pi\pi$ vertices, respectively. K is the π momentum in the dipion system; η_L^I, δ_L^I are the inelasticities and phase shifts of the dipion system; $f_{\mu\lambda\lambda}$ contains the specific t dependence of the various amplitudes, corrected for absorption as given by Dürr and Pilkuhn[2]; ψ is the rotation angle from the helicity frame to the Jackson frame, and (θ,ϕ) are the scattering angles of one π in the Jackson frame.

To account for the Pauli exclusion principle, we have assumed total spin flip at the nucleon vertex and corrected the predicted differential cross section accordingly. A maximum likelihood fit to the data with up to 4 partial waves was used to determine the η_L^I, δ_L^I in 40 MeV intervals of the $\pi\pi$ mass. The I=2 phase shifts and inelasticities were interpolated from other publications[3].

Because of the large number of parameters involved, and because of somewhat different assumptions used in the fitting at different $\pi\pi$ masses, we have divided the dipion mass distribution into 3 regions which were fitted separately. [ρ region (.6-.9) GeV; intermediate region (.9-1.08) GeV, f region (1.08-1.42) GeV]. The assumptions for the three regions are given in Table 1. The I=1 resonances ρ, g and the I=0 f^0 resonance are parametrized with a Breit-Wigner form; the phase shifts are a function of the central value m_0 and width Γ_0 of the resonances allowing us to have a smaller number of parameters to determine throughout the fit. We found that the N.P.T. as well as the correction for the Pauli exclusion principle have little effect on the phase shifts and inelasticities. In Fig. 1 the phase shifts and inelasticities of the I=0 partial waves as obtained from the fit are plotted as a function of the dipion mass. Only the central value and width of the ρ were fitted; for the f^0 and g^0 a few central values and widths were tried. We find that results for

TABLE I

Partial Wave	ρ Region	Intermediate Region	f Region
I = 0, S	Elastic → $\eta_0^0 = 1$ δ_0^0 variable	δ_0^0 variable η_0^0 variable	δ_0^0 variable η_0^0 variable
I = 1, P	Elastic → $\eta_1^1 = 1$ $\delta_1^{*1} = \arctan[m_0\Gamma(s)/(m_0^2-s)]$	η_1^1 variable δ_1^1 as in ρ region	η_1^1 variable δ_1^1 as in ρ region
I = 0, D	—	η_2^0 variable $\delta_2^0 = \arctan[m_0\Gamma(s)/(m_0^2 - s)]$	m_0 = 1.27 GeV Γ_0 = .20 GeV
I = 1, F	—	η_3^1 variable $\delta_3^1 = \arctan[m_0\Gamma'(s)/(m_0^2 - s)]$	m_0 = 1.68 GeV Γ_0 = .13 GeV

$$\Gamma(s) = \frac{2 m_0}{m_0+\sqrt{s}}\left(\frac{K}{K_0}\right)^{2L+1}\frac{D_L(K_0)}{D_L(K)}\cdot\Gamma_0, \quad \Gamma'(s) = \frac{m_0}{\sqrt{s}}\left(\frac{K}{K_0}\right)^{7}\cdot\Gamma_0$$

where: s=square of dipion mass, L = spin of the resonance, K, K_0 are momenta of the π in rest frame of dipion mass $\sqrt{s}$, m_0, respectively, $D_L(K)$ correction factor for different interaction radii.

*Γ_0, m_0 for the ρ were fitted only in the ρ region

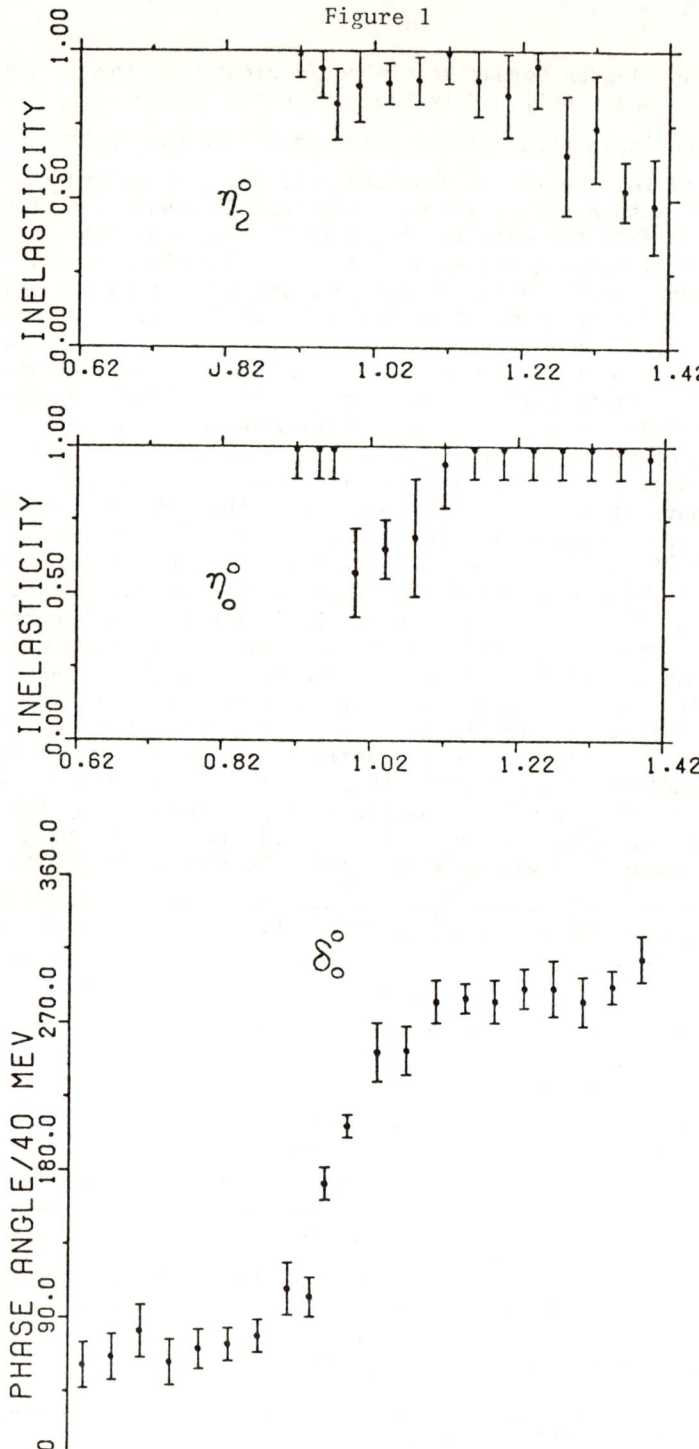

Figure 1

the I=0 partial wave inelasticities and δ_0^0 ($m_{\pi\pi}$) are not very sensitive to the exact parameters of the f and g mesons; the values used are listed in Table 1. For the ρ we find: m_0 = (.772 ± .004) GeV, Γ_0 = (.154 ± .015) GeV. In general our results for the inelasticities agree with those published by the Berkeley group.[4] Up to $m_{\pi\pi}$ = 1.04 GeV our δ_0^0 agrees with the values they have published, but above 1.04 GeV δ_0^0 continues to rise and agrees with results published by a CERN group.[5] We note that our preferred solution for δ_0^0 below 900 MeV corresponds to the "Down" solution. We also find the "Up" solution; it is 3.8 standard deviation less likely than the "Down" solution.

In Fig. 2 the differential cross section, the cosθ and Trieman-Yang angle distributions for the ρ and f regions are compared with the model predictions. In the ρ region the absolute predictions of the model differ only by 3% from the observed while in the f region the shapes agree but there is a discrepancy of about 20% in absolute normalization. The dipion mass distribution (not shown) also shows this normalization discrepancy in the f region. This effect is not well understood.

REFERENCES

1. B. Y. Oh, A. F. Garfinkel, R. Morse, W. D. Walker, J. D. Prentice, E. C. West, and T. S. Yoon, Phys. Rev. D1, 2494 (1970).
2. H. P. Dürr and H. Pilkuhn, Nuovo Cimento 40, 899 (1965).
3. J. P. Baton, G. Laurens and J. Reignier, Phys. Letters 33B, 525 and 528 (1970).
4. S. D. Protopopescu, M. Alston-Garnjost, A. Barbaro Galtieri, S. M. Flatte, J. H. Friedman, T. A. Lasinski, G. R. Lynch, M. S. Rabin, and F. T. Solmitz, Experimental Meson Spectroscopy 1972.
5. C. Grager, B. Hyams, C. Jones, P. Schlein, W. Blum, H. Dietl, W. Koch, E. Lorenz, G. Lutjens, W. Männer, J. Meissburger, W. Ochs, V. Stierlin, and P. Weilhammer, Experimental Meson Spectroscopy 1972.

FIGURE CAPTIONS

Fig. 1. δ_0^0, η_0^0, and η_2^0 as a function of dipion mass.

Fig. 2. dσ/dt, cosθ, and Trieman-Yang angle distributions for ρ^0 and f^0.

Figure 2

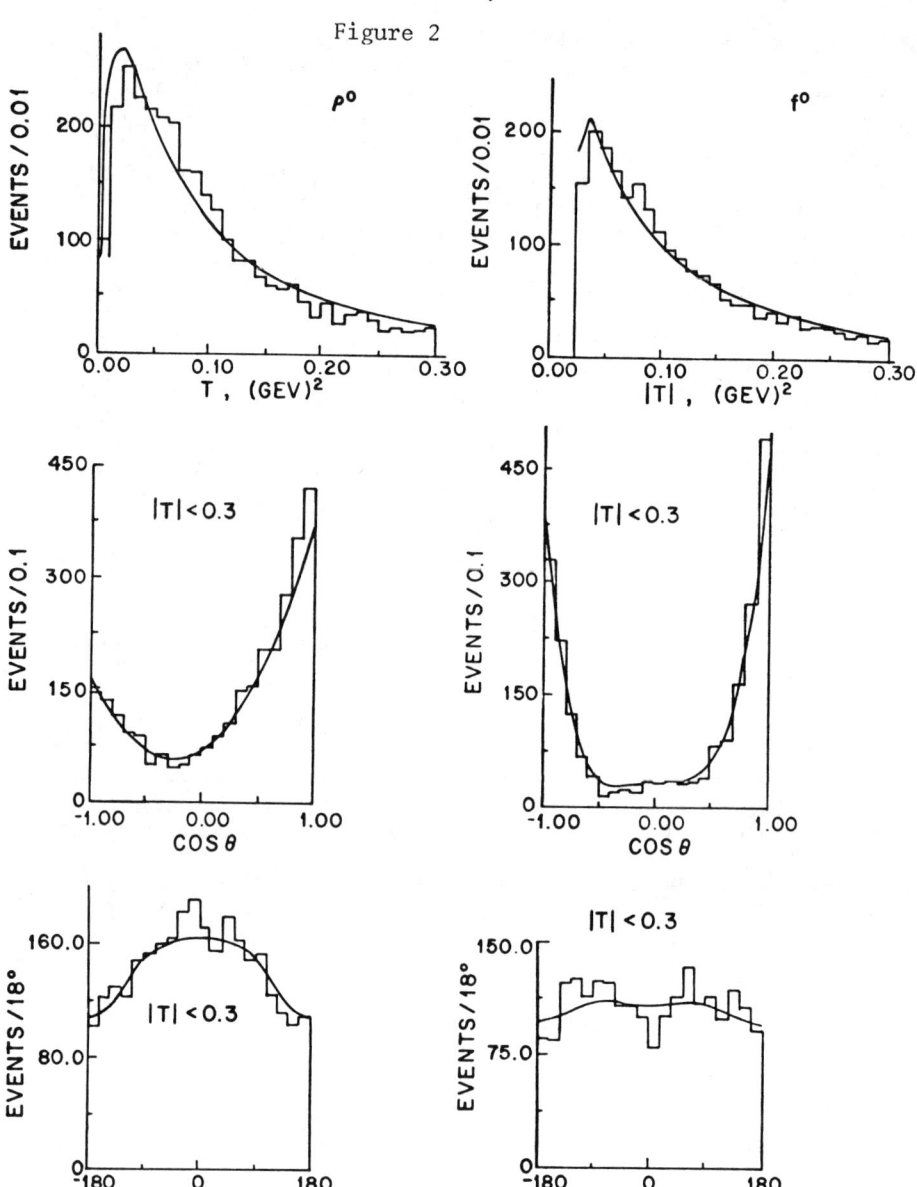

PRODUCTION OF ρ^0 AND f^0 IN 4 GEV/C $\pi^+ d$ INTERACTIONS

J. A. Charlesworth and R. L. Sekulin
Rutherford High Energy Laboratory, U.K.

M. J. Emms, J. B. Kinson, L. Riddiford, B. J. Stacey,
M. F. Votruba and P. L. Woodworth
University of Birmingham, U.K.

I. G. Bell, M. Dale and J. V. Major
University of Durham, U.K.

ABSTRACT

We have applied the Estabrooks and Martin[1] analysis to a sample of 5279 events produced in the reaction $\pi^+ n \to p\rho^0$, and have made a density matrix study, including a positivity analysis, of the $J = 0, 1, 2$ density matrix in the f^0 region, using a sample of 2385 events.

INTRODUCTION

We summarize the results of a preliminary study of ρ^0 and f^0 production in the reaction

$$\pi^+ d \to p_s p \pi^+ \pi^- \tag{1}$$

at 4 GeV/c. Our sample of 15485 events was obtained in an experiment in the CERN 2 m Bubble Chamber. Events corresponding to ρ^0 and f^0 production were selected by choosing events in the mass intervals 0.68-0.88 GeV and 1.20-1.34 GeV respectively.

The general features of reaction (1) are shown in the prism plot of Fig. 1 in which the mass combinations $M^2(\pi^+\pi^-)$ and $M^2(p\pi^-)$ are plotted parallel to the horizontal axes and the Van Hove angle ω is plotted parallel to the vertical axis. Strong bands due to forward ρ^0 and f^0 production can be seen in the upper half of the plot, and some backward production of these states is evident in the lower half.

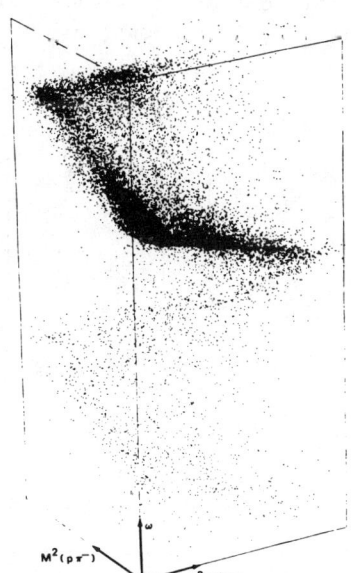

Fig. 1. Prism plot for $\pi^+ n \to p\pi^+\pi^-$ at 4 GeV/c.

We have studied the possible effects of a background due to incoherent N* production, and conclude that they are negligible in the ρ^0 and f^0 analyses described below.

ANALYSIS OF ρ^0 PRODUCTION

We have applied the model-dependent amplitude analysis of Estabrooks and Martin[1] - (EM) - to our data in the ρ^0 region. Details of this analysis have been discussed elsewhere[1] and we present here only the results of its application.

Using notation $\rho_{\lambda\lambda'}^{JJ'}$, for density matrix elements, and N_λ^J and U_λ^J for the (asymptotic) natural and unnatural parity exchange components in the production of the state $|J\lambda\rangle$, we show the results of the EM analysis on our data in Fig. 2.

In Fig. 2a the following features are observed:

(a) Production of the helicity zero component $\rho_{00}^{11} \frac{dN}{dt}$ dominates at lower t, as expected in one pion exchange models. A Williams model[2] form, $-t'e^{Bt}/(t-\mu^2)^2$, fits this component quite well below $|t| = 0.3(\text{GeV/c})^2$ with a slope $B = (7.2 \pm 0.6)(\text{GeV/c})^{-2}$;

(b) In contrast to the analysis[1] of the CERN-Munich data to the reaction $\pi^-p \to \rho^0 n$ at 17.2 GeV/c, U_1^1 is larger than N_1^1 over most of the $|t|$ range 0.1-$0.6(\text{GeV/c})^2$. This result depends only on the fact that ρ_{1-1}^{11} (which can be determined uniquely from the moments) is negative. The observed decrease in the proportion of natural parity exchange in going from higher to lower energy is no doubt partly due

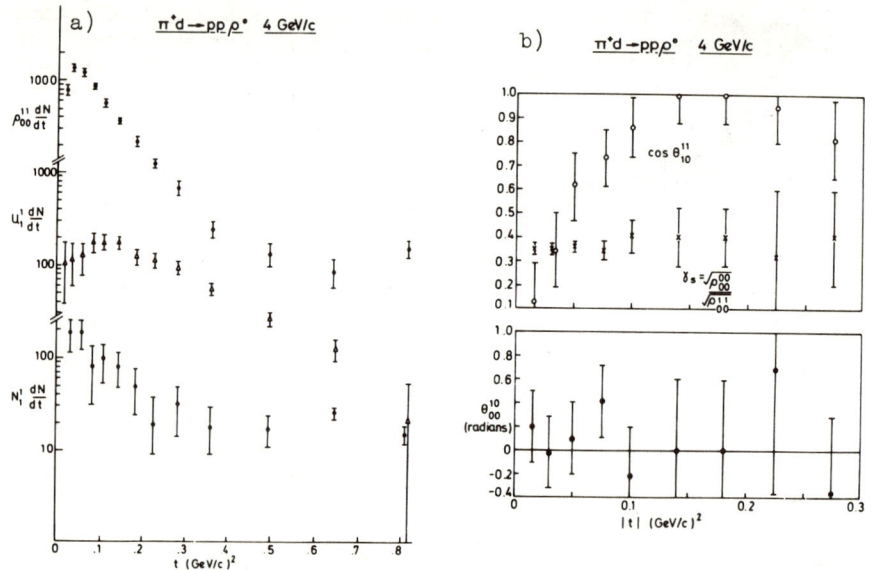

Fig. 2. Results of EM analysis of ρ^0 production at 4 GeV/c.

to the energy dependence of the A_2 pole contribution.

In Fig. 2b is shown the variation up to $|t| = 0.3(GeV/c)^2$ of the other parameters of the EM analysis: $\cos\theta_{10}^{11}$ (the phase angle between the amplitudes for production of the states $|11\rangle$ and $|10\rangle$), γ_s (the ratio of the moduli of the amplitudes for s and p wave production with helicity zero), and θ_{00}^{10} (the relative s-p wave decay phase in the helicity zero state). The values of these parameters are in qualitative agreement with those obtained from the 17.2 GeV/c data, θ_{00}^{10} being consistent with zero, and $\cos\theta_{10}^{11}$ varying rapidly at low t.

ANALYSIS OF f^0 PRODUCTION

Examination of the spherical harmonic moments of the $\pi^+\pi^-$ decay angular distribution in the f^0 region indicates that it is necessary to consider s p and d wave $\pi\pi$ production in this mass region. As in the case of mixed s and p wave production in the ρ^0 region, the information supplied by the moments is insufficient to determine uniquely all the J = 0, 1 and 2 density matrix elements, and we have made two approaches towards a solution of this problem:

A. <u>Analysis at low momentum transfer</u>. In the region $|t| < 0.07$ $(GeV/c)^2$, it is plausible that only the helicity zero states of the $\pi\pi$ system are occupied. With this assumption, the following conclusions may be made concerning the production of the $\pi\pi$ system[3]:
(a) The proportion of d-wave is $(34 \pm 10)\%$. The remaining $(66 \pm 10)\%$ corresponds to s and p wave production; it is not possible to further resolve these.
(b) The relative s-d-wave decay phase angle, θ_{00}^{20}, is close to zero i.e. the s and d waves are in phase. On the other hand, the p and d waves are found to be considerably different in phase. These results are independent of the relative amounts of s- and p- waves present.
B. <u>Positivity analysis of the J = 0, 1, 2 density matrix</u>. In order to study the f^0 over the whole t-range of its production, we have performed a model-independent analysis in which, by applying the positivity conditions of the type derived from the Schwartz inequalities to the expansion of the spherical harmonic moments in terms of density matrix elements, we have delimited the regions within which the density matrix elements, or physically interesting combinations of elements, must lie.

The results of this analysis are shown in Fig. 3. The following explanatory remarks and comments may be made:
(a) In Fig. 3a are shown 2-dimensional projections of the allowed regions of some density matrix elements as a function of t. In each case we have plotted 3 physically interesting combinations of density matrix elements which are constrained by the trace condition to add up to unity, and so have been plotted as distances from the sides of an equilateral triangle.

Thus the left-hand series of plots shows the variation with t

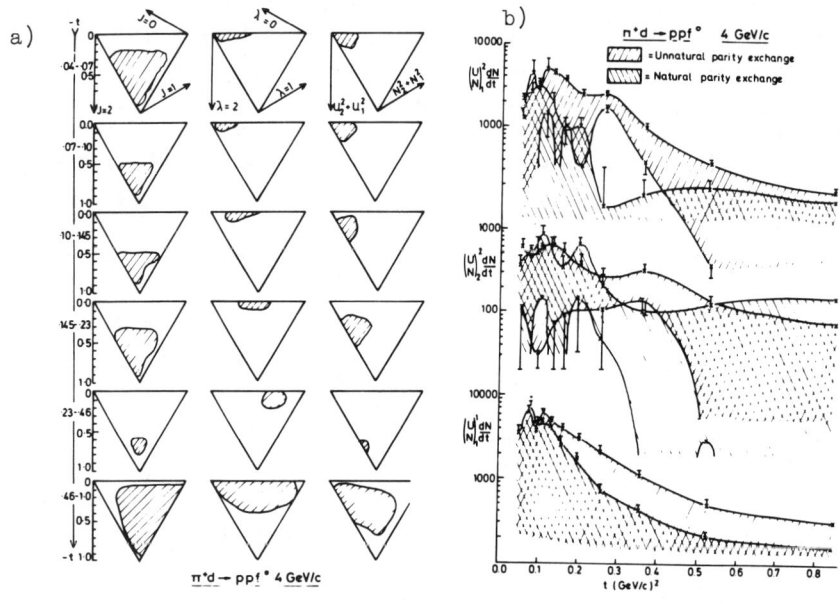
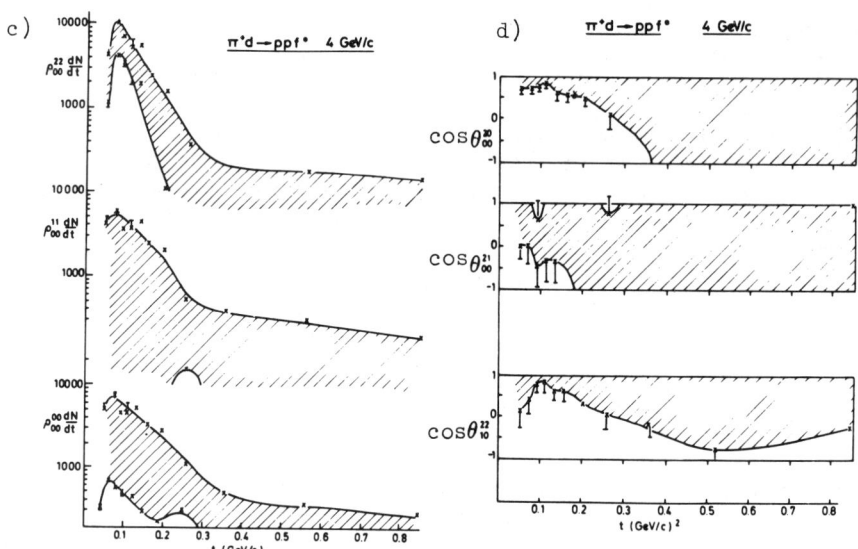

Fig. 3. Results of positivity analysis of f^0 production at 4 GeV/c. The shaded regions are the regions within which the Schwartz inequalities are satisfied, (see text).
The vertical bars indicate statistical errors.

of the relative proportions of J = 2, 1 and 0 production, constrained by the fact that the sum of these proportions must equal unity. It can be seen that, at some value of t, a non-zero contribution is required from each of these spin components.

In the centre series of plots is shown the analogous variation with t of the proportions of helicity 2, 1 and 0 production. One sees that the proportion of helicity zero production falls with increasing t, while the proportion of helicity one production rises; the proportion of helicity 2 production is small at all t except possibly the largest value considered.

In the right-hand series of plots, we compare the amounts of unnatural and natural parity exchange contributing to the production of the J = 2, λ = 1 and 2 states. It can be seen that there is a tendency for the amount of unnatural to exceed the amount of natural parity exchange, a result analogous to that found above for ρ^0 production.

(b) In Fig. 3c are shown as a function of t the one-dimensional projections of the limits on the unnormalized density matrix elements for helicity zero production in the J = 2, 1 and 0 states. The strongest constraints imposed by our analysis can be seen in the uppermost plot, corresponding to d-wave production; on the other hand, the relative productions of s- and p- waves, shown in the lower 2 plots, are not easily separated.

The helicity zero d-wave component can easily accommodate a Williams model type t-dependence, but the slope B required by the data, namely $B \gtrsim 9(GeV/c)^{-2}$, is greater than that obtained above in the case of ρ^0 production.

(c) In Fig. 3d are shown the limits on the unnormalized natural and unnatural parity exchange contributions to production of the states $|21\rangle$, $|22\rangle$ and $|11\rangle$ respectively. The tendency for the unnatural to exceed the natural parity exchange contribution, remarked upon above, is clear.

(d) The t dependence of the s-d and p-d relative decay phases is shown in the upper two plots of Fig. 3b. The conclusion drawn in the low t analysis above, that the s and d waves are close in phase is clearly also valid at low t in this general analysis. No strong statement can be made about the relative p-d phase however.

The lowest plot in Fig. 3 shows the variation with t of $\cos\theta_{10}^{22}$, the phase angle between the amplitudes for production of helicities 1 and 0 in the J = 2 state. One sees that the data can readily accommodate a variation of this angle similar to that found above for the analogous angle $\cos\theta_{10}^{11}$ in ρ^0 production.

REFERENCES

1. P. Estabrooks and A. D. Martin, Phys. Letters <u>41B</u>, 350 (1972).
   A. D. Martin, Invited talk at this conference.
2. P. K. Williams, Phys. Rev. <u>181</u>, 1963 (1969).
3. R. L. Sekulin, Rutherford Laboratory preprint RPP/H/106 (1973); Nuclear Physics, to be published.

PRELIMINARY RESULTS OF CHEW-LOW EXTRAPOLATIONS
USING $\pi^- p \to \pi^- \pi^+ n$ AT 4.5 GEV/C *

P. Jacques, S. Barish, J. Bensinger, M. Hearn,
W. Kononenko, E.M. O'Neill, W. Selove
University of Pennsylvania, Philadelphia, Pa. 19104

ABSTRACT

About 8,600 $\pi^-\pi^+ n$ events, from 110,000 2-prong events, have been analyzed using several forms of Chew-Low extrapolation. The results are compared on the basis of chi-square probabilities, stability against change in the upper limit in -t used, smoothness of $\sigma_{\pi\pi}$ as a function of $M_{\pi\pi}$, and agreement with the unitarity limit for $\sigma_{\pi\pi}$ at the ρ mass.

Using 8,600 events of the type $\pi^- p \to \pi^- \pi^+ n$ with $-t \leq 20\mu^2$ and a beam momentum of 4.5 Gev/c, Chew-Low extrapolations to the pion pole have been made to determine $\sigma_{\pi\pi}$ as a function of $M_{\pi\pi}$. No constraint on the value of $d\sigma/dt$ at t=0 has been applied. That is, $\sigma_{\pi\pi}$ is determined by the equations

$$\sigma_{\pi\pi}(M_{\pi\pi}) = -F(M_{\pi\pi}, -1)$$

where
$$F(M_{\pi\pi}, \delta) = \frac{\pi p^2 \mu^2 (\delta+1)^2 K}{f^2 M_{\pi\pi}^2 (M_{\pi\pi}^2/4 - \mu^2)^{\frac{1}{2}}} \frac{d^2 N}{dM_{\pi\pi} d\delta}$$

with
$$\delta = -t/\mu^2$$
μ = pion mass
p = beam momentum = 4.5 Gev/c
K = μb/event = .148

The data were analyzed in overlapping 40 Mev mass slices. The technique followed was to determine the value of $F(M_{\pi\pi}, \delta)$ for several values of δ in the physical region and then fit an assumed functional form to these values. $dN/d\delta$ was calculated by binning; i.e. $dN/d\delta \approx \Delta N/\Delta \delta$ where ΔN is the number of events in the range $\Delta \delta$. The value of the function at $\delta=-1$ gives the $\pi\pi$ cross section. Several forms were used for the extrapolation function, including the conformal mapping approach of Cutkosky, Deo and Ciulli.[2] The conformal mapping maps the complex δ plane into a new variable τ, where series expansions for F will have the lowest possible truncation error when terminated after a finite number of terms. It was found, however, that the X

*Supported by the United States Atomic Energy Commission

variable as used by Laurens[3] gave identical results to the conformal mapping.

Approximately 40 mass bins between 500 and 1300 Mev were used. For each extrapolation function tried, the quality of the fit was measured by the χ^2 probability in each mass bin; this probability should be uniformly distributed between 0 and 100% if the fit is good. In figure 1 χ^2 probability distributions for extrapolations using fits linear and quadratic in t and in X are shown. The fit linear in t is clearly bad, showing a large peak near zero probability, but there is no obvious way to differentiate among the others. However, the cross sections obtained from these are not consistent with each other, as can be seen in figure 2. The extrapolated $\pi\pi$ cross section is displayed, computed in overlapping 40 Mev mass bins, using fits quadratic in t and linear and quadratic in X.

In addition to the fact that the cross sections in figure 2 do not agree with each other, the p-wave unitarity limit, which is 116 mb at 760 Mev, is not reached, although the cross section should rise somewhat above that value due to the presence of some s-wave. These cross sections were also observed to change significantly when the upper limit on -t was reduced to $12\mu^2$ (there are 7,600 events with $-t \leq 12\mu^2$).

In an attempt to improve on this situation the fits were redone using the extended maximum likelihood technique[4], to avoid the loss in statistical power that occurs when the events are binned. This method did not bring the cross sections in figure 2 into agreement, but did produce smoother extrapolated cross sections and for this reason and the fact that the maximum likelihood fits were less sensitive to changes in the upper limit on -t we feel that the results obtained in this way are more trustworthy. Figure 3 shows the maximum likelihood cross section obtained from a quadratic fit in X, which still however does not reach the p-wave unitarity limit.

An alternate approach to the extrapolation problem is the use of form factors. We have used a Durr-Pilkuhn form factor[5] of the form

$$F_{NN\pi} = (1+R_N^2 Q_N^2)/(1+R_N^2 Q_{N_t}^2)$$
$$R_N = 2.86 \text{ Gev}^{-1}$$

for the nucleon vertex, and a p-wave Benecke-Durr form factor[5] of the form

$$F_{\rho\pi\pi} = (q^2/q_t^2) U_1(q_t R_\rho) / U_1(q R_\rho)$$
$$U_1(x) = (\frac{(2x^2+1)}{4x^2}\ln(4x^2+1) - 1)/(2x^2)$$
$$R_\rho = 2.31 \text{ Gev}^{-1}$$

for the meson vertex, as used by Wolf[5]. The maximum likelihood cross section, using a fit linear in t and using all events with $-t \leq 20\mu^2$, is shown in figure 4; essentially no change is produced when the fit is limited to events with $-t \leq 12\mu^2$. This method produces a cross section which is not inconsistent with the p-wave unitarity limit with some additional s-wave; also, the peak in the region of the f_o mass is of roughly the correct size (although the use of the p-wave meson vertex form factor is clearly incorrect at that mass).

Fig. 1a. χ^2 probability distribution for the fit linear in t.

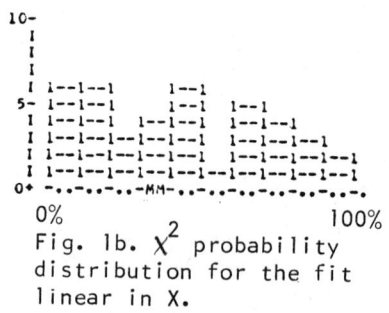

Fig. 1b. χ^2 probability distribution for the fit linear in X.

Fig. 1c. χ^2 probability distribution for the fit quadratic in t.

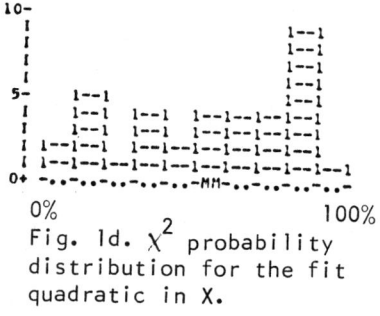

Fig. 1d. χ^2 probability distribution for the fit quadratic in X.

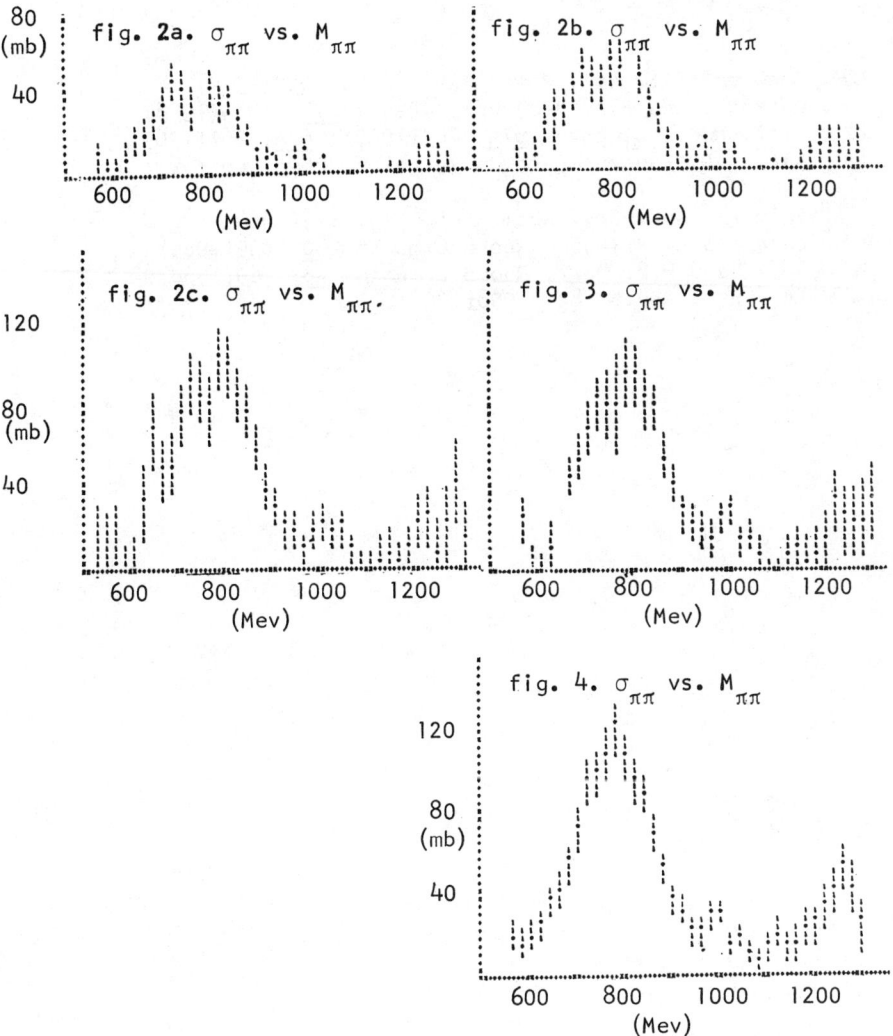

Extrapolated ππ total cross section as a function of ππ mass.

Fig. 2a. extrapolated using a fit quadratic in t
Fig. 2b.   "   "   "   " linear in X
Fig. 2c.   "   "   "   " quadratic in X
Fig. 3 .   "   "   "   " quadratic in X and fit
         using the extended maximum likelihood method.
Fig. 4 . extrapolated using a fit linear in t, with
         Durr-Pilkuhn and Benecke-Durr form factors.

REFERENCES

1. G.F. Chew and F.E. Low, Phys. Rev., 113, 1640(1959)
2. R.E. Cutkosky and B.B. Deo, Phys. Rev., 174, 1859(1968)
   S. Ciulli, Nuovo Cimento, 61A, 787(1969); 62A, 301(1969)
3. J.P. Baton, G. Laurens, and J. Reignier, Phys. Lett., 33B, 538(1970)
4. F.T. Solmitz, Ann. Rev. Nucl. Sci., 14, 375(1964)
5. H.P. Durr and H. Pilkuhn, Nuovo Cimento, 40, 899(1965)
   J. Benecke and H.P. Durr, Nuovo Cimento, 56, 269(1968)
6. G. Wolf, Phys. Rev., 182, 1598(1968)

A THEORETICAL CALCULATION OF LOW ENERGY ππ SCATTERING

K. S. Jhung and R. S. Willey
University of Pittsburgh, Pittsburgh, Pa. 1526

ABSTRACT

A theoretical calculation of the ππ s- and p-wave phase shifts below one GeV/c is presented. The calculation is based on the Padé approximant technique applied to the one loop approximation computed from the nonlinear σ-model lagrangian.

The starting point for low energy calculations of ππ scattering is the amplitude derived by Weinberg[1] using the ingredients of current algebra, PCAC, and a particular assumption about the nature of the $SU_2 \times SU_2$ symmetry breaking

$$M^{(0)}_{abcd} = \frac{1}{f_\pi^2}[\delta_{ab}\delta_{cd}(s-\mu^2) + \delta_{ac}\delta_{bd}(t-\mu^2) + \delta_{ad}\delta_{bc}(u-\mu^2)] \quad (1)$$

We note that this amplitude is purely real and grows rapidly with increasing energy. This implies violations of unitarity above the elastic threshold i.e. the Weinberg amplitude is a low energy theorem.

We consider field theories which incorporate the general principles used by Weinberg. In such theories, one can hope to extend the current algebra results above threshold by computing higher order corrections to the purely real Born terms. One such field theory model is the 'linear' σ-model[2] (LσM), characterized by an ordinary polynomial Lagrangian

$$\mathcal{L}_{L\sigma M}(\pi,\sigma,N).$$

The awkard point for the LσM is the nonexistence of the σ particle. We therefore turn our consideration to the 'nonlinear' σ-model (NLσM). In this case one has a Lagrangian

$$\mathcal{L}_{NL\sigma M}(\pi,N)$$

Which starts from the correct stable particles; the price one pays is that it is a nonpolynomial Lagrangian with derivative couplings. This raises special problems to which we will return.

First, there is a general problem for computing strong interaction processes starting from a Lagrangian. Perturbation theory doesn't converge in the presence of resonances. However, it has been realized[3] that this problem may be attacked with some hope of success by applying the Padé algorithm for summation of divergent series. The sequence of Padé approximants is known to converge in the potential theory case even when the potential is strong enough to form bound states.

Now we return to the problems specific to the NLσM. It is not renormalizable in the conventional sense. To deal with this problem we adopt a suggestion of Bessis and Zinn-Justin.[4] They suggest that the regularized, one-loop, NLσM amplitude should be obtained by first computing the one-loop amplitude in the (renormalizable) LσM, then expanding all renormalized Feynman integrals in powers of m_σ^{-1}, and dropping all terms which vanish in the limit $m_\sigma \to \infty$. This will leave ordinary Feynman integrals plus terms proportional to powers of m_σ^2 and $\ln m_\sigma^2$. To understand what is happening consider first the Born terms (zero loops) in the lim $m_\sigma \to \infty$.

$$M^{(0)}_{L\sigma M} \xrightarrow[M_\sigma \to \infty]{} M^{(0)}_{NL\sigma M} \qquad (2)$$

which is, in fact, just the Weinberg amplitude (1). Next, consider calculation of the imaginary part of the one-loop amplitude by iteration of the unitarity equation.

$$\lim_{m_\sigma \to \infty} \operatorname{Im} M^{(1)}_{L\sigma M} = \lim_{m_\sigma \to \infty} \Sigma |M^{(0)}_{L\sigma M}|^2$$
$$= \Sigma |M^{(0)}_{NL\sigma M}|^2 = \operatorname{Im} M^{(1)}_{NL\sigma M}. \qquad (3)$$

Thus the one-loop NLσM amplitude will differ from the $m_\sigma \to \infty$ limit of the one-loop LσM amplitude by at most a polynomial. Since the NLσM amplitude has an undetermined polynomial, the proposal of Bessis and Zinn-Justin is to define that polynomial by regularization with the LσM. In the limit $m_\sigma^2 \gg m_\pi^2$, s, individual LσM diagrams contribute terms of order m_σ^2, $m_\sigma \ln m_\sigma^2$, m_σ, $\ln m_\sigma^2$, 1, and 0. However, when all the diagrams are combined and all the chiral Ward identity constraints are imposed, the m_σ^2, $m_\sigma \ln m_\sigma$, and m_σ terms all cancel, and we are left with only one undetermined subtraction constant,

$$L \equiv \ln \frac{m_\sigma^2}{m_\pi^2}. \qquad (4)$$

The Feynman diagrams which contribute to the discontinuity of the one-loop NLσM amplitude are

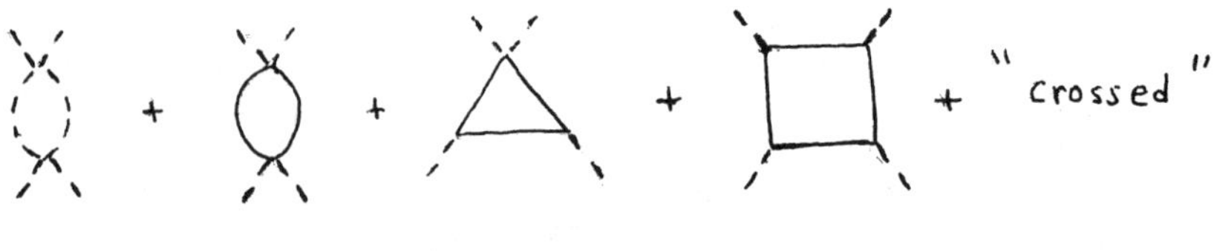

Fig. 1 One-loop Feynman diagrams for NLσM.

With the one-loop invariant amplitudes computed, we project out iso-spin and partial wave amplitudes, and form the [1,1] Padé approximant for the partial wave amplitudes.

$$T^{[1,1]}_{IJ}(s) = \frac{1}{f^2} T^{(0)}_{IJ}(s) \left[1 - \frac{1}{f^2} \frac{T^{(1)}_{IJ}(s)}{T^{(0)}_{IJ}(s)} \right]^{-1} \qquad (5)$$

These partial wave amplitudes satisfy elastic unitarity exactly, so one can compute the phase shifts from them.

One more point has to be considered. The Ward identities which are used to fix the renormalized constants, in the one-loop approximation, determine the effective πNN coupling to be

$$g_{eff} = \frac{m}{f_\pi} \quad \text{(one-loop)} \qquad (6)$$

In two-loop (and higher) order, we will obtain a (finite) renormalization of g_{eff}.

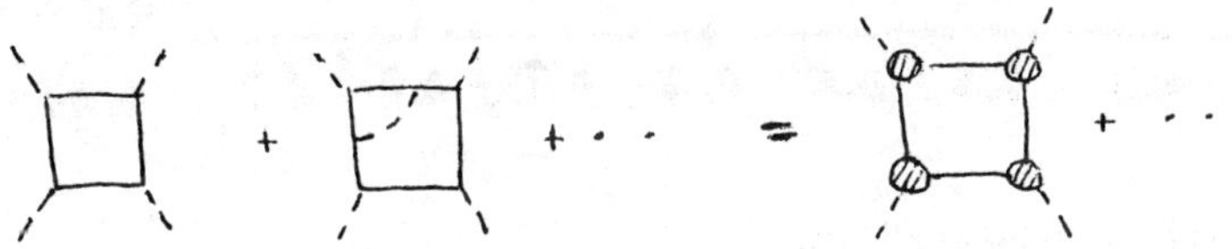

Fig. 2 Additional Feynman diagrams.

So we replace g_{eff} of (6) by

$$g = \frac{m}{f_\pi} g_A \quad \text{(Goldberger-Treiman relation)}$$

and treat g_A as a 'limited' parameter (because we don't consistently do a two loop calculation).

A set of calculated s- and p-wave phase shifts are shown in Fig. 3.

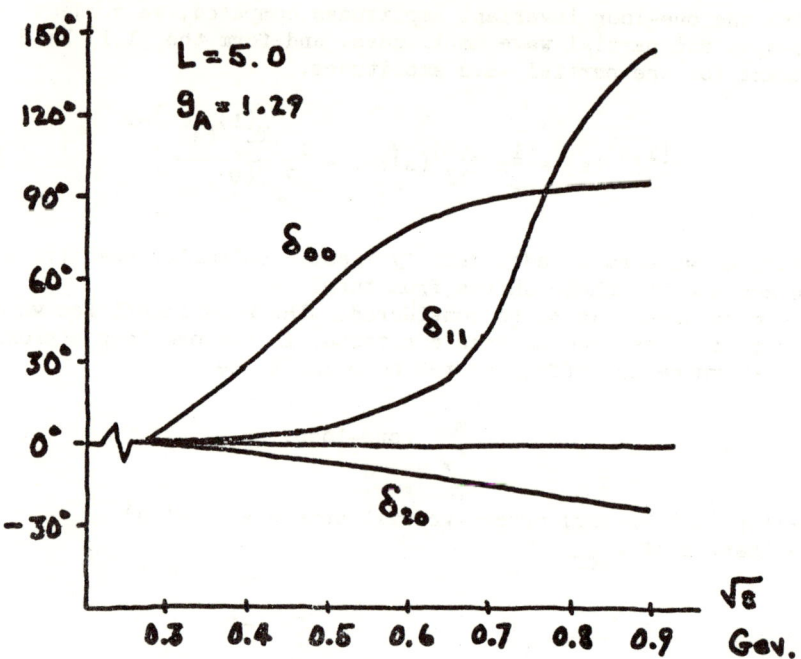

Fig. 3 Calculated phase shifts δ_{IJ}.

We make the following comments.
1) The s-wave phase shifts are sensitive to the constant L. Increasing L gives larger δ_{00}. δ_{20} is less sensitive than δ_{00}.
2) The p-wave phase shift is quite insensitive to L, but does depend strongly on g_A. With $g_A = 1.29$ one can produce a p-wave phase shift with an elastic Breit-Wigner shape with $m_\rho = 760$ MeV and $\Gamma_\rho = 140$ MeV.
3) All the considerations described here have been based on an $SU_2 \times SU_2$ model. Such a model is useful only for energies below the $K\bar{K}$ threshold.
4) One may think of this model as an extension of the Weinberg amplitude above threshold, into the resonance region. The predicted scattering lengths are little changed from Weinberg's values. They are $a_0 = 0.24\mu^{-1}$, $a_2 = -0.04\mu^{-1}$.

Preliminary results of these calculations were contributed to the XVI International Conference on High Energy Physics, Chicago-Batavia (1972). A detailed account will be published elsewhere.

REFERENCES

1. S. Weinberg, Phys. Rev. Lett. <u>17</u>, 616 (1966).
2. J. Schwinger, Ann. Phys. <u>2</u>, 407 (1957). M. Gell-Mann and

M. Levy, Nuovo Cimento 16, 705 (1960).
3. For a review of the Padé approximant and applications to particle physics see eg. J. L. Basdevant, F. de. Physik 20, 283 (1972).
4. D. Bessis and J. Zinn-Justin, Phys. Rev. D5, 1313 (1972).

RELATIVISTIC PROPAGATORS FOR π-π RESONANCES AND THE N-N INTERACTION

M. L. Nack, T. Ueda[*], and A. E. S. Green
Department of Physics and Astronomy
University of Florida, Gainesville, Fl. 32601

ABSTRACT

A relativistic resonance propagator is presented, whose form is designed to fit the full S-matrix and therefore the corresponding phase shift data, and to also satisfy a threshold scattering length relationship. This model is fitted to several recently reported π-π S and P waves phase shifts solutions. We find that the corresponding mass spectral functions lead to substantially different π-π system exchange contributions to the N-N interaction.

We present a summary of our construction of finite width propagators including the full δ_0^0 and δ_1^1 π-π phase shift data for use in the N-N interaction, and this is consistent with the close interrelationship of various strong interaction studies.[1,2]

We consider the elastic scattering partial wave S-matrix elements and due to the close relationship of the K-matrix unitarization[3,4] with the non-relativistic Breit-Wigner amplitude we use this unitary approximation to guide the choice of the form of our finite width amplitude and propagator

$$k A_\ell(s) \simeq B_{\ell\Gamma}(s) = \Delta_\Gamma^g(s) = \frac{\Gamma(s,\ell)}{(s_m-s)-i\,\Gamma(s,\ell)}, \quad (1)$$

where $s_m = m^2$, $\Delta^g(s,s_m) = g^2(s_m-s-io)^{-1}$ is our zero width propagator appearing in the zero width Born term $B_\ell = -\tfrac{1}{2}i\, S_{2\ell}$, and $B_{\ell\Gamma}$ now represents the full contribution of $S_{2\ell}, S_{4\ell}, \ldots$ due to our choice of $\Gamma(s,\ell) = g^2 f(s,\ell)$ where $g^2 = m\Gamma$. The general form of $f(s,\ell)$ is

$$f(s,\ell) = \theta(s-a)\,\theta(b-s)\,x(s,\ell), \\ x(s_m,\ell) = 1, \quad x(a,\ell) = 0 = x(b,\ell), \quad (2)$$

and we require it to satisfy as a stable particle limit $f(s,\ell) \to 1$ as $(a,b) \to (-\infty,\infty)$. The imposition of the

---
[*] On leave from Osaka University

scattering length requirement[1,2] $\tan \delta_\ell(k) \simeq a_\ell k^{2\ell+1}$ helps determine the following choice of $x(s,\ell)$ and consequently also a_ℓ;

$$x(s,\ell) = \left[\left(\frac{s-a}{s_m-a}\right)\left(\frac{b-s}{b-s_m}\right)\right]^{\ell+\frac{1}{2}}, \quad a_\ell = \frac{4^{\ell+\frac{1}{2}} m \Gamma}{(s_m-a)^{\ell+3/2}}\left(\frac{b-a}{b-s_m}\right)^{\ell+\frac{1}{2}}. \quad (3)$$

The extension of the theory to the propagation of a system of resonances indexed by ν on $\delta_\ell(s)$ is given by using Eqs. 1, 2, 3 and

$$B_{\ell\Gamma}(s) = \sum_\nu \Delta^g_{\Gamma_\nu}(s), \quad \Delta^g_{\Gamma_\nu}(s) = \frac{\Gamma_\nu(s,\ell)}{(s_\nu-s)-i\Gamma_\nu(s,\ell)}, \quad (4)$$

where $b_\nu = a_{\nu+1}$ is a continuity condition imposed on $\delta_\ell(s)$.

Using a dispersion relation[5] we obtain the normalized relationship between the spectral function (mass squared distribution) and the propagator;

$$\rho^g(s') = \frac{\mathrm{Im}\,\Delta^g_\Gamma(s')}{I^g}, \quad I^g = \int_a^\infty \mathrm{Im}\,\Delta^g_\Gamma(s')ds'. \quad (5)$$

An important characteristic of ρ^g is the mean mass; $\bar{m} = \langle\sqrt{s'}\rangle$. We then obtain $\Delta^g_\Gamma(s)$ in terms of $\delta_\ell(s)$, and the imaginary part of this equation and Eq. 5 result in

$$\delta_\ell(s) = \arcsin\sqrt{\rho^g(s)/\rho^g(s_m)}. \quad (6)$$

Using Eq. 6 we now fit the spectral functions ρ^g to the phase shifts over the (a, b) range of $\delta_\ell(s)$ to determine the parameters m, Γ, a, b. These parameters determine the scattering length by Eq. 3. Table 1 gives the five parameters associated with recently proposed phase shift solutions[6-11] along with values of $\bar{m}$. The corresponding fits are illustrated in Fig. 1.

Generalized OBEP of the N-N interaction (GOBEP) may be formulated using either the Nth order generalized field theory of Green[12] or equivalently the N/2-pole nucleon-meson vertex form factor of Ueda and Green[13]. As opposed to using a one parameter pole propagator, we pro-

Table I - Meson parameters (GEV)

δ_ℓ^I	Authors [ref.](year)	meson	m	$\bar{m}$	Γ	$\sqrt{a}$	$\sqrt{b}$	a_ℓ^I in $m_\pi^{-(2\ell+1)}$
δ_0^0	Baton et al.[6](1970)	ε	.720	.717	.200	.2774	1.000	.19
	Baton et al.[6](1970)	εd	.865	.773	.500	.2774	1.000	.42
	Flatte et al.[8](1972)	S	.890	.789	.400	.2774	1.000	.34
	Carroll et al.[7](1972)		1.270	1.281	.300	1.000	1.550	.29
δ_1^1	Baton et al.[6](1970)	ρ	.765	.803	.135	.2774	1.300	.021
	Scharenguivel et al.[9](1970)	ρ	.765	.857	.135	.2774	1.600	.017
	Moffat et al.[10](1972)	ε						.19
		ρ						.028
	Carrotte et al.[11](1970)	ε						.18

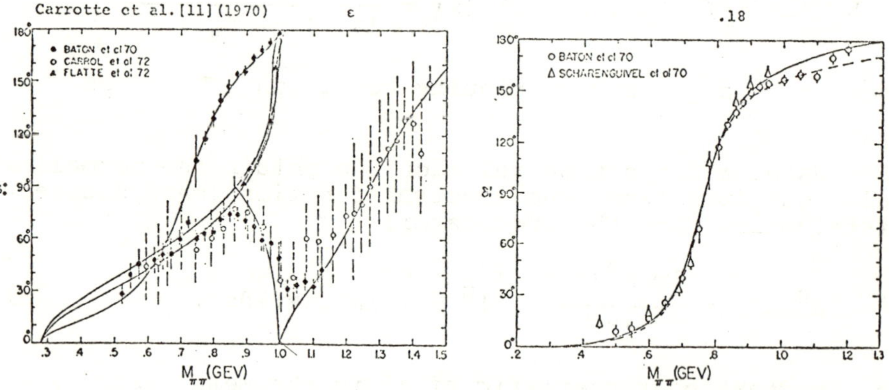

Fig. 1 Data and theoretical fits of the I=0
S and P wave phase shifts.

pose to use a four parameter (m,Γ,a,b) resonance propagator to give a more precise inclusion of the π-π δ_0^0 and δ_1^1 phase shifts in the N-N interaction.

The distributed mass potential in coordinate space is given by

$$J^N(r,\Lambda) = \int_a^b Y^N(r,\Lambda,t') \rho^g(t') dt'$$

$$= \frac{1}{2\pi^2} \int d^3k \, e^{i\underline{k}\cdot\underline{r}} F^N(k^2,\Lambda^2) \Delta_\Gamma(\underline{k}^2) ,$$

$$\Delta_\Gamma(\underline{k}^2) = \int_a^b \Delta(\underline{k}^2,t') \rho^g(t') dt' , \qquad (7)$$

where $\Delta_\Gamma(\underline{k}^2)$ would replace the pole propagators of δ_0^0 and

δ^1_1 normally used in a relativistic momentum space calculation.[14] For N = 4 we have

$$Y^4(r,\Lambda,m) = \frac{1}{r\tau^4}\left\{e^{-mr} - e^{-\Lambda r}\left[1 + \tfrac{1}{2}\Lambda r\tau + \tfrac{1}{8}\Lambda r(1+\Lambda r)\tau^2 + \tfrac{1}{48}\Lambda r(3+3\Lambda r+\Lambda^2 r^2)\tau^3\right]\right\} \quad (8)$$

where $\tau = 1-(m/\Lambda)^2$ and $t' = m^2$. This is the so-called quadrupole regulated potential of Ueda and Green.[13,15,16]

Our work shows a close relationship between the Baton down (εd) solution of the low energy part of δ^0_0 and the S*, and this is illustrated in Fig. 1 by the dashed reflection of εd through $\frac{\pi}{2}$ since ρ^g depends only on $\sin^2\delta = \sin^2(\pi-\delta)$. Consequently the main implications of the differing π-π δ^0_0 solutions on the N-N problem can be illustrated by comparing the ε and S* solutions, and this can be seen in Fig. 2 by examining a coordinate space difference function of the two potentials

$$D_{\varepsilon S*} = g^2_\varepsilon J^4_\varepsilon - g^2_{S*} J^4_{S*} . \quad (9)$$

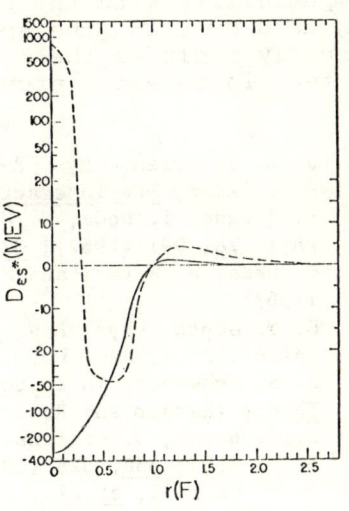

Fig. 2 D versus r.

To show this realistically we take $g^2_\varepsilon = 14$ and adjust g^2_{S*} so that the difference vanishes at 1F. Such a cancellation brings the difference into the same order of magnitude as the residual static terms in current GOBEP which survive after the major cancellation between the static repulsive contributions of the ω meson and the π-π static attractive contribution of the scalar isoscalar meson, i.e. the ε or S*. It is the scalar-vector cancellation which makes the N-N interaction so complicated not only by bringing out relativistic spin and velocity dependent terms which survive the cancellation but also by amplifying the influence of the width and shape of the π-π phase shifts. The solid curve compares this paper's distributed S* and ε, and the dashed curve compares this paper's distributed S* with the more sharply distributed ε of our past work.[16] The large values of $D_{\varepsilon S*}$, especially in the core region of 0 to 1.0 F indicate the different input possibilities of the π-π δ_0^0 data into the N-N problem.

It was noted several years ago[15] that a GOBEP solution of the N-N phase shift data favored the εd over the ε solution of the π-π δ_0^0 phase shift. In this conference the S* seems to be emerging as the favored solution, and in Fig. 1 we have shown a close similarity between the εd and S* solutions. We can now expect to obtain compatibility with the S* and a GOBEP solution of the N-N data by using the spectral function of Eq. 5, which properly includes the low mass contribution of the S*, in the finite width propagator of Eq. 7.

References

1. J. L. Petersen, Phys. Reports 2, 155 (1971).
2. H. Pilkuhn, The Interaction of Hadrons (Wiley, 1967).
3. S. Sawada, T. Ueda, W. Watari and M. Yonezawa, Prog. Theor. Phys. 28, 991 (1962); 32, 380 (1964). M. Kikugawa, S. Sawada, T. Ueda, W. Watari and M. Yonezawa, Prog. Theor. Phys. 37, 88 (1967).
4. S. N. Gupta, Phys. Rev. 117, 1146 (1960); C. W. Bock and R. D. Haracz, Phys. Rev. D5, 39 (1972); Phys. Rev. D6, 1373 (1972).
5. S. S. Schweber, An Introduction to Relativistic Quantum Field Theory (Harper and Row, 1962); see Sect. 17b, Eq. 76.
6. J. P. Baton, G. Laurens, and J. Reignier, Phys. Letters 33B, 525 (1970); 33B, 528 (1970).
7. J. T. Carroll, et al., Phys. Rev. Letters 28, 318 (1972).
8. S. M. Flatte, et al., Phys. Letters 38B, 232 (1972).
9. J. Scharenguivel, et al., Purdue University Report, 1970 (unpublished).
10. J. W. Moffat and B. Weisman, Phys. Rev. D6, 238 (1972).
11. J. B. Carrotte and R. C. Johnson, Phys. Rev. D2, 1945 (1970).
12. A. E. S. Green, Phys. Rev. 75, 1926 (1948; A. E. S. Green and and T. Sawada, Nucl. Phys. B2, 276 (1967); Rev. Mod. Phys. 39, 594 (1967).
13. T. Ueda and A. E. S. Green, Phys. Rev. 174, 1304 (1968); Nucl. Phys. B10, 289 (1969).
14. A. Gersten, R. H. Thompson, and A. E. S. Green, Phys. Rev. D3, 2076 (1971).
15. R. W. Stagat, F. Riewe and A. E. S. Green, Phys. Rev. Letters, 24, 631 (1970); Phys. Rev. C3, 552 (1971).
16. A. E. S. Green, F. Riewe, M. L. Nack, and L. D. Miller, proceedings of symposium on "Present Status and Novel Developments in the Nuclear Many-Body Problem", Rome, 9/72 (to be published).

ISOSPIN 2 ππ PHASE SHIFTS FROM AN EXPERIMENT $\pi^+ p \to \pi^+\pi^+ n$ AT 12.5 GeV/c

G. Grayer[*], B. Hyams, C. Jones and P. Weilhammer
CERN, Geneva, Switzerland

W. Blum, H. Dietl, W. Koch, E. Lorenz, G. Lütjens,
W. Männer, J. Meissburger and U. Stierlin
Max-Planck-Institut für Physik
und Astrophysik, Munich, Germany

W. Hoogland
Zeeman Laboratorium, University of
Amsterdam[**], Amsterdam, The Netherlands

ABSTRACT

The I = 2 s- and d-wave ππ phase shifts δ_s^2 and δ_d^2 have been determined in the ππ mass region from 500-1500 MeV, using a sample of 17,500 events $\pi^+ p \to \pi^+\pi^+$ from a spark chamber experiment. For this purpose the moments of the ππ angular distribution in the s-channel helicity frame have veen written as bilinear expressions of OPE amplitudes modified by absorption and have been fitted to the measured values of the moments in four mass bins of width 0.25 GeV/c². The value of δ_s^2 in the ρ-mass region ($\delta_s^2 \simeq -15.5°$) corresponds to an elastic cross-section of $\simeq 7.5$mb.

The values of δ_d^2 are about a factor of 10 smaller than the corresponding values of δ_s^2 for the whole mass region considered.

INTRODUCTION

We have determined that I = 2 s-wave and d-wave phase shifts δ_s^2 and δ_d^2 from a sample of 17,500 events

$$\pi^+ p \to \pi^+\pi^+ n \qquad (1)$$

with a forward-produced $\pi^+\pi^+$ system, obtained in a wire chmber experiment at the CERN Proton Synchrotron. For a discussion of the apparatus as well as for the methods used to extract the spherical harmonic moments of the ππ angular distribution, we refer to other publications[1,2].

---

[*] Now at M.P.I.
[**] The work in Amsterdam is part of the joint research programs of F.O.M. and Z.W.O.

METHOD AND RESULTS

Our apparatus yields a clean sample of reaction (1) events. Background due to events with additional π^0's can be estimated to be $(3 \pm 1)\%$, while the contamination by events produced by particles other than π^+ (essentially protons) is less than 1%. Since the acceptance falls when the $\pi^+\pi^+$ invariant mass increases, we have restricted our analysis to mass values below 1.5 GeV/c².

The invariant mass spectrum and the t-channel spherical harmonic moments of the $\pi^+\pi^+$ system as a function of mass are shown in Figs. 1 and 2, respectively (only even moments have to be considered, because of the symmetry of the $\pi^+\pi^+$ system). The following observations can be made:

i) All mass dependences are smooth. No resonance structure is seen, in agreement with the expectation that such effects are absent in exotic channels.

ii) $\langle Y_2^0 \rangle$ is substantially different from zero and positive above 0.5 GeV/c², indicating the presence of a d-wave component even at relatively low $\pi\pi$ masses, with δ_d^2 having the same sign as δ_s^2.

iii) The m = 1 moments are significantly different from zero, indicating non-negligible contributions of helicity 1 amplitudes. The m = 2 moments, however, are consistent with zero in both t- and s-channel helicity frames.

The t-dependence of the moments is shown in Fig. 3 for events in the mass interval 0.75-1.25 GeV/c².

Following the absorption model formulated by Williams[3] which has been successfully used[4] to describe the reaction $\pi^- p \to \pi^- \pi^+ n$, we determined δ_s^2 and δ_d^2 from the s-channel moments by fitting them to bilinear expressions of the s-wave and d-wave production amplitudes. These amplitudes are:

$$S = \gamma\, A_s\, \frac{(-t')^{1/2}}{t - \mu^2}\, e^{B(t-\mu^2)} \tag{2a}$$

$$D_0 = \gamma\, \sqrt{5}\, A_d\, \frac{(-t')^{1/2}}{t - \mu^2}\, e^{B(t-\mu^2)} \tag{2b}$$

$$D_1^- = D_1 - D_{-1} = \gamma\, \frac{\sqrt{30}}{M_{\pi\pi}}\, A_d \left\{ \frac{2(-t')}{t - \mu^2} + c \right\} e^{B(t-\mu^2)} \tag{2c}$$

$$D_1^+ = D_1 + D_{-1} = -\gamma\, \frac{\sqrt{30}}{M_{\pi\pi}}\, A_d\, c\, e^{B(t-\mu^2)}, \tag{2d}$$

where we have made the usual assumption that only spin flip contributes, and that the production amplitudes are coherent in phase. We have further assumed that the helicity-two amplitudes may be neglected. A_s and A_d are given by $e^{i\delta} \sin \delta$ (the elasticity parameters η_s and η_d being assumed unity). The normalization constant γ follows from the Chew-Low formula. For reaction (1):

$$\gamma^2 = \frac{M_{\pi\pi}^2}{4\pi \, m_p^2 \, p_{lab}^2} \frac{g^2}{4\pi} \frac{8\pi}{q_\pi}, \tag{3}$$

where μ is the pion mass
$M_{\pi\pi}$ the mass of the $\pi\pi$ system
P_{lab} the beam momentum in the lab system
$g^2/4\pi = 2 \times 14.6$ the $p n \pi_{1/2}$ coupling constant, and
$q_\pi = [(1/4) \, m_{\pi\pi}^2 - \mu^2]^{1/2}$

Equations (2) reduce to the Williams amplitudes for $c = 1$, in which case the $m = 1$ moments are forced to change sign at $-t = \mu^2$. In this analysis, however, c has been fixed to a value of 0.5 in accordance with the observed behaviour of the $m = 1$ moments in reaction (1), which indicates a cross-over nearer to $t = 0$.

Our results for δ_s^2 and δ_d^2, however, are rather insensitive to the chosen value of c. Using $c = 1$, for instance, does not lead to significantly different values.

Solutions for the $\pi^+\pi^+$ phase shifts were obtained in the mass intervals 0.5 GeV/c^2–1.50 GeV/c^2 in bins of width 0.25 GeV/c^2, by performing a χ^2 fit to the moments $N\langle Y_0^0 \rangle$, $N\langle Y_2^0 \rangle$, $N\langle \text{Re } Y_2^1 \rangle$, $N\langle Y_4^0 \rangle$, $N\langle \text{Re } Y_4^1 \rangle$ and $N\langle \text{Re } Y_4^2 \rangle$, in the region $t_{min} < |t| < 0.2$ (GeV/c)^2.

The results of this fit in which δ_s^2 and δ_d^2 for each mass interval, and B were free parameters, are given in Table 1. The quoted errors are statistical only.

Table 1: Fitted values of δ_s^2 and δ_d^2

Mass interval (GeV/c^2)	δ_s^2 (degrees)	δ_d^2 (degrees)	B $[\text{GeV/c}]^{-2}$
0.50 – 0.75	(−13.8 ±0.6)	(−1.0 ±0.15)	
0.75 – 1.00	(−19.0 ±0.6)	(−1.9 ±0.2)	−2.8 ±0.3
1.00 – 1.25	(−23.5 ±0.6)	(−2.8 ±0.2)	
1.25 – 1.50	(−25.8 ±0.6)	(−2.9 0.3)	

The sign of δ_s^2 was taken to be negative as indicated, for example, by the analysis of $\pi^-\pi^0$ scattering data[5].

REFERENCES

1. G. Grayer et al., High statistics study of the reaction $\pi^-p \to \pi^-\pi^+n$: apparatus, method of analysis, and general features of results at 17 GeV/c, to be published.
2. G. Grayer et al., Nuclear Instrum. Methods **99**, 579 (1972).
3. P.K. Williams, Phys. Rev. **D1**, 1312 (1970).
4. G. Grayer et al., Contributed talk to the 4th Int. Conf. on High-Energy Collisions, Oxford (1972).
5. J. Baton et al., Phys. Letters **33** B, 528 (1970).

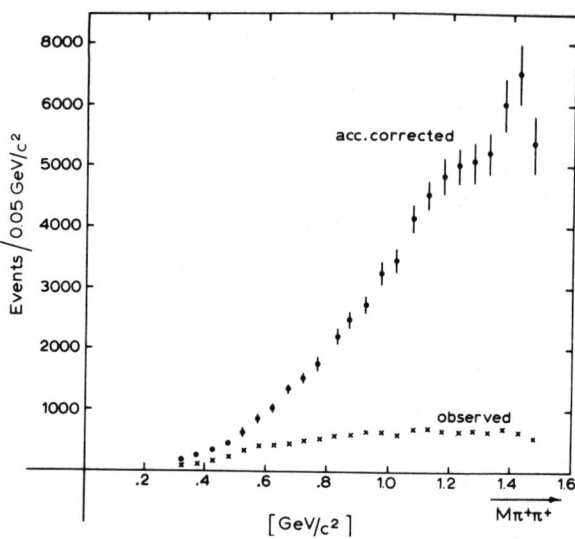

Fig. 1 $\pi^+\pi^+$ invariant mass spectrum

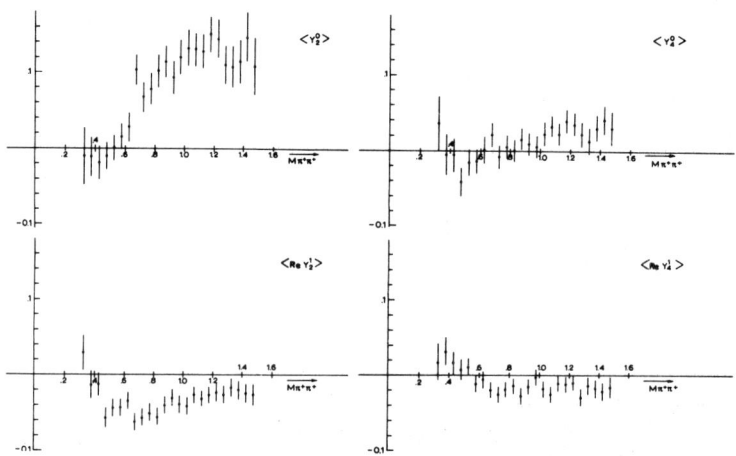

Fig. 2 $\pi^+\pi^+$ moments in the t-channel as function of mass for t-interval $t_{min} < |t| < 0.2$ (GeV/c)^2

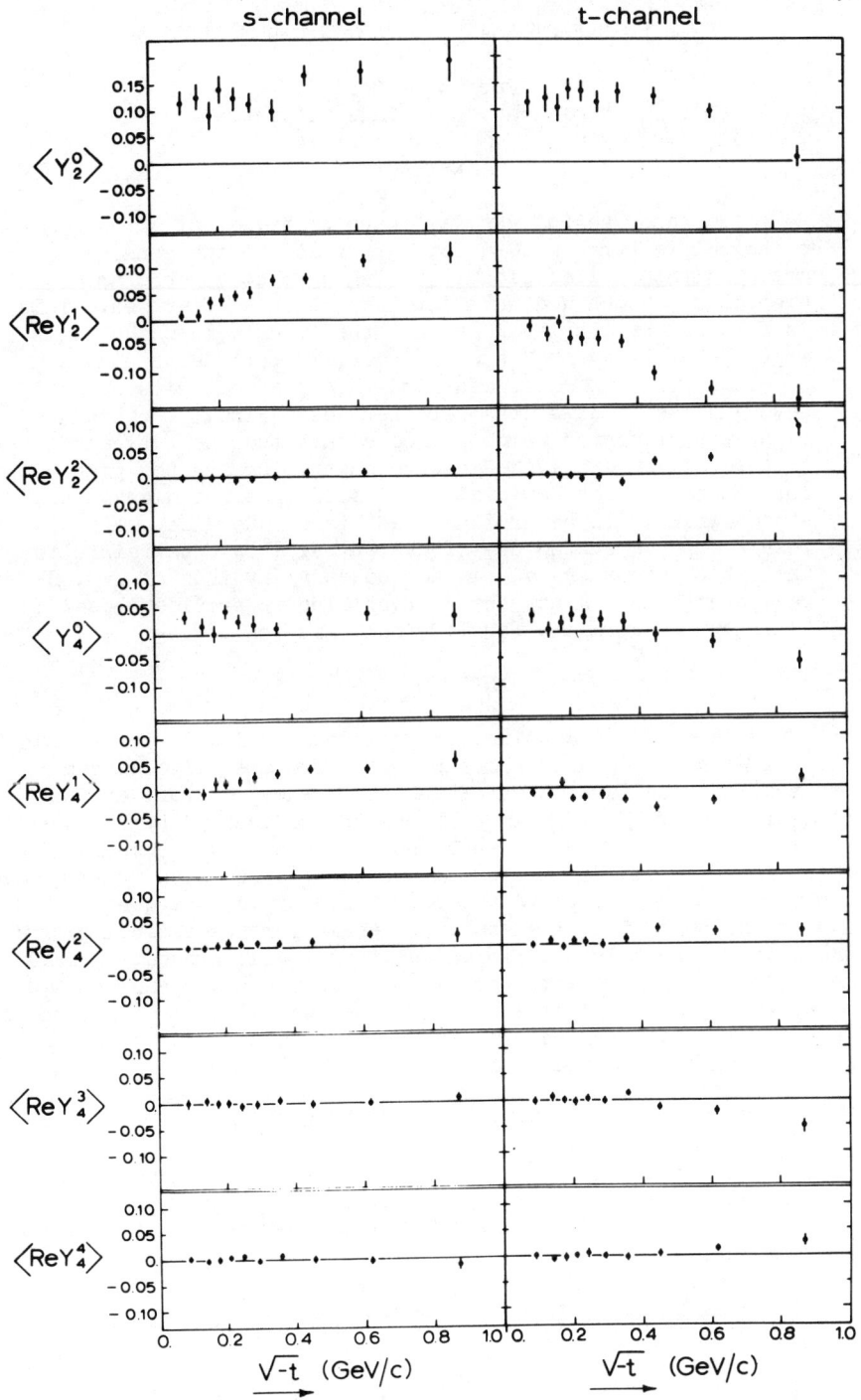

Fig. 3 $\pi^+\pi^+$ moments as a function of t in mass interval $0.75 < M_{\pi\pi} < 1.25$ GeV/c^2

PION FORM FACTOR AND INELASTIC π-π SCATTERING

Robert L. Goble
University of Utah, Salt Lake City, Utah 84112

ABSTRACT

The paper begins with a lightning review of the principal features of π-π scattering and of the generally accepted theoretical picture (using current algebra and smoothness) which accounts for most of these features. This is followed by a brief look into the future, in which it is argued that P-wave π-π scattering will continue to play a prominent role in experiment and theory at energies above 1 GeV, just as it does at lower energies. Finally, calculations are presented which indicate that the theoretical picture of low energy π-π P-wave scattering and the pion form factor is not noticeably modified by the inelastic states which will be prominent at higher energies. This result depends on the use of current algebra constraints for inelastic scattering and is not necessarily true of I=0, S-wave scattering, since the current algebra provides less stringent constraints for I=0 inelastic scattering.

INTRODUCTION

It is late; the limousines are waiting; so I will try to be brief. I wish to touch on three points. Two are quite general, appropriate to the last paper of the conference: first, an instantaneous review of the experimental and theoretical picture of π-π scattering that has been presented here; second, two comments about the future which did not appear in the panel discussion this morning. These will provide the context for my third topic, the question of how valid is the neglect of inelastic states in our theoretical picture of low energy π-π scattering. To answer this I will present a small calculation taken from some work in progress with Jim Ball at Utah on the larger topic of the title. (For those who wish to go out now and sit in the limousine, I should say that the calculation supports the neglect, at least for the P-waves.)

INSTANT REVIEW OF THE CONFERENCE

To some extent this conference is a celebration. After years of struggle and confusion, experiments now give a clear qualitative picture of π-π scattering below 1.1 GeV. Here are five features we all agree on:
1) Low energy phase shifts are small.
2) The parameters of the ρ are known.

3) δ_0^2 is small and negative.
4) We have the "down" solution for δ_0^0; i.e., no rapid variation below .9 GeV.
5) There is rapid variation in the I=0, S-wave near the K-$\bar{\text{K}}$ threshold.

Of course there are still enormous problems:
1) What really are the low energy amplitudes? (Do they agree with current algebra?)
2) What really are the ϵ and S* (K$\bar{\text{K}}$ threshold) parameters?
3) What do the amplitudes do above 1.1 GeV?
4) How good are the extrapolations? What is absorption doing? etc.

These were the main topics of the conference and I have nothing to add to them.

The theoretical situation is exactly parallel to the experimental. We have a simple qualitative picture for π-π scattering up to the ρ region which we can agree with, but detailed applications of this picture and extensions or modifications of it are still shrouded in controversy. There are two basic ingredients: the low energy current algebra constraints and minimal energy dependence consistent with the existence of resonances and elastic unitarity.[1] An elegant modern formulation of these assumptions was given yesterday morning by Pennington.[2] The picture provides a connection between the mass and width of the ρ, and says that δ_0^0 should be slowly varying and that δ_0^2 should be small; thus it gives a theoretical interpretation of the experimental features 1-4 on my list of things we agree on. What matters in the current algebra contribution to this picture is that there be a zero in the amplitudes somewhere around s=0, and that the slope have the current algebra value (given in terms of the pion decay constant F_π). At the energies where present experiments are reliable, the actual location of the zero (which depends on the pion mass) makes no difference; the slope, however, is crucial.

A LOOK AT THE FUTURE

There are two basic possibilities. One is, perhaps, to solve some of problems 1-4. The second is to go to higher energies and to look at inelastic channels. This, as was apparent in this morning's panel discussion, will be very messy. Two things which have not been mentioned at this conference may help. One is that there soon will be excellent high energy e^+e^- annihilation experiments running at SLAC and DESY. These will provide information (with good statistics) related to scattering in I=1, P-wave states. The second is that there are many as yet poorly formulated theoretical suggestions that we are fast approaching an asymptotic world. Perhaps these ideas can be used to tie down some features of the high energy amplitudes in a way analogous to the way current algebra constrains low energy amplitudes. Here are two possibilities: we may be able to learn things just from multiplicities; also, a propos of the e^+e^- annihilations, the SLAC deep inelastic data suggests that heavy photons have peculiar and possibly simple couplings to hadrons.

DO INELASTIC CHANNELS AFFECT THE LOW ENERGY PICTURE?

Inelastic channels play an important role in π-π scattering above 1 GeV and will determine the nature of any high energy simplicity. One is led naturally to wonder how inelastic channels affect the connection that the current algebra slope has with resonance parameters and the other qualitative features of π-π scattering at moderate energies. We can make a simple model to examine this question by considering a two-channel problem, π-π and K-$\bar{\text{K}}$ scattering in the I=1, P-wave. We assume that the scattering matrix satisfies the following three properties: It satisfies elastic unitarity exactly (with both π-π and $\bar{\text{KK}}$ intermediate states); it has no left-hand singularities except for a single far distant pole; it has a single resonance. If we further require that the matrix approach the current algebra limit as s is small, and that the resonance be at the ρ mass, we have a two-channel model which incorporates the basic theoretical ideas discussed earlier. (The requirement that it be a single distant left-hand pole simplifies the algebra without appreciably affecting the conclusions. Nearby left-hand singularities could, of course, give a pronounced effect.) These properties, plus time-reversal invariance, completely determine a four-parameter form for the scattering matrix, which is a two-channel generalization of the amplitude in reference 1. This form is a simple relativistic N/D with N the single pole; it corresponds to the infinite sum of bubble graphs. The four parameters can be chosen as follows: 1) the mass of the resonance is fixed at the ρ mass; 2&3) the low energy ππ→ππ and ππ→$\bar{\text{KK}}$ amplitudes are fixed at the current algebra values; 4) there is one free parameter, which, however, is essentially irrelevant to the π-π scattering amplitude once 1-3 are fixed and can be varied enormously without changing anything. The resulting π-π phase shift is plotted as the solid line in Figure 1. The other lines in Figure 1 differ only in the choice of the low energy ππ-$\bar{\text{KK}}$ amplitude: the dashed line corresponds to choosing it to be zero, thus reproducing the old 1-channel calculation; the dash-dotted line corresponds to setting it to be three times the current algebra value (here the value of the fourth parameter matters slightly and a typical value has been arbitrarily chosen). Clearly it is reasonable to neglect the inelastic $\bar{\text{KK}}$ states at energies up to the ρ mass provided that the amplitude is constrained to be within the current algebra value at low energies. If low energy ππ→$\bar{\text{KK}}$ scattering were much larger, however, the low energy shape of the π-π amplitude could be altered. Similar considerations apply to ππ→πω and ππ→ other 4π states since current algebra restricts the low energy ππ→4π amplitudes (here I am assuming that the ω is basically a 3π state). However, the situation for I=0, s-wave ππ and $\bar{\text{KK}}$ scattering is much less clear. As mentioned before, the important quantity in determining the shape of the amplitudes at moderate energies is the low energy slope, not the location of the zero. But the I=0, $\pi\pi \to K\bar{K}$ amplitude is an even amplitude and the slope is unconstrained by current algebra. (The calculation is just like that in the Adler-Weissberger relation.) Consequently there is no reason to expect the slope to be small so the shape of a 1-channel current algebra model phase shift such as that of reference 1 or the one which Ray Willey just presented[3] could be substantially modified.

A final remark about form factors: the idea of ρ dominance gives a connection between the magnitude of the pion form factor and the slope of the low energy P-wave $\pi\pi$ amplitude, given by current algebra.[4] Again this connection will not be seriously modified by inelastic states subject to current algebra constraints, but would be noticeably changed by a state with a larger low energy coupling.

REFERENCES

1. This picture is assumed one way or another by most workers in the field. A more detailed statement of the form given here is in L.S. Brown and R.L. Goble, Phys. Rev. Letters 20, 346 (1968), and L.S. Brown and R.L. Goble, Phys. Rev. D4, 723 (1971). For other versions see references 2 and 3.
2. M.R. Pennington, this conference. This formulation has the advantage of making explicit the connection between different partial wave amplitudes.
3. R. Willey, this conference.
4. G.J. Gounaris and J.J. Sakurai, Phys. Rev. Letters 21, 244 (1968).

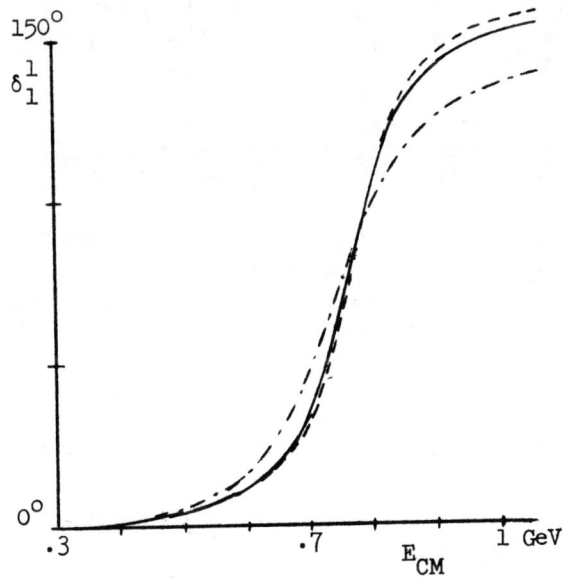

Figure 1.

Three parametrizations of the P-wave π-π phase shift, δ_1^1, as a function of the total CM energy. The parameters are described in the text above.

PHENOMENOLOGY OF ππ SCATTERING

J.L. Basdevant
Faculté des Sciences de Paris, France

C.D. Froggatt
Glasgow University, Scotland

J.L. Petersen
Nordita, Copenhagen, Denmark

ABSTRACT

We present the set of ππ amplitudes below 1100 MeV consistent with all present experimental information and the requirements of crossing, unitarity and analyticity. The corresponding ππ scattering lengths lie on a universal curve and we discuss how further experiments can resolve these final ambiguities in their values. We illustrate in what sense the Roy equations cannot resolve the 'up-down' ambiguity.

In this paper, we discuss the set of ππ amplitudes consistent with crossing analyticity, unitarity and experimental information[1] below 1.1 GeV. Our method is to construct unitary solutions of the exact s and p partial wave equations for ππ scattering recently studied by Roy[2]. We refer to our previous papers[3] for details and for a discussion of the driving terms - i.e. the contributions from higher partial waves and effects above 1100 MeV.

We select all solutions of the Roy equations which are compatible with the following experimental information.

(i) For the p wave, we fix the mass and width of the rho meson M = 765 MeV and Γ = 135 MeV but allow the scattering length a_1 to be arbitrary.

(ii) The isoscalar s wave phase shift δ_0^0 in the mass region $500 < M_{\pi\pi} < 900$ MeV lies in the between-down band[1].

(iii) Inelasticity due to 4π production below 1 GeV is negligible, but there is a strong cusp or S* effect causing δ_0^0 to accelerate rapidly through 180° and a sharp onset of inelasticity at the $K\bar{K}$ threshold.

(iv) The I = 2 s wave phase shift has a rather smooth behaviour with a value in the range $\delta_0^2 = -15° \pm 5°$ at the rho mass[1].

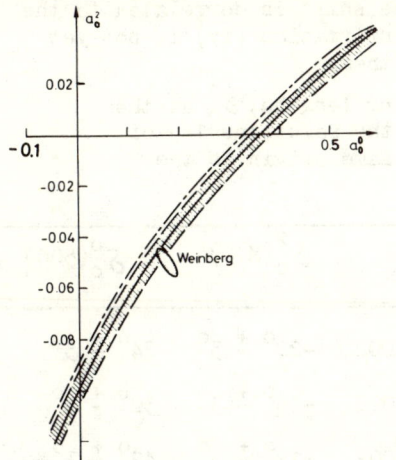

Fig. 1. Allowed region in the (a_0^0, a_0^2) plane.

After imposing the information (i) - (iii) above, there are only solutions of the Roy equations on a universal curve or narrow allowed band of finite extent in the s-wave scattering length (a_0^0, a_0^2) plane, as shown in figure 1. The full and dot-dashed lines correspond to two extreme possibilities for the S^* effect. The surrounding shaded band indicates the uncertainty in our results obtained by varying the rho width by 25 MeV, the driving terms within their estimated errors and the s-wave phase shifts from those of Baton et al[4] to those of Protopopescu et al[5]. For either set of data, we obtain solutions along the whole universal curve. This was illustrated in ref[3] for the phase shifts of Baton et al[4]. In fig 2 we show the three corresponding typical solutions for the Berkeley phase shifts [5]. These three curves illustrate the spread of allowed behaviour below 500 MeV for a given set of experimentally determined s wave phase shifts above 500 MeV, with the present statistical accuracy. The crucial parameters characterizing these solutions are given in table I.

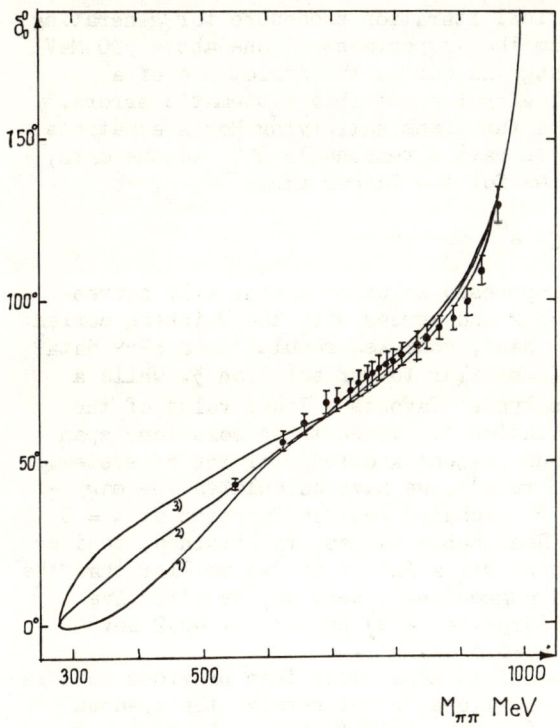

Fig. 2. Typical solutions for δ_0^0 fitted to the energy dependent phase shifts of ref. 5.

In our solutions, the I = 2 s wave phase shift is correlated to the behaviour of δ_0^0 but the experimental information (iv) is not yet precise enough to discriminate between them.

Table I. Values of the scattering lengths, δ_0^2 at the rho mass and δ_0^0 at 500 MeV for the solutions 1-3 of figure 2. Errors shown have the same origin as the bands in figure 1.

a_0^0	a_0^2	a_1^1	$\delta_0^2(M_\rho)$	$\delta_0^0(500)$
-0.05	-0.115 ± 0.005	0.034 ± 0.002	$-22° \pm 3°$	$24° \pm 5°$
0.16 ± 0.02	-0.048 ± 0.003	0.035 ± 0.002	$-16° \pm 3°$	$30° \pm 5°$
0.60	0.043 ± 0.004	0.041 ± 0.002	$-12° \pm 3°$	$42° \pm 3°$

Pennington and Protopopescu[6] have claimed that imposing the Berkeley I = 0 data in the Roy equations leads to a well defined scattering length: $a_0^0 = 0.15 \pm 0.07$. Their criterion to define a solution is that their numerical iteration procedure for generating consistent $\pi\pi$ amplitudes from the 'experimental' one above 500 MeV must converge. The small range is due to the strict use of a particular data set, not allowing for possible systematic errors. However, by first finding the functions satisfying Roy's equations and then selecting those which have a reasonable χ^2 to the data, we find a set of $\pi\pi$ amplitudes for the larger range

$$-0.05 < a_0^0 < 0.6$$

The Pennington and Protopopescu solution essentially corresponds to our central solution 2 and agrees with the Weinberg current algebra value. On the other hand, some Ke4 results[7] and $\pi^0\pi^0$ data[8] favour a higher value for a_0^0 similar to our solution 3, while a recent low energy πN data analysis[9] favours a lower value of the type corresponding to our solution 1. These three solutions span the differences allowed by our present knowledge of the $\pi\pi$ system.

By using the Olsson sum rule[10], we have calculated the magnitude of the effective Regge ρ exchange residue function at t = 0 for each of our solutions. Reasonable values are obtained, tending to favour a residue function nearly a factor of two smaller than the Lovelace-Veneziano value and suggesting a zero of the effective residue function (for the absorptive part) nearer $t = -0.2$ GeV^2 than the 'nonsense' value $t = -0.6$ GeV^2 [11].

We obtain a larger class of $\pi\pi$ amplitudes than previous authors and, in particular, the Roy equations do not resolve the up-down ambiguity as claimed in ref. 6. This is illustrated in figure 3 where we present two typical solutions, with scattering lengths

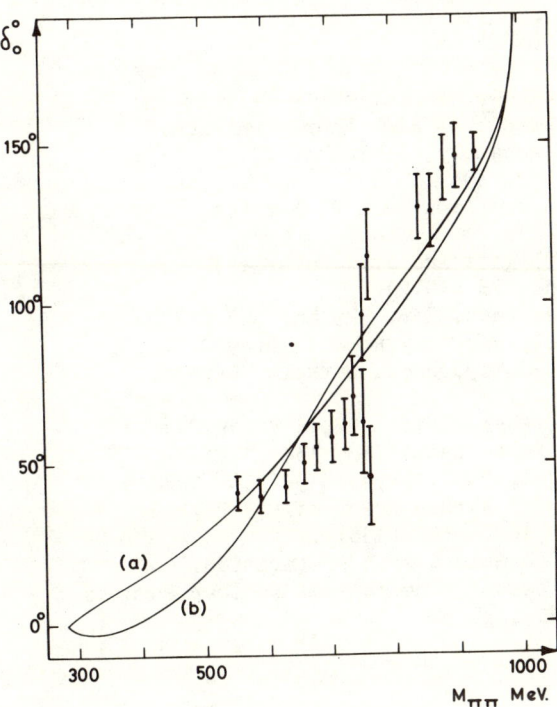

Fig. 3. Typical solutions for δ_0^0 fitted to the 'up' phase shifts of ref. 5. We also plot the 'down' phase shifts at $M\pi\pi$ = 750 and 760 MeV.

(a) a_0^0 = 0.12, a_0^2 = -0.051 and
(b) a_0^0 = -0.145, a_0^2 = -0.137, which are required to fit the energy independent Berkeley 'up' phase shifts[5] with $\chi^2 \leq 40$. The 'up' solutions lie on the same universal curve but extended further to the left. Although we can fit the Berkeley data with our forms such that χ^2 is of the order of one per data point, we should point out that it is impossible for an analytic function to reproduce the narrow effect at 750 to 800 MeV in detail without using more parameters than seems physically reasonable. In particular, a narrow resonance sitting on the smooth background will tend to push the phase shift quickly through 180° [12]. Crossing on the other hand (as expressed by the Roy equations) is unable to rule out such a behaviour.

Finally, we discuss how future experimental information could select between our various 'between-down' solutions. Our isoscalar s wave amplitudes for a particular set of experimental data differ mostly below the rho (see fig 2) and more accurate data in this region are required. We note that the errors on $\delta_0^0(500)$, the isoscalar phase shift at 500 MeV, for the three solutions of table I are predominantly due to the experimental uncertainty (between ref 4 and ref 5) on δ_0^0 in the rho region. A knowledge of δ_0^0 from 500 to 900 MeV with a true uncertainty of ± 3° would allow us to select a_0^0 with an uncertainty of 0.05 for the lower values of a_0^0 and of 0.15 for the higher values of a_0^0. Further accuracy in the rho region alone is then unlikely to improve the result significantly. An important check on such a result would come from measuring δ_0^2 with a similar accuracy and using our predicted correlations between δ_0^0 and δ_0^2 as for instance in the s-p wave interference term in $\pi^+\pi^-$ scattering. The final confirmation would come from an accurate Ke4 experiment with known (small) systematic errors.

REFERENCES

1. R. Diebold, Meson Resonances, Rapporteur's Talk at the XVI International Conference on High Energy Physics, National Accelerator Laboratory, September 1972.
2. S.M. Roy, Phys. Lett. 36B, 353 (1971)
   J.L. Basdevant, J.C. Le Gouillou and H. Navelet, Nuovo Cim. 7A, 363 (1972).
3. J.L. Basdevant, C.D. Froggatt and J.L. Petersen, Phys. Lett. 41B, 173 and 178 (1972).
4. J.P. Baton et al., Phys. Lett. 33B, 525 and 528 (1970).
5. S.D. Protopopescu et al., Berkeley Report LBL-970.
6. M.R. Pennington and S. Protopopescu, Berkeley Reports LBL-963 and 1323.
7. A. Zylbersztejn et al., Phys. Lett. 38B, 457 (1972).
8. J.R. Bensinger et al., Phys. Lett. 36B, 134 (1971).
9. H. Nielsen and G.C. Oades, Nucl. Phys. B49, 586 (1972)
   F. Elvekjar, University of Aarhus Preprint (1972).
10. M.G. Olsson, Phys. Rev. 162, 1338 (1967).
11. J.L. Basdevant and C. Schomblond - in preparation.
12. J.M. Blatt and V.F. Weisskopf, Theoretical Nuclear Physics (J. Wiley and Sons, 1952), p.398

COMMENTS ON "PHENOMENOLOGY OF $\pi\pi$ SCATTERING"*

M. R. Pennington
Lawrence Berkeley Laboratory, University of California
Berkeley, California 94720

S. D. Protopopescu
Brookhaven National Laboratory, Upton, Long Island, N. Y. 11973

ABSTRACT

We discuss how our aims and methods of applying Roy's equations to a particular set of data differ from those of Basdevant, Froggatt and Petersen and emphasize that there is no disagreement between us concerning s-wave scattering lengths.

We feel that some of the points made in our papers[1,2] on the application of Roy's equations to the data were misunderstood by the authors of the previous paper in these proceedings[3] (hereafter referred to as BFP). We hope that the following comments may clarify the problem.

Firstly and most obviously our aims were different, ours being to check the consistency of a particular set of data[4] with Roy's equations and with reasonable estimates of the high energy (HE) $\pi\pi$ amplitude. We did this in the following way. We first estimated the $\pi\pi$ amplitude above 1 GeV from Regge and resonance phenomenology with sizeable errors. These estimates, together with the Berkeley phase shifts,[4] are our input. In order to apply Roy's equations to the data we have next to determine the two subtraction terms, which we chose to be the s-wave scattering lengths a_0^0 and $2a_0^0 - 5a_0^2$.

However, these two subtraction terms are not independent, but are related by the Olsson sum rule for $2a_0^0 - 5a_0^2$ (or, equivalently by equating the Froissart-Gribov representation for a_1^1 with Roy's equation for a_1^1). We therefore have one free parameter which we take to be a_0^0, for convenience. We then chose values for a_0^0 between 0 and 0.4 and demanded agreement between the input and output s and p waves in Roy's equations from threshold to where the data begin (i.e. 500 MeV). For this we used an iteration procedure.[1] So for each $a_0^0 \in (0, 0.4)$ we have a consistent set of phase shifts for the s and p waves from threshold to 500 MeV, but as yet we have no guarantee that these phase shifts join on with the data beyond that energy. This we next ensured by requiring input-output agreement for just the $I = 0$ s-wave from 520 to 700 MeV within the errors in the data. These errors only include the statistical uncertainty in the phase shifts, e.g. δ_0^0 has an error of $\pm 4^\circ$, and we did not allow for possible systematic deviations.

---

* This work was supported by the U.S. Atomic Energy Commission.

This appreciably reduces the allowed range of a_0^0 to 0.15 ± 0.07. For details of this procedure see Ref. (1).

Having fixed a_0^0 and $2a_0^0 - 5a_0^2$ in this way, we computed[2] f_0^0, f_1^1, f_0^2 in the entire energy region upto 1050 MeV. It is highly non-trivial that for Re f_0^0 beyond 720 MeV we obtained reasonable input-output agreement. In fact there is absolutely no guarantee that the output real part is within the unitarity bound. We should stress that we did not vary any parameters in our HE estimates or use any fitting procedure. In particular, our agreement around 950 MeV depends crucially on our estimate of a large δ_0^1 under the f_0 resonance. We then compute Re f_1^1 for which no input-output consistency has been required anywhere the data exists, i.e. beyond 500 MeV. We find near perfect agreement.[2] It is easy to check that appreciably changing our idea of the HE $\pi\pi$ amplitude would destroy this agreement and, for example, shift the mass of the output ρ resonance (although such changes would alter Re f_0^0 much more than Re f_1^1).

The agreement we do obtain, which we must emphasize is highly non-trivial, shows that our estimates of the HE $\pi\pi$ amplitude and our values for the subtraction terms are correct given the input data with its particular error band. The only feature which makes the Berkeley data[4] different from earlier data (aside from the S^*, of course) is the smallness of its statistical errors.

In contrast, BFP looked for general solutions to Roy's equations and unitarity generally compatible with the data. They demanded exact input-output agreement from the outset in the entire energy region. They also allowed a greater variation in the I = 0 s-wave phase shifts than we did and so correspondingly found a larger range of scattering lengths. Indeed looking at Fig. 2 of the preceding paper[3] and remembering we used the Berkeley band of phase shifts down to 500 MeV with a total uncertainty of $\pm 4°$ for δ_0^0 we see immediately from the results of BFP why we obtained values for a_0^0 close to their solution 2 and ruled out the extremes of 1 and 3.

Similar remarks also apply to our distinguishing between "up" and "down" solutions. We, in fact, constrained the output real part for f_0^0 to have a rapid variation between 750 and 800 MeV by our choice of subtraction terms in this case,[2] so as to agree with the "up" solution--see Fig. 5 of Ref. (2). However looking beyond that energy and remembering the $\pm 4°$ uncertainty, the down solution of Fig. 2 of Ref. (2) is clearly favored. The same conclusion can be drawn from the results of BFP,[3] although the "up" solution they use appears to be different from ours,[2] by comparing figures 2,3 and remembering the errors.

There is therefore no disagreement between our results and those of BFP. The smaller range of values we quote for a_0^0 and our ability to distinguish "down" from "up" solutions hinges on the fact that we do not allow for systematic deviations in the phase shifts and take the quoted statistical uncertainty as being the total error. This fixes our solutions to be close to solution 2 of BFP. Although our aims and methods differ from those of BFP there is no disagreement between their resulting scattering lengths and ours.

REFERENCES

1. M. R. Pennington and S. D. Protopopescu, Phys. Rev. D7, 1429 (1973).
2. M. R. Pennington and S. D. Protopopescu, Phys. Rev. D (May 1973).
3. J. L. Basdevant, C. D. Froggatt and J. L. Petersen, in these proceedings.
4. S. D. Protopopescu et al., Phys. Rev. D7, 1279 (1973).

STRUCTURE IN THE MOMENTUM TRANSFER DISTRIBUTION OF
$\pi^- + p \to \rho + N$ at 2.3 GeV/c

S. Hagopian and V. Hagopian
The Florida State University, Tallahassee, Fla. 32306

W. Selove
University of Pennsylvania, Philadelphia, Pa. 19104

ABSTRACT

The data from $\pi^- + p \to \rho^0 + n$ and $\pi^- + p \to \rho^- + p$ at 2.3 GeV/c is consistant with a dip in $d\sigma/dt$ at $t \sim -0.6$ $(GeV/c)^2$ for ρ^- production and a break at $t \sim -0.6$ $(GeV/c)^2$ for ρ^0 production, in agreement with theoretical predictions and some other experiments.

EXPERIMENTAL RESULTS

The experiment was performed in the Princeton-Pennsylvania Accelerator 15-inch bubble chamber and was measured by the University of Pennsylvania Flying Spot Digitizer.[1] We present results from 3,615 events of the type $\pi^- + p \to \rho^0 + n$ and 2,180 events of the type $\pi^- + p \to \rho^- + p$ (0.58 μb/event). The ρ region is defined as $660 \leq M(\pi\pi) \leq 860$ MeV. The four-momentum transfer distribution $d\sigma/dt$ for ρ production has in addition to a forward peak several dips and bumps some of which seem to be energy independent. Of particular current interest is a dip at $t \sim -0.6$ which is predicted by the Regge pole model with wrong signature nonsense zero for the ω trajectory in ρ^- production.[2] Figure 1a shows $d\sigma/dt$ for ρ^- and ρ^0 production at 2.3 GeV/c. A clear dip is seen in $d\sigma/dt$ for ρ^0 production at $t \sim -0.5$ to -0.6 $(GeV/c)^2$. A break is seen in $d\sigma/dt$ for ρ^0 production at about the same t value. This is in agreement with experiments at other energies.[3]

The quantity $\rho_{00}\, d\sigma/dt$ is also predicted by the absorption model[4] to have a dip in the neighborhood of $t \sim -0.6$ $(GeV/c)^2$, where ρ_{00} is the density matrix element calculated in the helicity frame. Figure 1b shows "ρ_{00}" $d\sigma/dt$, where "ρ_{00}" is the helicity frame matrix element defined:
"ρ_{00}" $= 1/3(1 + 2(\rho_{00} - \rho_{11}))$. This includes the s wave effect:
"ρ_{00}" $= \rho_{00}^{p\ wave} + 1/3\, \rho_{00}^{s\ wave}$. Again in the figure 1b a break is seen near $t \sim -0.6$ $(GeV/c)^2$. A dip or break in $\rho_{00}\, d\sigma/dt$ seems to occur in a number of experiments. This effect is discussed in the review talk by J. A. J. Matthews in these proceedings.

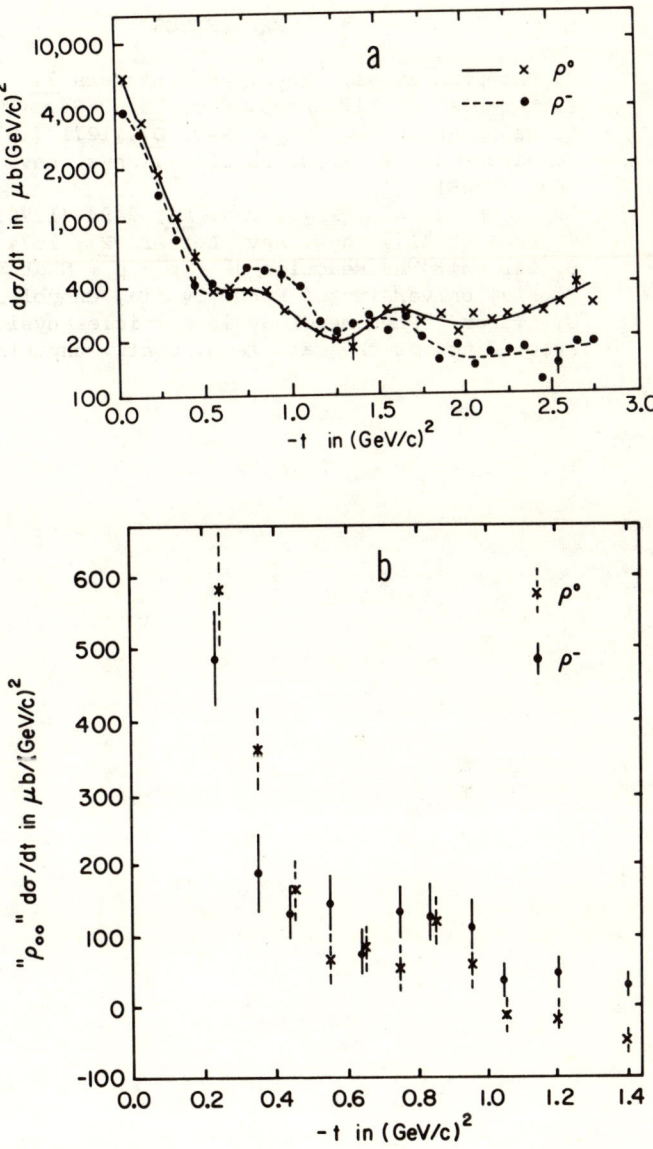

Figure 1 The reaction $\pi^- + p \to \rho + N$ at 2.3 GeV/c, (a) $d\sigma/dt$ vs t. (b) "ρ_{00}" $d\sigma/dt$ vs t.

REFERENCES

1. S. Hagopian et al., Phys. Rev. Letters $\underline{24}$, 1445 (1970).
   S. Hagopian et al., Phys. Rev. $\underline{D5}$, 2684, (1972).
   S. Hagopian et al., Phys. Rev. $\underline{D7}$, 1271 (1973).
2. V. Barger and R. J. N. Phillips, Phys. Rev. Letters $\underline{20}$, 564, (1968).
3. B. Y. Oh, et al., Phys. Rev. $\underline{D1}$, 2494 (1970); D. J. Crennel et al., Phys. Rev. Letters $\underline{27}$, 1674 (1971); S. Barish, "The Reaction $\pi^- + p \to \rho + N$ at 4.5 BeV/c, Thesis, University of Pennsylvania, unpublished (1972).
4. G. C. Fox, "Phenomenology in Particle Physics," Proceedings of the Cal. Tech. Conf., unpublished (1971).

$\pi^- p \to \pi^- \pi^+ n$ AMPLITUDE ANALYSIS AND EXTRAPOLATION TO THE π EXCHANGE POLE

P. Estabrooks * and A.D. Martin +
CERN-Geneva, Switzerland

ABSTRACT

We solve analytically for the $\pi^- p \to \pi^- \pi^+ n$ amplitudes in the ρ region at 17.2 GeV/c and find two acceptable solutions. We discuss the extrapolation of s and t channel amplitudes to the π pole and conclude that the former should be used for $\pi\pi$ phase shift analyses.

It has been noted [1] that an amplitude analysis of the reaction $\pi^- p \to \pi^- \pi^+ n$ (with S and P wave dipion production dominant) is possible under the assumption that exchanges with the quantum numbers of the A_1 can be neglected. Here we wish to point out that a twofold ambiguity is inherent in such an analysis. To see this, we express the observables in terms of the production amplitudes (S, P_0, $P_\pm$ in the notation of Ref. 2) as follows [1]:

$$\sigma \equiv \frac{d\sigma}{dt} = (|S|^2 + |P_0|^2) R + |P_+|^2 + |P_-|^2 \qquad (1)$$

$$\alpha \equiv (\rho_{00} - \rho_{11})\sigma = |P_0|^2 R - \tfrac{1}{2}(|P_+|^2 + |P_-|^2) \qquad (2)$$

$$\beta \equiv \rho_{1-1}\sigma = \tfrac{1}{2}(|P_+|^2 - |P_-|^2) \qquad (3)$$

$$\gamma_{10} \equiv \sqrt{2}\,\mathrm{Re}\,\rho_{10}\sigma = |P_-||P_0|\cos\varphi \qquad (4)$$

$$\gamma_{0S} \equiv \mathrm{Re}\,\rho_{0S}\sigma = |P_0||S| R \cos\Delta \qquad (5)$$

$$\gamma_{1S} \equiv \sqrt{2}\,\mathrm{Re}\,\rho_{1S}\sigma = |P_-||S| \cos(\varphi - \Delta) \qquad (6)$$

where $R = 1$ for t channel quantities and $R = t/(t-t_{min})$ for s channel quantities. R accounts for the small π pole contribution to s channel nucleon helicity non-flip amplitudes so that S, P_0 denote the nucleon flip amplitudes.

---

* Supported by the National Research Council of Canada.

\+ On leave of absence from the University of Durham, England.

From these six observables we can solve for the six quantities $|P_0|$, $|P_\pm|$, $\gamma_s \equiv |S|/|P_0|$, φ and Δ. To do this we reduce the equations to a cubic equation in $x \equiv |P_0|^2$.

$$-3R^3 x^3 + R^2(B+3A)x^2 + (3R\gamma_{10}^2 - R\gamma_{15}^2 - \gamma_{os}^2 - AB)Rx + (A\gamma_{os}^2 - RB\gamma_{10}^2 + 2R\gamma_{10}\gamma_{os}\gamma_{15}) = 0, \quad (7)$$

where $A = \alpha + \beta$ and $B = \sigma + 2\alpha$. In Fig. 1 we show the amplitudes for the two allowed solutions as determined from the 17.2 GeV CERN-Munich [3] s channel density matrix elements in the ρ mass region ($700 \leq M_{\pi\pi} \leq 850$ MeV). The other solution is unphysical; $|P_0|^2 < 0$. The physical solutions have similar values of $|P_0|^2$, but solution 1 is characterized by $|\cos\varphi| \approx 1$ and solution 2 by $|\cos\Delta| \approx 1$. Only the relative sign of φ and Δ can be determined from Eq. (1) to (6). The values shown for Δ for solution 2 correspond to $\sin\varphi > 0$ while, for solution 1, $|\cos\varphi| \approx 1$ and so only $|\Delta|$ is shown. $\gamma_s \cos\Delta$ is remarkably constant in each solution. The more erratic behaviour of γ_s for solution 1 is mainly due to a sizeable $\gamma_s \sin\Delta$ component.

Similar properties are found [2] when the analysis is repeated in 20 MeV $\pi\pi$ mass bins in the range $500 \leq M_{\pi\pi} \leq 980$ MeV. Moreover, for both solutions, $|P_0|$, γ_s and Δ suitably extrapolated to the π exchange pole give $\pi\pi$ phase shifts consistent with elastic unitarity. However, only solution 1 is in agreement with the shape of the $\pi^0\pi^0$ mass spectrum observed [4] in the reaction $\pi^- p \to \pi^0 \pi^0 n$.

For the 150 MeV mass bin about the ρ, solution 1 appears to have anomalously large average values of $|\Delta|$ and γ_s. This feature is understandable since, in this mass region, the P wave phase shift is rapidly varying while the S wave is approximately constant. Therefore, care must be taken in extracting $\pi\pi$ phase shifts from large mass bins across which the amplitudes are rapidly varying.

To investigate whether it is better to perform the phase shift analysis in the s or t channel, we have analysed the moments in both channels in the ρ mass region. In what follows, we study only solution 1 although the discussion applies equally well to solution 2. If P_0 were pure π pole exchange, we would expect

$$G_\pi \equiv \sqrt{\frac{2R}{-t}} \left(\frac{\mu^2 - t}{M_{\pi\pi}}\right) |P_0| \quad (8)$$

to show an exponential decrease in t. From Fig. 2, we see that this form is an appreciably better description in the s channel

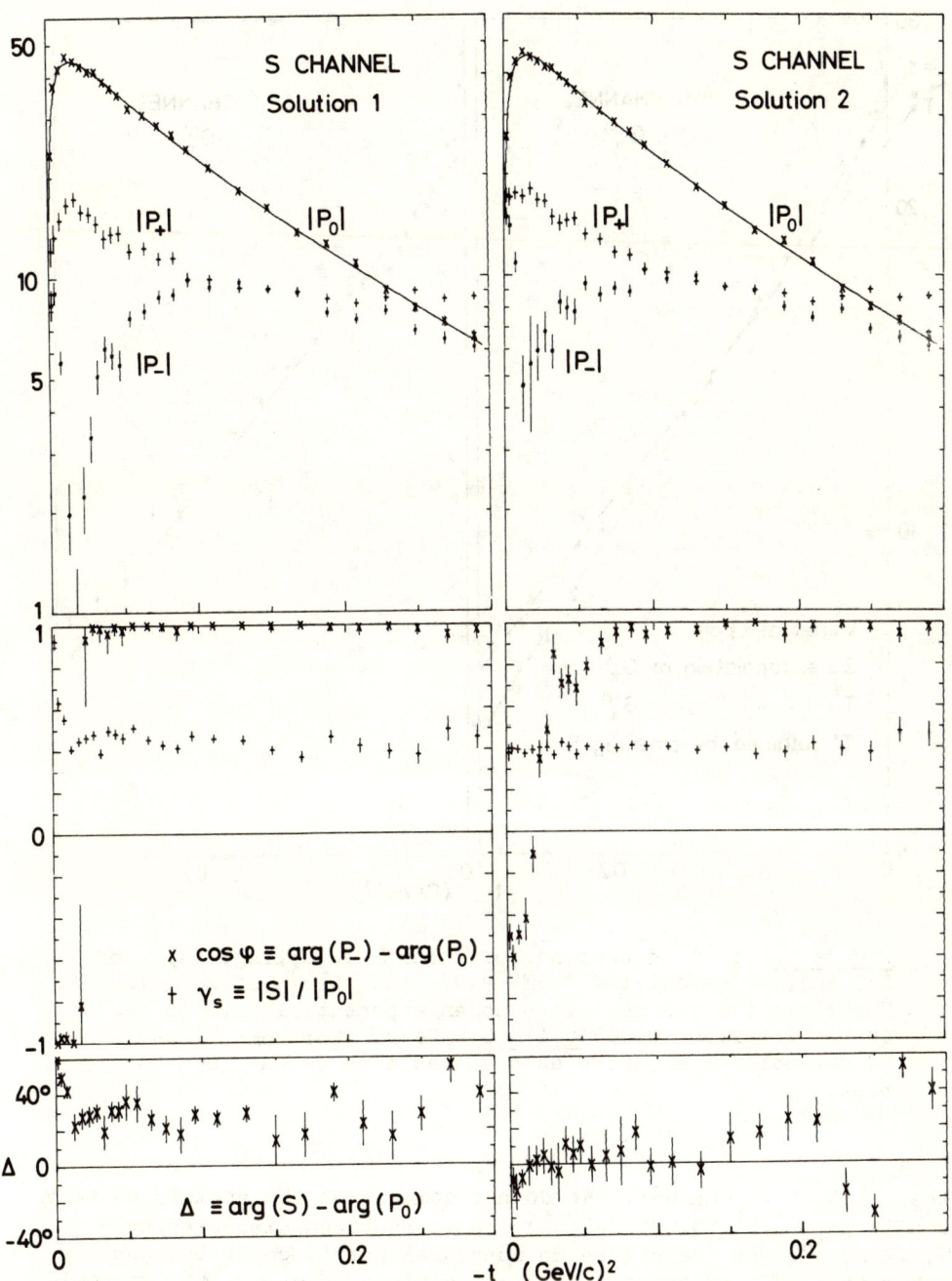

Figure 1 : Results of an s channel amplitude analysis of the 17.2 GeV $\pi^-p \to \pi^-\pi^+n$ data [3]. The curves are the best fits to $|P_0|$ using an exponential form of G_π of Eq. (8) in the interval $0.005 \leq |t| \leq 0.2$ $(GeV/c)^2$.

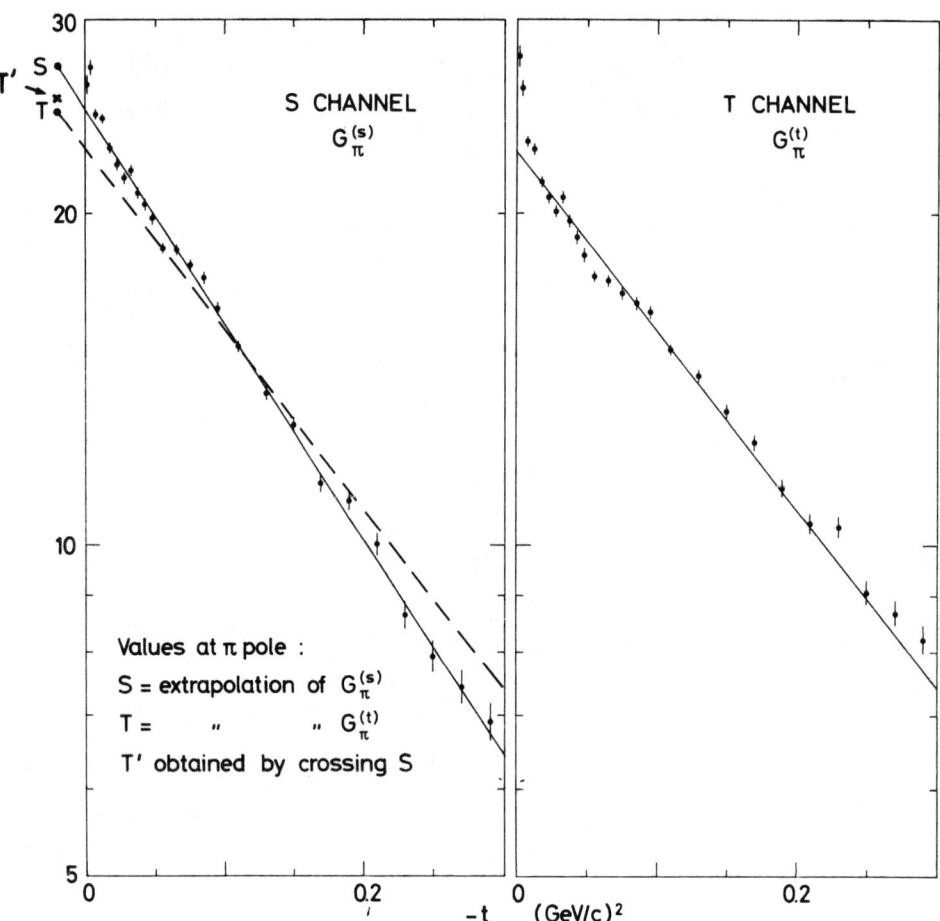

Figure 2 : t dependence of the π coupling, G_π of Eq. (8), as calculated from $P_0^{(s)}$ and $P_0^{(t)}$ of solution 1. The lines are the best fits to an exponential form of G_π in the interval $0.005 \leq |t| \leq 0.2$ GeV^2. For comparison, the t channel fit is shown as a dashed line on the s channel plot.

than in the t channel. We do not comment on the anomaly at very small $|t|$ ($|t| \leq 0.005$ GeV/c^2) as it has been discussed by W. Männer [5]. If there were no unnatural parity contributions other than π exchange, then, extrapolating to $t = \mu^2$ we would have

$$\left.\frac{G_\pi^{(s)}}{G_\pi^{(t)}}\right|_{t=\mu^2} = \cos\chi \Big|_{t=\mu^2} = \sqrt{\frac{M_{\pi\pi}^2}{M_{\pi\pi}^2 - 4\mu^2}}, \qquad (9)$$

as compared to the points S and T in Fig. 2. Here χ is the s-t crossing angle. From Eq. (9) and the point S, we would expect T to be at T'. The extrapolation of G_π is more stable in the s than in the t channel to changes of the t interval fitted. For instance, if data in the range $0.005 < |t| < 0.1$ GeV^2 are used, then S, and consequently T', are unchanged but T is raised to essentially T'. The difference between T and T', which indicates the presence of non π exchange contributions to P_0 can be understood in terms of the contamination of $P_0^{(t)}$ arising from the destructive non π contribution (C in the notation of Ref. 6) to $P_-^{(s)}$ since:

$$P_0^{(t)} = P_0^{(s)} \cos\chi + P_-^{(s)} \sin\chi. \qquad (10)$$

The existence of C is necessary to explain the sign change in $\text{Re}\rho_{10}^{(s)}$ near $t = -\mu^2$. For small $|t|$, $\sin\chi \approx (2\sqrt{-t'})/M_{\pi\pi}$, so the contamination in $P_0^{(t)}$ increases rapidly with decreasing $M_{\pi\pi}$. The increase [6] of C with decreasing $M_{\pi\pi}$ further enhances the T-T' discrepancy at low values of $M_{\pi\pi}$.

ACKNOWLEDGEMENTS

We thank C. Michael and the members of the CERN-Munich collaboration, in particular Wolfgang Männer, for useful discussions.

REFERENCES

1. P. Estabrooks and A.D. Martin, Phys.Letters 41B, 350 (1972).
2. P. Estabrooks et al., CERN preprint TH.1661 (1973).
3. G. Grayer et al., Proc. of IV Int.Conf. on H.E. Collisions, Oxford (1972); we use data in a slightly larger mass bin.
4. W. Apel et al., Phys.Letters 41B, 542 (1972).
5. W. Männer, contribution to Int. Conf. on $\pi\pi$ scattering and related topics, Tallahassee, March (1973).
6. A.D. Martin and P. Estabrooks, Proc. of VIII Rencontre de Moriond (1973).

Index of Authors

Alston-Garnjost, M.	1
Anderson, J. C.	312
Ayed, R.	307
Ayres, D. S.	284, 302
Baillon, P.	260
Barbaro-Galtieri, A.	1
Bareyre, P.	307
Barish, S.	322
Basdevant, J. L.	346
Beier, E. W.	26
Bell, I. G.	317
Bensinger, J.	322
Blum, W.	37, 117, 206, 337
Borgeaud, P.	307
Carnegie, R. K.	298
Cashmore, R. J.	144
Charlesworth, J. A.	317
Dale, M.	317
David, M.	307
Diaz, J.	312
DiBianca, F. A.	312
Diebold, R.	284, 302
Dietl, H.	37, 117, 206, 337
Emms, M. J.	317
Engler, A.	312
Ernwein, J.	307
Estabrooks, P.	37, 357
Feltesse, J.	307
Fickinger, W.	312
Field, R. D.	153
Flatté, S. M.	1
Friedman, J. H.	1
Froggatt, C. D.	346
Goble, R. L.	342
Grayer, G.	37, 117, 206, 337
Green, A. E. S.	332
Greene, A. F.	284, 302
Gutay, L. J.	255
Hagopian, S.	354
Hagopian, V.	354
Hearn, M.	322
Hoogland, W.	337
Hyams, B.	37, 117, 206, 337
Jacques, P.	322

Jones, C. 37, 117
                                                    206, 337
Jhung, K. S. 327
Kane, G. L. 247
Kimel, J. D. 274
Kinson, J. B. 317
Koch, W. 37, 117
                                                    206, 337
Kononenko, W. 322
Kraemer, R. W. 312
Kramer, S. L. 284, 302
Laurens, G. 270
Lemoigne, Y. 307
Lorenz, E. 37, 117
                                                    206, 337
Lütjens, G. 37, 117
                                                    206, 337
Lynch, G. R. 1
Major, J. V. 317
Männer, W. 37, 117
                                                    206, 337
Martin, A. D. 37, 357
Marty, P. 307
Matison, M. J. 1
Matthews, J. A. J. 188
Meissburger, J. 37, 117
                                                    206, 337
Morgan, D. 346
Nack, M. L. 332
Ochs, W. 117, 206
O'Neill, E. M. 322
Pawlicki, A. J. 284, 302
Pennington, M. R. 89, 351
Petersen, J. L. 346
Protopopescu, S. D. 351
Rabin, M. S. 1
Reya, E. 274
Riddiford, L. 317
Robinson, D. K. 312
Schlein, P. 117
Sekulin, R. L. 317
Selove, W. 322, 354
Solmitz, F. T. 1
Stacey, B. J. 317
Stierlin, U. 37, 117
                                                    206, 337
Stirling, A. V. 307
Sullivan, C. A. 312
Toaff, S. 312
Ueda, T. 332
Vasavada, K. V. 255

Villet, G.	307
Votruba, M. F.	317
Wagner, F.	206
Walker, W. D.	80
Weilhammer, P.	37, 117, 206, 337
Weisser, F.	312
Wicklund, A. B.	284, 302
Willey, R. S.	327
Williams, P. K.	135
Woodworth, P. L.	317

AIP Conference Proceedings

		L.C. Number	ISBN
No. 1	Feedback and Dynamic Control of Plasmas (Princeton 1970)	70-141596	0-88318-100-2
No. 2	Particles and Fields - 1971 (Rochester 1971)	71-184662	0-88318-101-0
No. 3	Thermal Expansion - 1971 (Corning 1971)	72-76970	0-88318-102-9
No. 4	Superconductivity in d- and f- Band Metals (Rochester 1971)	74-188879	0-88318-103-7
No. 5	Magnetism and Magnetic Materials - 1971 (2 parts) (Chicago 1971)	59-2468	0-88318-104-5
No. 6	Particle Physics (Irvine 1971)	72-81239	0-88318-105-3
No. 7	Exploring the History of Nuclear Physics (Brookline 1967,1969)	72-81883	0-88318-106-1
No. 8	Experimental Meson Spectroscopy - 1972 (Philadelphia 1972)	72-88226	0-88318-107-X
No. 9	Cyclotrons - 1972 (Vancouver 1972)	72-92798	0-88318-108-8
No.10	Magnetism and Magnetic Materials - 1972 (Denver 1972)	72-623469	0-88318-109-6
No.11	Transport Phenomena - 1973 (Brown University Conference)	73-80682	0-88318-110-X
No.12	Experiments on High Energy Particle Collisions - 1973 (Vanderbilt Conference)	73-81705	0-88318-111-8
No.13	π-π Scattering - 1973 (Tallahassee Conference)	73-81704	0-88318-112-6

QC
793.5
M428
I 56
1973

FEB 3 1975